T0236382

Lecture Notes in Computer Science 9116

Commenced Publication in 1973
Founding and Former Series Editors:
Gerhard Goos, Juris Hartmanis, and Jan van Leeuwen

Editorial Board

David Hutchison
  Lancaster University, Lancaster, UK
Takeo Kanade
  Carnegie Mellon University, Pittsburgh, PA, USA
Josef Kittler
  University of Surrey, Guildford, UK
Jon M. Kleinberg
  Cornell University, Ithaca, NY, USA
Friedemann Mattern
  ETH Zurich, Zürich, Switzerland
John C. Mitchell
  Stanford University, Stanford, CA, USA
Moni Naor
  Weizmann Institute of Science, Rehovot, Israel
C. Pandu Rangan
  Indian Institute of Technology, Madras, India
Bernhard Steffen
  TU Dortmund University, Dortmund, Germany
Demetri Terzopoulos
  University of California, Los Angeles, CA, USA
Doug Tygar
  University of California, Berkeley, CA, USA
Gerhard Weikum
  Max Planck Institute for Informatics, Saarbrücken, Germany

More information about this series at http://www.springer.com/series/7412

Jesús Ariel Carrasco-Ochoa
José Francisco Martínez-Trinidad
Juan Humberto Sossa-Azuela
José Arturo Olvera López
Fazel Famili (Eds.)

Pattern Recognition

7th Mexican Conference, MCPR 2015
Mexico City, Mexico, June 24–27, 2015
Proceedings

 Springer

Editors

Jesús Ariel Carrasco-Ochoa
National Institute of Astrophysics, Optics,
 and Electronics
Puebla
Mexico

José Francisco Martínez-Trinidad
National Institute of Astrophysics, Optics,
 and Electronics
Puebla
Mexico

Juan Humberto Sossa-Azuela
National Polytechnic Institute of Mexico
Ciudad de México
Mexico

José Arturo Olvera López
Autonomous University of Puebla
Puebla
Mexico

Fazel Famili
University of Ottawa
Ottawa, ON
Canada

ISSN 0302-9743   ISSN 1611-3349   (electronic)
Lecture Notes in Computer Science
ISBN 978-3-319-19263-5   ISBN 978-3-319-19264-2   (eBook)
DOI 10.1007/978-3-319-19264-2

Library of Congress Control Number: 2015939168

LNCS Sublibrary: SL6 – Image Processing, Computer Vision, Pattern Recognition, and Graphics

Springer Cham Heidelberg New York Dordrecht London
© Springer International Publishing Switzerland 2015
This work is subject to copyright. All rights are reserved by the Publisher, whether the whole or part of the material is concerned, specifically the rights of translation, reprinting, reuse of illustrations, recitation, broadcasting, reproduction on microfilms or in any other physical way, and transmission or information storage and retrieval, electronic adaptation, computer software, or by similar or dissimilar methodology now known or hereafter developed.
The use of general descriptive names, registered names, trademarks, service marks, etc. in this publication does not imply, even in the absence of a specific statement, that such names are exempt from the relevant protective laws and regulations and therefore free for general use.
The publisher, the authors and the editors are safe to assume that the advice and information in this book are believed to be true and accurate at the date of publication. Neither the publisher nor the authors or the editors give a warranty, express or implied, with respect to the material contained herein or for any errors or omissions that may have been made.

Printed on acid-free paper

Springer International Publishing AG Switzerland is part of Springer Science+Business Media
(www.springer.com)

Preface

MCPR 2015 was the seventh Mexican Conference on Pattern Recognition. This edition was jointly organized by the Computer Science Department of the National Institute for Astrophysics Optics and Electronics (INAOE) of Mexico and the Center for Computing Research of the National Polytechnic Institute of Mexico (CIC-IPN), under the auspices of the Mexican Association for Computer Vision, Neurocomputing and Robotics (MACVNR), which is a member society of the International Association for Pattern Recognition (IAPR). MCPR 2015 was held in Mexico-City, Mexico, during June 24–27, 2015.

MCPR 2015 attracted not only Mexican researchers but as in previous years; it also included worldwide participation. MCPR provided to the Pattern Recognition community a space for scientific research exchange, sharing of expertise and new knowledge, and establishing contacts that improve cooperation between research groups in pattern recognition and related areas in Mexico and the rest of the world.

Three invited speakers gave keynote addresses on various topics in pattern recognition:

- Prof. Xiaoyi Jiang, Department of Computer Science, University of Münster, Germany.
- Prof. Martin Hagan, School of Electrical and Computer Engineering, Oklahoma State University, USA.
- Prof. Daniel P. Lopresti, Department of Computer Science and Engineering, Lehigh University, USA.

These prestigious researchers also presented enlightening tutorials during the conference. To all of them, we express our sincere gratitude and appreciation for these presentations.

We received contributions from 16 countries. In total 63 manuscripts were submitted, out of which 29 were accepted for publication in these proceedings and for presentation at the conference. Each of these submissions was strictly peer-reviewed by at least two members of the Program Committee, all of them are specialists in Pattern Recognition, who prepared an excellent selection dealing with ongoing research.

The selection of papers was extremely rigorous in order to maintain the high quality standard of the conference. We would like to thank the members of the Program Committee for their efforts and the quality of the reviews. Their work allowed us to offer a conference program of high standard. We also extend our gratitude to all authors who submitted their papers to the conference and our regrets to those we turned down.

The authors of selected papers have been invited to submit extended versions of their papers for a Special Issue of the Intelligent Data Analysis Journal published by IOS Press.

Finally, our thanks go to the National Council of Science and Technology of Mexico (CONACYT) for providing a key support to this event.

June 2015   Jesús Ariel Carrasco-Ochoa
                                              José Francisco Martínez-Trinidad
                                              Juan Humberto Sossa-Azuela
                                              José Arturo Olvera-López
                                              Fazel Famili

Organization

MCPR 2015 was sponsored by the Computer Science Department of the National Institute of Astrophysics, Optics and Electronics (INAOE) and the Center for Computing Research of the National Polytechnic Institute of Mexico (CIC-IPN).

General Conference Co-chairs

Fazel Famili	School of Electrical Engineering and Computer Science, University of Ottawa, Ottawa, Canada
Jesús Ariel Carrasco-Ochoa	Computer Science Department, National Institute of Astrophysics, Optics and Electronics (INAOE), Mexico
José Francisco Martínez-Trinidad	Computer Science Department, National Institute of Astrophysics, Optics and Electronics (INAOE), Mexico
Juan Humberto Sossa-Azuela	Center for Computing Research of the National Polytechnic Institute of Mexico (CIC-IPN), Mexico
José Arturo Olvera-López	Autonomous University of Puebla (BUAP), Mexico

Local Arrangement Committee

Cerón Benítez Gorgonio
Cervantes Cuahuey Brenda Alicia
López Lucio Gabriela
Meza Tlalpan Carmen

Scientific Committee

Asano, A.	Kansai University, Japan
Batyrshin, I.	Mexican Petroleum Institute, Mexico
Benedi, J.M.	Universidade Politécnica de Valencia, Spain
Borges, D.L.	Universidade de Brasília, Brazil
Castelan, M.	CINVESTAV, Mexico
Chen, Chia-Yen	National University of Kaohsiung, Taiwan
Facon, J.	Pontifícia Universidade Católica do Paraná, Brazil
Gatica, D.	Idiap Research Institute, Switzerland
Gelbukh, A.	CIC-IPN, Mexico
Goldfarb, L.	University of New Brunswick, Canada
Gomes, H.	Universidade Federal de Campina Grande, Brazil
Graña, M.	University of the Basque Country, Spain
Heutte, L.	Université de Rouen, France
Hurtado-Ramos, J.B.	CICATA-IPN, Mexico

Igual, L.	University of Barcelona, Spain
Jiang, X.	University of Münster, Germany
Kampel, M.	Vienna University of Technology, Austria
Klette, R.	University of Auckland, New Zealand
Kober, V.	CICESE, Mexico
Koster, W.	Universiteit Leiden, The Netherlands
Laurendeau, D.	Université Laval, Canada
Lazo-Cortés, M.S.	Universidad de las Ciencias Informáticas, Cuba
Lopez-de-Ipiña-Peña, M.K.	Universidad del País Vasco, Spain
Lorenzo-Ginori, J.V.	Universidad Central de Las Villas, Cuba
Mayol-Cuevas, W.	University of Bristol, UK
Menezes, P.	University of Coimbra-Polo II, Brazil
Mora, M.	Catholic University of Maule, Chile
Morales, E.	INAOE, Mexico
Nolazco, J.A.	ITESM-Monterrey, Mexico
Pina, P.	Instituto Superior Técnico, Portugal
Pinho, A.	University of Aveiro, Portugal
Pinto, J.	Instituto Superior Técnico, Portugal
Pistori, H.	Dom Bosco Catholic University, Brazil
Raducanu, B.	Universitat Autònoma de Barcelona, Spain
Raposo-Sanchez, J.M.	Instituto Superior Técnico, Portugal
Real, P.	University of Seville, Spain
Rojas, R.	Free University of Berlin, Germany
Roman-Rangel, E.F.	University of Geneva, Switzerland
Ross, A.	West Virginia University, USA
Rueda, L.	University of Windsor, Canada
Ruiz-Shulcloper, J.	CENATAV, Cuba
Sanchez-Cortes, D.	Idiap Research Institute, Switzerland
Sanniti di Baja, G.	Istituto di Cibernetica, CNR, Italy
Sang-Woon, K.	Myongji University, South Korea
Sansone, C.	Università di Napoli, Italy
Sappa, A.	Universitat Autònoma de Barcelona, Spain
Schizas, C.	University of Cyprus, Cyprus
Sousa-Santos, B.	Universidade de Aveiro, Portugal
Spyridonos, P.	University of Ioannina, Greece
Sucar, L.E.	INAOE, Mexico
Valev, V.	University of North Florida, USA
Vaudrey, T.	University of Auckland, New Zealand
Vitria, J.	University of Barcelona, Spain
Zagoruiko, N.G.	Russian Academy of Sciences, Russia
Zhi-Hua, Z.	Nanjing University, China

Additional Referees

Alvarez-Vega, M.
Arco, L.
Dias, P.
Escalante-Balderas, H.J.
Feregrino-Uribe, C.
García-Borroto, M.
Gomez-Gil, M.P.
Hernández-Rodríguez, S.
Kleber, F.
Matsubara, E.

Montes Y Gómez, M.
Morales-Reyes, A.
Neves, A.J.R.
Pedro, S.
Pereira, E.
Reyes-García, C.A.
Sánchez-Vega, J.F.
Villaseñor-Pineda, L.
Tao, J.

Sponsoring Institutions

National Institute of Astrophysics, Optics and Electronics (INAOE)
Center for Computing Research of the National Polytechnic Institute of Mexico
(CIC-IPN)
Mexican Association for Computer Vision, Neurocomputing and Robotics
(MACVNR)
National Council of Science and Technology of Mexico (CONACYT)

Contents

Image Processing and Analysis

Robotics and Computer Vision

Natural Language Processing and Recognition

Applications of Pattern Recognition

Pattern Recognition and Artificial Intelligent Techniques

Recommendation of Process Discovery Algorithms Through Event Log Classification

Damián Pérez-Alfonso[1]([✉]), Osiel Fundora-Ramírez[2], Manuel S. Lazo-Cortés[3], and Raciel Roche-Escobar[1]

[1] University of Informatics Sciences, Habana, Cuba
{dalfonso,roche}@uci.cu
[2] XETID, Habana, Cuba
ofundora@xetid.cu
[3] National Institute for Astrophysics Optics and Electronics, Puebla, México
mlazo@inaoep.mx

Abstract. Process mining is concerned with the extraction of knowledge about business processes from information system logs. Process discovery algorithms are process mining techniques focused on discovering process models starting from event logs. The applicability and effectiveness of process discovery algorithms rely on features of event logs and process characteristics. Selecting a suitable algorithm for an event log is a tough task due to the variety of variables involved in this process. The traditional approaches use empirical assessment in order to recommend a suitable discovery algorithm. This is a time consuming and computationally expensive approach. The present paper evaluates the usefulness of an approach based on classification to recommend discovery algorithms. A knowledge base was constructed, based on features of event logs and process characteristics, in order to train the classifiers. Experimental results obtained with the classifiers evidence the usefulness of the proposal for recommendation of discovery algorithms.

Keywords: Process discovery · Process mining · Classification

1 Introduction

The execution of business processes on information systems is recorded on event logs. Process mining involves discovery, conformance and enhancement of process starting from event logs. A process discovery algorithm is a function that maps an event log onto a process model, such that the model is "representative" for the behavior seen in the event log [1]. Several algorithms have been developed for discovering process models that reflect the actual execution of processes. These models allow business analysts to make performance evaluations, anomaly identification, compliance checking, among others analysis.

Noise, duplicate tasks, hidden tasks, non-free choice constructs and loops are typical problems for discovery algorithms [2]. These problems are related to unstructured processes, commonly present in real environments [3]. Thus,

© Springer International Publishing Switzerland 2015
J.A. Carrasco-Ochoa et al. (Eds.): MCPR 2015, LNCS 9116, pp. 3–12, 2015.
DOI: 10.1007/978-3-319-19264-2_1

algorithms performance depends on event log characteristics and their associated process.

Several executions of different discovery algorithms could be required trying to obtain a quality model, thus becoming a time consuming and error-prone task. Selecting the right algorithms is a hard task due to the variety of variables involved. Several techniques execute different discovery algorithms for an event log and evaluate their resulting models using quality metrics [4]. Nevertheless, this empirical evaluation approach is computationally expensive and time consuming.

On the other hand, a recommendation technique is based on regression, but it requires reference models [5]. Reference models are not commonly available in contexts where process discovery is required. If there are reference models, it is unwise to assume that they reflect the actual execution of processes.

Studies that attempt to establish the algorithms with better performance under certain conditions have been published using the aforementioned empirical evaluation techniques [6]. However, the impact of each condition on model quality is not clearly defined yet. Therefore, the actual use of these studies remains limited.

The aim of this paper is to evaluate the usefulness of an approach based on classification to recommend discovery algorithms [7]. A knowledge base is constructed considering event log features such as: control-flow patterns, invisible tasks and infrequent behavior (noise). The recommendation procedure based on classification is tested over the knowledge base with different classifiers.

The paper is structured as follows: in the next section concepts and approaches related to process discovery and recommendation of discovery algorithms are presented. In Sect. 3, phases of the recommendation procedure, followed by a description of the creation process of the knowledge base are presented. In Sect. 4, experimental results of the classification based recommendation and their respective analysis are provided. A set of current techniques and approaches to assess and recommend discovery algorithms are discussed in Sect. 5. Finally, the last section is devoted to conclusions and outlines for future work.

2   Recommendation of Process Discovery Algorithms

A process discovery algorithm constructs a process model starting from an event log. An event is the occurrence of an activity of a process and a trace is a non-empty finite sequence of events recorded during one execution of such process. So, an event log is a multi-set of traces belonging to different executions of the same process.

Obtaining a quality model is the main goal of a process discovery algorithm. There are various metrics and approaches for estimating process model quality, though there is a consensus on the following quality criteria [1]:

– *Fitness:* The model should allow the behavior present in the event log.
– *Precision:* The model should not allow a behavior unrelated to the one stored in the log.

– *Generalization:* The model should generalize the behavior present in the log.
– *Simplicity:* The model should be as simple as possible. Also referred as structure, is influenced by the vocabulary of modeling language.

These criterion compete among them due to the inverse relationship between generalization and precision. A too general model could lead to allow much more behavior than the one presents in the log, it is also known as underfitting model. On the contrary, a too precise or overfitting model is undesirable. The right balance between overfitting and underfitting is called *behavioral appropriateness.* The *structural appropriateness* of a model refers to its ability to clearly reflect the performance recorded with the minimal possible structure [8]. A quality model requires both, behavioral appropriateness and structural appropriateness [9].

In order to obtain a quality model, a discovery algorithm should tackle several challenges related to event logs characteristics. Heterogeneity of data sources from real environments, can lead to difficult cases for discovery algorithms [10]. Infrequent traces and data recorded incompletely and/or incorrectly can induce wrong interpretations of process behavior. Moreover, data provided by parallel branches and ad-hoc changed instances generate complex sequences on event logs, this creates traces that are harder to mine.

Process structure is another source of challenges for discovery algorithms. Presence of control-flow patterns like non-free choices, loops and parallelism affect the discovery algorithms. For example, algorithms such as α, $\alpha^{\dagger}$, $\alpha^{\#}$ and α^* do not support non-free choices [11]. On the other hand, DWS Mining and α^{++} can deal with non-free choice but cannot support loops [2].

In order to identify which discovery algorithm allows obtaining suitable models for particular situations, a set of techniques for algorithms evaluation have been developed. Performance of these algorithms is determined through evaluation of quality of obtained models. Defined quality metrics are grouped under two main methods [12]. One method compares the discovered model with respect to the event log and is called *model-log.* The other method, called *model-model,* assesses similarity between discovered model and a reference model of process.

Evaluation frameworks allow end users to compare the performance of discovery algorithms through empirical evaluation with quality metrics [4,13]. Moreover, recommending a discovery algorithm for a given event log, based on empirical evaluation, involves time and resource consumption for each of the algorithms chosen as a possible solution. So, alternative approaches, based on classification, have been proposed to recommend process discovery algorithms [5,7].

3   Classification of Event Logs for Recommendation of Process Discovery Algorithms

In this section we evaluate the usefulness of an approach based on classification to recommend discovery algorithms [7]. *Classification* is the problem concerning the construction of a procedure that will be applied to a continuing sequence of cases, in which each *new case* must be assigned to one of a set of *pre-defined classes* on the basis of *observed attributes or features* [14]. An event log on which is necessary to recommend a discovery algorithm is considered as a *new case* to be classified. The recommended algorithm is the *pre-defined class* to be assigned to an event log based on its *observed features.*

Taking into account the challenges for discovery algorithms the classification mechanism for recommendation of discovery algorithms consider the following factors:

1. The event log is the main information source that is available in all environments for process characterization.
2. The peculiarities of event log. must be considered in addition to process characteristics.
3. The results obtained by quality metrics on discovered models provide information about performance of process discovery algorithm facing event logs and process characteristics.

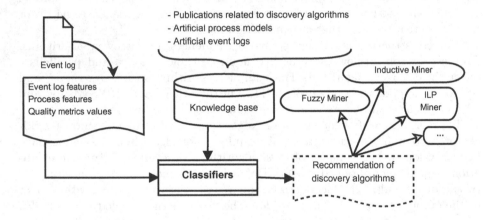

Fig. 1. Recommendation of discovery algorithms through event log classification

The stages for classification of event logs in order to recommend process discovery algorithms can be observed in Fig. 1. It shows that staring point is the new case to be classified. This new case is composed of event log and the process features that affect discovery algorithms and desired values for quality metrics on each quality criterion. The discovery algorithm that could discover a model for that log with the desired values on quality metrics is the *class*. The classifiers are trained in a knowledge base composed by cases with the same structure of aforementioned case, but labeled with the corresponding discovery algorithm. The discovery algorithm selected as the class for the new case is recommended to be applied on the new event log in order to obtain a process model with the specified quality values.

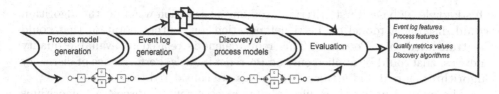

Fig. 2. Generation of artificial cases for the knowledge base

3.1 Building the Knowledge Base

A major challenge for this classification problem is the building of a knowledge base. In order to construct the knowledge base for this problem the following features were selected to build the cases:

1. The event log features.
2. The process characteristics.
3. The discovery algorithm used to obtain the process model.
4. The quality metrics obtained values based on the discovered model.

The sequence of phases followed to generate the artificial cases for the knowledge base is presented in Fig. 2. The outcome of each phase is used as input to the following.

The goal of the first and second phases is to obtain the event log and process features that affect discovery algorithms. In these two phases was used the *Process log generator* tool [15]. Using this tool several process models were generated in a random way. 67 of these process models that combine loops, non-free choice and invisible tasks were selected for the second phase. These process features were considered due to their impact on process discovery algorithms [6].

In order to generate the event logs a factorial complete experimental design was performed. For this factorial design five classes were considered: noise (C_1), noise interval (C_5), loops (C_3), parallelism (C_4) and invisible tasks (C_5). Noise could appear on different proportion of traces on event logs, on this cases 5 different proportions were used: 0, 25, 50, 75, 100. The same stand for noise interval: 0, 25, 50, 75 and 100 were the distribution used. Control flow patterns (C_3, C_4) and invisible tasks (C_5) were considered on a boolean manner. Formula 1 was used to calculate the required number of event logs to combine the aforementioned criteria.

$$F = C_1 * C_2 * C_3 * C_4 * C_5$$
$$F = 5 * 5 * 2 * 2 * 2 = 200$$

(1)

Considering results from Formula 1, 201 event logs were generated combining the five features already stated. One third of these event logs has 500 traces each one. The second third has 1000 traces on each event log, while the last third has 1500 traces on each one.

On the third phase, discovery algorithms are applied on the generated event logs. Even there are several discovery algorithms, only five where selected for this knowledge base. First criterion used on the selection was that the algorithm could discover a model on Petri net notation, or another notation that could be translated to a Petri net. This requirement is related to available quality metrics, that could be applied only on Petri net models. Performance of discovery algorithms on real and artificial event logs are considered too.

Therefore, published results about assessment of discovery algorithms [2,4,6,11] lead us to select the Heuristic Miner [16], ILP Miner [17], Inductive Miner [18], Genetic Miner [19] and Alpha Miner [20]. All these algorithms are available as plug-ins on the process mining framework ProM [21]. So, ProM 6.3 was used to obtain five process model from each event log using every selected algorithm.

The main goal of the last phase in the generation of artificial cases for the knowledge base is to evaluate the performance of discovery algorithms. One quality metric was selected for each quality dimension or criterion. So, were selected *fitness* [22], ETC [23], ARC Average [22] and Behavioral Generalization [24]. All these metrics belong to *model-log* method and are implemented in CoBeFra [25], a benchmarking tool. Therefore, using the generated event logs and the discovered models, the values for these quality metrics were obtained using CoBeFra.

Once all the phases were executed the information obtained were used to create the cases. One case was created for each discovered model. Each case is represented as a vector $c_i = \{at_i, aa_i, and_i, xor_i, l_i, it_i, nd_i, ni_i, f_i, p_i, g_i, s_i, DA\}$. In this vector i refers to the ordinal number of the discovered model. The at_i and aa_i variables stand for amount of traces and amount of activities in the event log used to discover the i model. Moreover, and_i, xor_i, l_i and it_i refers to the amount of parallelism, exclusive choice, loops and invisible tasks respectively, in the process related to the i model. The noise distribution and interval on the event log used to discover the i model are represented as nd_i and ni_i. Variables f_i, p_i, g_i and s_i express the values of the quality metrics obtained on the i model, related to fitness, precision, generalization and simplicity, respectively. Last but not least, DA is the discovery algorithm used to create the i model and this is the class that labeled the case.

Following the aforementioned description, a knowledge base was constructed with 795 cases. A ProM plug-in was developed to visualize and manage the knowledge base. This plug-in allows integration with other techniques in ProM.

4   Testing Classifiers

In order to find suitable classifiers for the knowledge base built, a set of well-known classifiers were trained and assessed. Before training, the data set was normalized to values between 0 and 1. Results for each classifier training are presented in Table 1, expressed in terms of Incorrectly Classified Instances (ICI) and Mean Absolute Error (MAE). Classifiers implementation on WEKA [26] where used in all cases, with default configuration values.

Based on the results presented on Table 1, seven classifiers were selected: Classification Via Regression, Multilayer Perceptron, Simple Logistic, Logistic, J48, Filtered Classifier and MultiClass Classifier. A ProM plug-in was developed

Table 1. Results of classifiers training

Classifiers	Cross-validation		Use training set		Percentage split	
	ICC	MAE	ICC	MAE	ICC	MAE
Classification via regression	231	0.2956	175	0.2644	80	0.3042
Multilayer perceptron	259	0.3329	196	0.2822	89	0.332
Simple logistic	264	0.3547	257	0.3421	87	0.3529
Logistic	286	0.3411	266	0.3314	103	0.348
PART	309	0.357	216	0.3282	136	0.3761
J48	317	0.3475	273	0.3278	103	0.359
OneR	325	0.4521	289	0.4263	103	0.4389
Bayes net	334	0.3724	334	0.3671	97	0.3454
Filtered classifier	343	0.3631	201	0.3508	135	0.3612
MultiClass classifier	343	0.3797	334	0.3752	126	0.3813
IBK	452	0.5317	280	0.4186	158	0.5389
Naive bayes	452	0.4234	474	0.4135	168	0.4254
K*	495	0.4329	525	0.4329	208	0.4334

Table 2. Execution times for empirical evaluation and classification of new event logs

Event log	Case vector	Empirical evaluation	Classification
1	$c_1 = \{500, 26, 2, 4, 1, 0, 0, 50, 1, 1, 1, 1, ?\}$	12 h 22 m 43 s	10 s
2	$c_2 = \{1000, 17, 1, 2, 1, 2, 25, 75, 1, 1, 1, 1, ?\}$	1 h 30 m 8 s	9 s
3	$c_3 = \{1500, 18, 2, 1, 2, 2, 0, 75, 1, 1, 1, 1, ?\}$	25 h 26 m 53 s	15 s
4	$c_4 = \{500, 25, 4, 1, 0, 0, 100, 25, 1, 1, 1, 1, ?\}$	37 h 10 m 23 s	40 s
5	$c_5 = \{500, 25, 6, 9, 2, 2, 25, 25, 1, 1, 1, 1, ?\}$	14 h 24 m 48 s	10 s

to integrate the WEKA implementation of these classifiers into ProM. Using this classification plug-in, the classifiers could be trained in ProM with the previously mentioned knowledge base. With the trained classifiers, the plug-in enables the recommendation of discovery algorithms through classification of new cases.

Five new event logs were generated to assess the recommendation provided by the classification plug-in developed. Empirical evaluation of discovery algorithms on these event logs were used as reference for this assessment. Starting from the features of these event logs, five new cases were prepared (Table 2). Each case has the structure $c_i = \{at_i, aa_i, and_i, xor_i, l_i, it_i, nd_i, ni_i, f_i, p_i, g_i, s_i, DA\}$.

Recommendation through classification means a significant time improvement with respect to empirical evaluation, as can be seen in Table 2. Besides, for each event log (with the exception of event log 4) the class proposed by six of the seven classifiers match with the discovery algorithm with best results on empirical evaluation.

Heuristics Miner was the only algorithm with quality values distinct from 0 in empirical evaluation of event log 4. This result only matches with the class obtained by three classifiers, because other four proposed Alpha Miner and the remaining propose Inductive Miner. Discovery algorithms were impacted by the high level of noise distribution in this event log ($nd_4 = 100$). Models obtained from this kind of logs are incomprehensible and have very low values on quality metrics. This situation could be the explanation for mismatching of classifiers with event log 4. Nevertheless, further experimentation with highly noisy event logs is required to prove this hypothesis.

5 Related Work

Evaluation frameworks allow end users to compare the performance of discovery algorithms through empirical evaluation [13]. But, using this framework as a recommendation mechanism is not suitable due to the cost involved on empirical assessments of discovery algorithms. Following the *model-log* method, another evaluation framework, that includes a parameter optimization step, has been proposed [4]. Nevertheless, the negative examples generation created serious performance problems in the experiments with complex event logs [4].

Other proposed solution is based on selecting reference models of high quality and building from these a regression model to estimate the similarity of other process models [5]. Created serious performance problems However, this approach needs reference models for the evaluation and prediction, a requirement that severely limits its application. In multiple real-world environments, where discovery algorithms need to be applied, the process models are not described or are inconsistent and/or incomplete. Besides, this solution assumes that the actual execution of the processes keeps a close relationship with their reference models. But, inexact results can be expected in contexts where features of the actual logs differ from logs artificially generated by the reference models. Furthermore, the construction of a regression model from process model features discards issues such as noise and lack of information on event logs. These issues have a significant impact on the performance of discovery algorithms.

6 Conclusions and Future Work

Event logs features and process characteristics affect the performance of process discovery algorithms. Classical approaches that select discovery algorithms based on empirical assessments are computationally expensive and time consuming.

This paper evaluates the recommendation of discovery algorithms as a classification problem. For this purpose, a knowledge base, with artificially generated cases was built. Cases combine features of event logs and process characteristics with impact on performance of discovery algorithms. Besides, each case contains the values of one quality metric from each quality criterion.

Two ProM plug-ins developed allow to train seven well known classifiers over the knowledge base built. Recommendation of these classifiers match entirely, on four from five event logs, with the discovery algorithm with best quality values on empirical evaluation. In all cases recommendation through classification was obtained in a significant lower time than through empirical evaluation.

Experimentation with highly noisy logs and multiple classifier systems is suggested as future work. Besides, research is required to apply the proposed approach on event logs from real environments. In this context, low level patterns such as indirect successions and repeated events could be used to extract process characteristics from real event logs.

References

1. van der Aalst, W.M.P.: Process Mining. Discovery, Conformance and Enhancement of Business Processes. Springer, Heidelberg (2011)
2. De Weerdt, J., De Backer, M., Vanthienen, J., Baesens, B.: A multi- dimensional quality assessment of state-of-the- art process discovery algorithms using real- life event logs. Inf. Syst. **37**, 654–676 (2012)
3. Desai, N., Bhamidipaty, A., Sharma, B., Varshneya, V.K., Vasa, M., Nagar, S.: Process trace identification from unstructured execution logs. In: 2010 IEEE International Conference on Services Computing (SCC), pp. 17–24 (2010)
4. Ma, L.: How to evaluate the performance of process discovery algorithms. Master thesis, Eindhoven University of Technology, Netherlands (2012)
5. Wang, J., Wong, R.K., Ding, J., Guo, Q., Wen, L.: Efficient selection of process mining algorithms. IEEE Trans. Serv. Comput. **99**(1), 1–1 (2012)
6. vanden Broucke, S.K.L.M., Delvaux, C., Freitas, J., Rogova, T., Vanthienen, J., Baesens, B.: Uncovering the Relationship Between Event Log Characteristics and Process Discovery Techniques. In: Lohmann, N., Song, M., Wohed, P. (eds.) BPM 2013 Workshops. LNBIP, vol. 171, pp. 41–53. Springer, Heidelberg (2014)
7. Pérez-Alfonso, D., Yzquierdo-Herrera, R., Lazo-Cortés, M.: Recommendation of process discovery algorithms: a classification problem. Res. Comput. Sci. **61**, 33–42 (2013)
8. van der Aalst, W.M.P., Rubin, V., Verbeek, H., Van Dongen, B., Kindler, E., Günther, C.: Process mining: a two-step approach to balance between underfitting and overfitting. Softw. Syst. Model. **9**(1), 87–111 (2010)
9. Rozinat, A., van der Aalst, W.M.P.: Conformance testing: measuring the fit and appropriateness of event logs and process models. In: Bussler, C.J., Haller, A. (eds.) BPM 2005. LNCS, vol. 3812, pp. 163–176. Springer, Heidelberg (2006)
10. Ly, L.T., Indiono, C., Mangler, J., Rinderle-Ma, S.: Data transformation and semantic log purging for process mining. In: Ralyté, J., Franch, X., Brinkkemper, S., Wrycza, S. (eds.) CAiSE 2012. LNCS, vol. 7328, pp. 238–253. Springer, Heidelberg (2012)
11. van Dongen, B.F., Alves de Medeiros, A.K., Wen, L.: Process mining: overview and outlook of petri net discovery algorithms. In: Jensen, K., van der Aalst, W.M.P. (eds.) Transactions on Petri Nets and Other Models of Concurrency II. LNCS, vol. 5460, pp. 225–242. Springer, Heidelberg (2009)
12. De Weerdt, J., De Backer, M., Vanthienen, J., Baesens, B.: A critical evaluation study of model-log metrics in process discovery. In: Muehlen, M., Su, J. (eds.) BPM 2010 Workshops. LNBIP, vol. 66, pp. 158–169. Springer, Heidelberg (2011)

13. Rozinat, A., Medeiros, A.K.A.d., Günther, C.W., Weijters, A.J.M.M., van der Aalst, W.M.P.: Towards an evaluation framework for process mining algorithms. BPM Center Report (2007)
14. Michie, D., Spiegelhalter, D.J., Taylor, C.C., Campbell, J. (eds.): Machine Learning, Neural and Statistical Classification. Ellis Horwood, Upper Saddle River (1994)
15. Burattin, A., Sperduti, A.: PLG: a framework for the generation of business process models and their execution logs. In: Muehlen, M., Su, J. (eds.) BPM 2010 Workshops. LNBIP, vol. 66, pp. 214–219. Springer, Heidelberg (2011)
16. Weijters, A., van der Aalst, W.M.P., de Medeiros, A.K.A.: Process mining with the heuristics miner-algorithm. Technische Universiteit Eindhoven. Technical report WP 166 (2006)
17. van der Werf, J.M.E.M., van Dongen, B.F., Hurkens, C.A.J., Serebrenik, A.: Process discovery using integer linear programming. In: van Hee, K.M., Valk, R. (eds.) PETRI NETS 2008. LNCS, vol. 5062, pp. 368–387. Springer, Heidelberg (2008)
18. Leemans, S.J.J., Fahland, D., van der Aalst, W.M.P.: Discovering block-structured process models from event logs - a constructive approach. In: Colom, J.-M., Desel, J. (eds.) PETRI NETS 2013. LNCS, vol. 7927, pp. 311–329. Springer, Heidelberg (2013)
19. De Medeiros, A., Weijters, A., van der Aalst, W.M.P.: Genetic process mining: an experimental evaluation. Data Min. Knowl. Discov. 14(2), 245–304 (2007)
20. van der Aalst, W.M.P., Weijters, A.J.M.M., Maruster, L.: Workflow mining: discovering process models from event logs. IEEE Trans. Knowl. Data Eng. 16(9), 1128–1142 (2004)
21. Verbeek, H.M.W., Buijs, J.C.A.M., van Dongen, B.F., van der Aalst, W.M.P.: XES, XESame, and ProM 6. In: Soffer, P., Proper, E. (eds.) CAiSE Forum 2010. LNBIP, vol. 72, pp. 60–75. Springer, Heidelberg (2011)
22. Rozinat, A., Aalst, W.M.P.: Conformance checking of processes based on monitoring real behavior. Inf. Syst. 33(1), 64–95 (2008)
23. Adriansyah, A., Munoz-Gama, J., Carmona, J., van Dongen, B.F., van der Aalst, W.M.P.: Alignment Based Precision Checking. In: La Rosa, M., Soffer, P. (eds.) BPM Workshops 2012. LNBIP, vol. 132, pp. 137–149. Springer, Heidelberg (2013)
24. vanden Broucke, S.K.L.M., De Weerdt, J., Baesens, B., Vanthienen, J.: Improved artificial negative event generation to enhance process event logs. In: Ralyté, J., Franch, X., Brinkkemper, S., Wrycza, S. (eds.) CAiSE 2012. LNCS, vol. 7328, pp. 254–269. Springer, Heidelberg (2012)
25. De Weerdt, J., Baesens, B., Vanthienen, J.: A comprehensive benchmarking framework (CoBeFra) for conformance analysis between procedural process models and event logs in ProM. In: Proceedings of the IEEE Symposium on Computational Intelligence and Data Mining (CIDM 2013), part of the IEEE Symposium Series in Computational Intelligence 2013 (2013)
26. Hall, M., Frank, E., Holmes, G., Pfahringer, B., Reutemann, P., Witten, I.H.: The WEKA data mining software: An update. SIGKDD Explor. Newsl. 11(1), 10–18 (2009). 07535

A New Method Based on Graph Transformation for FAS Mining in Multi-graph Collections

Niusvel Acosta-Mendoza[1,2]([✉]), Jesús Ariel Carrasco-Ochoa[2],
José Fco. Martínez-Trinidad[2], Andrés Gago-Alonso[1],
and José E. Medina-Pagola[1]

[1] Advanced Technologies Application Center (CENATAV), 7a ♯ 21406 e/ 214
and 216, Siboney, Playa, CP: 12200 Havana, Cuba
{nacosta,agago,jmedina}@cenatav.co.cu
[2] Instituto Nacional de Astrofísica, Óptica Y Electrónica (INAOE),
Luis Enrique Erro No. 1, Sta. María Tonantzintla, CP: 72840 Puebla, Mexico
{nacosta,ariel,fmartine}@ccc.inaoep.mx

Abstract. Currently, there has been an increase in the use of frequent approximate subgraph (*FAS*) mining for different applications like graph classification. In graph classification tasks, FAS mining algorithms over graph collections have achieved good results, specially those algorithms that allow distortions between labels, keeping the graph topology. However, there are some applications where multi-graphs are used for data representation, but FAS miners been designed to work only with simple-graphs. Therefore, in this paper, in order to deal with multi-graph structures, we propose a method based on graph transformations for FAS mining in multi-graph collections.

Keywords: Approximate graph mining · Approximate multi-graph mining · Graph-based classification

1 Introduction

In Data Mining, frequent pattern identification has become an important topic with a wide range of applications in several domains of science, such as: biology, chemistry, social science and linguistics, among others [1]. Therefore different techniques for pattern extraction, where frequent subgraph mining algorithms have been developed.

From these techniques, in the last years, the most popular strategies are the approximate approaches, because in practice there are specific problems where exact matching is not applicable with a positive outcomes [2–4]. Therefore, it is important to tolerate certain level of variability: semantic distortions, vertices or edges mismatching during frequent pattern search it implies to evaluate the similarity between graphs considering approximate matching. In this sense, several algorithms have been developed for frequent approximate subgraph (*FAS*) mining, which use different approximate graph matching techniques allowing the detection

© Springer International Publishing Switzerland 2015
J.A. Carrasco-Ochoa et al. (Eds.): MCPR 2015, LNCS 9116, pp. 13–22, 2015.
DOI: 10.1007/978-3-319-19264-2_2

of frequent subgraphs when some minor (non-structural) variations in the sub-graphs are permitted [3,5]. Different heuristics and graph matching approaches have been used as basis for developing FAS mining algorithms, for example: edit distance [6], heuristic over uncertain graphs [4,7], and heuristics based on semantic substitutions [3,5,8–10], being this last approach the idea followed in this paper. However, these algorithms perform FAS mining on simple-graph collections, and they were not designed to deal with other structures as multi-graphs. Neverthe-less, several authors argue that the nature of the phenomenon in some application can be better modeled through multi-graphs [11–13].

Analyzing the aforementioned algorithms, some problems are detected: (1) the reported algorithms for mining FASs from graph collections, which allow semantic distortions and keep the graph topology, were not designed to deal with multi-graphs; (2) the algorithms reported in [4,7], the sub-isomorphism tests are computationally expensive because the occurrences of the candidates in the collection are not stored. These tests are more expensive when approx-imate matching is used than for exact matching; (3) the algorithms proposed in [5,7], which compute representative patterns, were not designed to deal with semantic distortions between vertex and edge label sets in graph collections; (4) the algorithms introduced in [8–10] were not designed for dealing with graph collections. Thus, in this paper, we focus on the approximate graph mining app-roach, which allows semantic distortions between vertex and edge label sets, keeping the graph topology, over multi-graph collections.

The organization of this paper is the following. In Sect. 2, some basic concepts and notations are provided. Our proposed method for processing multi-graphs is introduced in Sect. 3. Several experiments are presented in Sect. 4. Finally, our conclusions and future work directions are discussed in Sect. 5.

2   Background

In this section, the background and notation needed to understand the following sections, as well as the FAS mining problem are presented. Notice that most of definitions are presented in a way that they allow treating both types of undirected graphs: simple-graphs and multi-graphs.

Definition 1 (Labeled Graph). A *labeled graph* with the domain of labels $L = L_V \cup L_E$, where L_V and L_E are the label sets for vertices and edges respectively, is a 5-tuple, $G = (V, E, \phi, I, J)$, where V is a set whose elements are called *vertices*, E is a set whose elements are called *edges*, $\phi : E \to V \times V$ is the *incidence function* (the edge e, through the function $\phi(e)$, connects the vertices u and v if $\phi(e) = \{u, v\}$), $I : V \to L_V$ is a *labeling function* for assigning labels to vertices and $J : E \to L_E$ is a *labeling function* for assigning labels to edges.

Definition 2 (Subgraph and Supergraph). Let $G_1 = (V_1, E_1, \phi_1, I_1, J_1)$ and $G_2 = (V_2, E_2, \phi_2, I_2, J_2)$ be two graphs, G_1 is a *subgraph* of G_2 if $V_1 \subseteq V_2$, $E_1 \subseteq E_2$, ϕ_1 is a restriction of ϕ_2 to V_1, I_1 is a restriction of I_2 to V_1, and J_1 is a restriction of J_2 to E_1 (a restriction of a function is the result of trimming its domain). In this case, the notation $G_1 \subseteq G_2$ is used, and it is also said that G_2 is a *supergraph* of G_1.

Definition 3 (Multi-edge and Loop). An edge $e \in E$, where $\phi(e) = \{u, v\}$ and $u \neq v$, is a *multi-edge* if there is $e' \in E$ such that $e \neq e'$ and $\phi(e) = \phi(e')$; otherwise, e is a *simple-edge*. If $|\phi(e)| = 1$, e is called a *loop*, i.e. $\phi(e) = \{u\} = \{v\}$.

In this way, the set of edges E of a graph can be partitioned into three disjoint subsets $E^{(s)}$, $E^{(m)}$, and $E^{(l)}$ containing simple-edges, multi-edges, and loops, respectively.

Definition 4 (Simple-graph and Multi-graph). A graph is a *simple-graph* if it has no loops and no multi-edges, i.e. it has only simple-edges, $E = E^{(s)}$, being $E^{(m)} = \emptyset$ and $E^{(l)} = \emptyset$; otherwise, it is a *multi-graph*.

Definition 5 (The Operator $\oplus$). Let $G_1 = (V_1, E_1, \phi_1, I_1, J_1)$ and $G_2 = (V_2, E_2, \phi_2, I_2, J_2)$ be two graphs, where for each $v \in V_1 \bigcap V_2$ $I_1(v) = I_2(v)$, and for each $e \in E_1 \bigcap E_2$ $\phi_1(e) = \phi_2(e)$ and $J_1(e) = J_2(e)$. In this case, it is said that G_1 and G_2 are *mutually compatible* graphs. Thus, the *sum* of G_1 and G_2 is a supergraph of G_1 and G_2 denoted by $G_1 \oplus G_2 = (V_3, E_3, \phi_3, I_3, J_3)$, where $V_3 = V_1 \bigcup V_2$; $E_3 = E_1 \bigcup E_2$; for each $v \in V_1$ $I_3(v) = I_1(v)$ and for each $v \in V_2$ $I_3(v) = I_2(v)$; for each $e \in E_1$ $\phi_3(e) = \phi_1(e)$ and $J_3(e) = J_1(c)$, and for each $e \in E_2$ $\phi_3(e) = \phi_2(e)$ and $J_3(e) = J_2(e)$. We will use the notation $\bigoplus_i G_i$ for denoting the successive sum of several graphs G_i.

Definition 6 (Isomorphism). Given two graphs G_1 and G_2, a pair of functions (f, g) is an *isomorphism* between these graphs if $f : V_1 \to V_2$ and $g : E_1 \to E_2$ are bijective functions, where:

1. $\forall u \in V_1 : f(u) \in V_2$ and $I_1(u) = I_2(f(u))$
2. $\forall e_1 \in E_1$, where $\phi_1(e_1) = \{u, v\}$: $e_2 = g(e_1) \in E_2$, and $\phi_2(e_2) = \{f(u), f(v)\}$ and $J_1(e_1) = J_2(e_2)$.

If there is an isomorphism between G_1 and G_2, it is said that G_1 and G_2 are *isomorphic*.

Definition 7 (Sub-isomorphism). Given three graphs G_1, G_2 and G_3. If G_1 is isomorphic to G_3 and $G_3 \subseteq G_2$, then it is said that there is a *sub-isomorphism* between G_1 and G_2, denoted by $G_1 \subseteq_s G_2$, and it is also said that G_1 is *sub-isomorphic* to G_2.

Definition 8 (Similarity). Let Ω be the set of all possible labeled graphs in L, the *similarity* between two graphs $G_1, G_2 \in \Omega$ is defined as a function $sim : \Omega \times \Omega \to [0, 1]$. The higher the value of $sim(G_1, G_2)$ the more similar the graphs are, and if $sim(G_1, G_2) = 1$ then there is an isomorphism between these graphs.

Definition 9 (Approximate Isomorphism and Approximate Sub-isomorphism). Let G_1, G_2 and G_3 be three graphs, let $sim(G_1, G_2)$ be a similarity function among graphs, and let τ be a similarity threshold, there is an *approximate isomorphism* between G_1 and G_2 if $sim(G_1, G_2) \geq \tau$. Also, if there is an approximate isomorphism between G_1 and G_2, and $G_2 \subseteq G_3$, then there is an *approximate sub-isomorphism* between G_1 and G_3, denoted as $G_1 \subseteq_A G_3$.

Definition 10 (Maximum Inclusion Degree). Let G_1 and G_2 be two graphs, let $sim(G_1, G_2)$ be a similarity function among graphs, since a graph G_1 can be enough similar to several subgraphs of another graph G_2, the *maximum inclusion degree* of G_1 in G_2 is defined as:

$$maxID(G_1, G_2) = \max_{G \subseteq G_2} sim(G_1, G), \tag{1}$$

where $maxID(G_1, G_2)$ means the maximum value of similarity at comparing G_1 with all of the subgraphs G of G_2.

Definition 11 (Approximate Support). Let $D = \{G_1, \ldots, G_{|D|}\}$ be a graph collection, let $sim(G_1, G_2)$ be a similarity function among graphs, let τ be a similarity threshold, and let G be a graph. Thus, the *approximate support* (denoted by *appSupp*) of G in D is obtained through equation (2):

$$appSupp(G, D) = \frac{\sum_{\{G_i | G_i \in D, G \subseteq_A G_i\}} maxID(G, G_i)}{|D|} \tag{2}$$

Using (2), G is a *FAS* in D if $appSupp(G, D) \geq \delta$, for a given support threshold δ, a similarity function among graphs $sim(G_1, G_2)$ and a similarity threshold τ. The values of δ and τ are in $[0, 1]$ because the similarity is defined in $[0, 1]$.

FAS mining consists in, given a support threshold δ and a similarity threshold τ, finding all the FAS in a collection of graphs D, using a similarity function *sim*.

3   Method Based on Transformation for Processing Multi-graphs

Following the idea used in [11,14] for transforming a multi-graph into a simple-graph[1], we propose a method for transforming a multi-graph collection into a simple-graph collection without losing any topological or semantic information.

The proposed method for transforming a multi-graph into a simple-graph comprises two steps: (1) transforming each loop into a simple-edge, and (2) transforming each edge (simple-edge or multi-edge) into two simple-edges. Then, each loop is changed by adding a new vertex with an special label (κ) and a simple-edge with the loop label. Later, each edge e which connects two vertices $u \neq v$ is changed by a new vertex with a special label (ϱ) and two simple-edges (e_1 and e_2) both with the label of e. The simple-edge e_1 connects the new vertex with u and e_2 connects the new vertex with v. The transformation from a multi-graph into a simple-graph is formally defined in Definition 12.

---

[1] It is important to highlight that in [11,14] the transformation is different to the one introduced in this paper. Furthermore, these transformations were not proposed for mining frequent patterns in multi-graphs.

Definition 12 (Multi-graph Simplification). Let $G = (V, E, \phi, I, J)$ be a connected multi-graph, and let κ and ϱ be two different vertex labels that will be used in two kind of vertices for representing all loops and all edges, respectively. The *multi-graph simplification* of G is a graph defined as:

$$G' = \bigoplus_{e \in E} G'_e, \tag{3}$$

where the graph G'_e is defined as follows:

- If e is an edge and $\phi(e) = \{u, v\}$, the graph G'_e is defined as $G'_e = (V_1, E_1, \phi_1, I_1, J_1)$, where $V_1 = \{u, v, w\}$, $E_1 = \{e_1, e_2\}$, $E_1 \cap E = \emptyset$, $\phi_1(e_1) = \{u, w\}$, $\phi_1(e_2) = \{w, v\}$, I_1 is a restriction of I to V_1 with $I_1(w) = \varrho$, $J_1(e_1) = J_1(e_2) = J(e)$, and $w \notin V$ is a new vertex.
- If e is a loop and $\phi(e) = \{v\}$, the graph G'_e is defined as $G'_e = (V_2, E_2, \phi_2, I_2, J_2)$, where $V_2 = \{v, w\}$, $E_2 = \{e'\}$, $e' \notin E$, $\phi_2(e') = \{v, w\}$, I_2 is a restriction of I to V_2 with $I_2(w) = \kappa$, $J_2(e') = J(e)$, and $w \notin V$ is a new vertex.

Based on Definition 12, an algorithm for transforming a multi-graph into a simple-graph, called *M2Simple*, can be introduced. First, by traversing the edges in the input multi-graph, for each edge e, according to Definition 12, the graph G'_e can be calculated and added to G'. Thus, the complexity of this algorithm is $O(k * n^2 * d)$, where k is the largest number of edges between two vertices, n is the number of vertices in the graph with the largest amount of vertices, and d is the number of graphs. Notice that for building simplifications only vertices with label κ and ϱ were added, including the simple-edges connecting such vertices.

For doing compatible a simple-graph with the original multi-graph collection, a reversing process is required. Then, there is a transformation from a simple-graph to a multi-graph. In Definition 13, the required condition that a simple-graph must fulfill in order to be transformed to a multi-graph is introduced.

Definition 13 (Returnable Graph). Let κ and ϱ be the special labels used in the Definition 12. The simple-graph $G' = (V', E', \phi', I', J')$ is *returnable* to a multi graph if it fulfills the following conditions:

1. Each vertex $v \in V'$ with $I'(v) = \varrho$ has exactly two incident edges e_1 and e_2, such that $J'(e_1) = J'(e_2)$, and
2. Each vertex $v \in V'$ with $I'(v) = \kappa$ has exactly one incident edge.

The transformation from a simple-graph into a multi-graph is formally defined in the Definition 14.

Definition 14 (Graph Generalization). Let $G' = (V', E', \phi', I', J')$ be a returnable connected graph and let κ and ϱ be the special labels used in the Definitions 12 and 13. Let V'_ϱ be the set of all of $v \in V'$ such that $I'(v) = \varrho$, and let V'_κ be the set of all of $v \in V'$ such that $I'(v) = \kappa$. Thus, the *generalization* of G' is a graph defined as:

$$G = \bigoplus_{w \in V'_\varrho \cup V'_\kappa} G_w, \tag{4}$$

where the graph G_w is defined as follows:

- If $I'(w) = \varrho$, by the first returnable condition there are exactly two incidents edges e_1 and e_2, such that $\phi'(e_1) = \{u, w\}$ and $\phi'(e_2) = \{w, v\}$, then G_w is defined as $G_w = (V_1, E_1, \phi_1, I_1, J_1)$, where $V_1 = \{u, v\}$, $E_1 = \{e\}$, $e \notin E'$, $\phi_1(e) = \{u, v\}$, I_1 is a restriction of I' to V_1, and $J_1(e) = J'(e_1) = J'(e_2)$.
- If $I'(w) = \kappa$, by the second returnable condition there is exactly one incident edge e_2, such that $\phi'(e_2) = \{v, w\}$, then G_w is defined as $G_w = (V_2, E_2, \phi_2, I_2, J_2)$, where $V_2 = \{v\}$, $E_2 = \{e\}$, $e \notin E'$, $\phi_2(e) = \{v\}$, I_2 is a restriction of I' to V_2, and $J_2(e) = J'(e_2)$.

An algorithm for transforming a returnable simple-graph into a multi-graph, called *S2Multi* can be introduced. Based on Definition 14, this algorithm comprises two steps: first, the algorithm traverses the vertices with label ϱ for adding edges to G. Then, the algorithm traverses the vertices with label κ for adding loops to G. Taking into account the way in which G is built, through the proposed simplification process (see Definition 12), the complexity of this algorithm is $O((n + k * n^2) * d)$, where n, k and d are the same as used for complexity of the M2Simple algorithm. Notice that, for building generalizations, only vertices with label κ and ϱ were removed, including the simple-edges connecting such vertices.

The correctness of our method, which is based on transformations, is proved trough the following theorems. The proofs of these theorems were omitted due to space limitations.

Theorem 1. *Let $G_1 = (V_1, E_1, \phi_1, I_1, J_1)$ and $G_2 = (V_2, E_2, \phi_2, I_2, J_2)$ be two isomorphic connected (returnable) graphs. Then, the graphs, $G'_1 = (V'_1, E'_1, \phi'_1, I'_1, J'_1)$ and $G'_2 = (V'_2, E'_2, \phi'_2, I'_2, J'_2)$, obtained from the simplifications (generalizations) of G_1 and G_2 are also isomorphic.*

Theorem 2. *Let $G = (V, E, \phi, I, J)$ be the simplification (generalization) of the connected graph $G_1 = (V_1, E_1, \phi_1, I_1, J_1)$, and let κ and ϱ be the special labels; then G_1 is isomorphic to the generalization (simplification) of G.*

Theorem 3. *Let $G'_1 = (V'_1, E'_1, \phi'_1, I'_1, J'_1)$ and $G'_2 = (V'_2, E'_2, \phi'_2, I'_2, J'_2)$ be two multi-graphs, and let $G_1 = (V_1, E_1, \phi_1, I_1, J_1)$ and $G_2 = (V_2, E_2, \phi_2, I_2, J_2)$ be two simple-graphs, such that G_1 and G_2 are simplifications of G'_1 and G'_2, respectively; then G'_2 is sub-isomorphic to $G'_1 \Leftrightarrow G_2$ is sub-isomorphic to G_1.*

Corollary 1. *Let $D = \{G_1, \ldots, G_N\}$ be a collection of N simple-graphs, let $D' = \{G'_1, \ldots, G'_N\}$ be a collection of N multi-graphs, where G_i is the simplification of G'_i, for each $1 \leq i \leq N$. Let G be the simplification of a multi-graph G'. Thence, the support of G in D is the same that the support of G' in D'.*

Corollary 2. *Let D and D' be the collections of simple-graphs and multi-graphs, respectively used in Corollary 1. Let G be the simplification of a multi-graph G'. Thence, if G is FAS in $D \Leftrightarrow G'$ is FAS in D'.*

Finally, it is important to highlight that the transformation process of multi-graph collections into simple-graph collections allows us applying any algorithm for traditional FAS mining.

4   Experiments

In this section, with the aim of studying the performance of the proposed method as well as its usefulness for image classification tasks, two experiments are performed. All our experiments were carried out using a personal computer with an Intel(R) Core(TM) i7 with 64 GB of RAM. The M2Simple and S2Multi algorithms was implemented in python and executed on GNU/Linux (UBUNTU).

In our first experiment, the performance of our transformation algorithms (M2Simple and S2Multi) is evaluated. For this evaluation, three kinds of synthetic multi-graph collections were used. These synthetic collections were generated using the PyGen[2] graph emulation library. For building these collections, first, we fix $|D| = 5000$ and $|E| = 800$, varying $|V|$ from 100 to 500, with increments of 100 (see Table 1(a)). Next, we fix $|V| = 500$ and $|D| = 5000$ varying $|E|$ from 600 to 1000, with increments of 100 (see Table 1(b)). Finally, we vary $|D|$ from 10000 to 50000, with increments of 10000, keeping $|V| = 500$ and $|E| = 800$ (see Table 1(c)). Notice that, we assign a descriptive name for each synthetic collection, for example, $D5kV600E2k$ means that the collection has $|D| = 5000$, $|V| = 600$ and $|E| = 2000$.

In Table 1, the performance results, in terms of runtime, of both proposed transformation algorithms (M2Simple and S2Multi) is shown. These results were achieved by transforming each multi-graph collection into a simple-graph collection using M2Simple. Then, each obtained simple-graph collection was transformed into a multi-graph using S2Multi. In this table, the first column shows the collection and the other two columns show the runtime in seconds of both proposed transformation algorithms.

According to the complexity of the proposed transformation processes presented in Sect. 3, remarked by the results shown in Table 1, we can conclude that our transformation algorithms are more sensitive to the number of edge variations

Table 1. Runtime results in seconds of our transformation algorithms over several synthetic collections.

(a) **Varying $|V|$ from 1000 to 5000 with $|D| = 5000$ and $|E| = 1000$.**

Collection	M2Simple	S2Multi
D5kV100E800	16.77	22.34
D5kV200E800	18.00	23.83
D5kV300E800	18.60	25.21
D5kV400E800	18.93	25.63
D5kV500E800	20.06	27.27

(b) **Varying $|E|$ from 1000 to 5000 with $|D| = 5000$ and $|V| = 1000$.**

Collection	M2Simple	S2Multi
D5kV500E600	15.95	21.18
D5kV500E700	17.97	24.64
D5kV500E800	20.06	27.27
D5kV500E900	22.54	31.27
D5kV500E1k	24.88	32.74

(c) **Varying $|D|$ from 10000 to 50000 with $|E| = 1000$ and $|V| = 1000$.**

Collection	M2Simple	S2Multi
D10kV500E800	41.28	55.64
D20kV500E800	81.45	109.33
D30kV500E800	122.47	163.48
D40kV500E800	157.54	252.94
D50kV500E800	199.83	268.52

---

[2] PyGen is available in http://pywebgraph.sourceforge.net.

than to the number of vertex variations in this kind of experiments. This is because the required runtime grows faster when the amount of edges increases than increasing the number of vertices. Furthermore, M2Simple receives many more vertices and edges than S2Multi for the same multi-graph collection, since S2Multi creates an additional vertex and an additional edge for each transformed edge or loop. On the other hand, the number of graph of the collection is an important variable to take into account, because it affects the performance of both proposed algorithms when it has high values. Finally, the most important fact to take into account is that the transformations are performed in a runtime less than 5 min over collections with high dimensions.

In addition, three graph collections representing images generated with the Random image generator of Coenen[3] are used. For obtaining the Coenen image collections, we randomly generate 1000 images with two classes. These images were randomly divided into two sub-sets: one for training with 700 (70 %) images and another for testing with 300 (30 %) images. The first two collections, denoted by *C-Angle* and *C-Distance*, are represented as simple-graph collections, using a quad-tree approach [15] and the angles and the distances between regions as edge labels, respectively. These two collections have 21 vertex labels, 24 edge labels and 135 as the average graph size. The last collection, denoted by *C-Multi*, was also built using a quad-tree approach, but it uses both (angles and distances) values as labels for two multi-edges, respectively. This collection comprises 21 vertex labels, 48 edge labels and the average graph size is 270, where all edges are multi-edges. These collections are used for assessing the performance of our method based on graph transformation and also for showing the usefulness of the identified patterns on a multi-graph collection. The algorithm for FAS mining used in our method is VEAM [3].

In order to show the usefulness of the patterns computed in multi-graphs through our proposed method, in Table 2 we show the classification results by using the patterns computed after applying our method in the multi-graph collections. For these experiments, the classifier J48graft taken from Weka v3.6.6 [16] using the default parameters, was used. This table shows the results of the accuracy and F-measure achieved over the three graph collections representing the image collection. The first column shows the used support threshold values. The other three consecutive columns show the accuracy results using the collection specified in the top of these columns, and in the other three consecutive columns the F-measure results are shown. The results achieved using VEAM over C-Angle and C-Distance collections represent the solution to the problem if our proposal were not available since for applying VEAM the images would be represented as simple-graphs using only one edge between vertices as in C-Angle or C-Distance.

The classification results achieved (see Table 2) over the image collection show the usefulness of the patterns computed by our proposal, where in most of these cases, the best classification results are obtained by using our method. Furthermore, the representation of the images as multi-graphs allows us to obtain better classification results than using simple-graphs.

---

[3] www.csc.liv.ac.uk/~frans/KDD/Software/ImageGenerator/imageGenerator.html.

Table 2. Classification results (%) achieved over the different graph collection representing the image collection using the J48graft classifier with similarity threshold $\tau = 0.4$ and different support threshold values.

Support	Accuracy			F-measure		
	C-Angle	C-Distance	C-Multi	C-Angle	C-Distance	C-Multi
0.8	78.33	78.33	78.33	68.90	68.90	68.90
0.7	77.67	76.33	**78.67**	74.13	72.37	**77.46**
0.6	79.67	91.33	**92.33**	77.15	90.85	**91.76**
0.5	89.00	**93.33**	93.33	88.17	92.59	**93.00**
0.4	89.67	91.27	**91.65**	89.27	90.13	**91.59**
Average	*82.87*	*86.12*	*86.86*	*79.52*	*82.97*	*84.54*

5 Conclusions

In this paper, a new method based on graph transformation for FAS mining in multi-graph collections is proposed. In our method, we first transform a multi-graph collection into a simple-graph collection, then the mining is performed over this collection using a traditional FAS miner and later, an inverse transformation allow us returning the mining results (simple-graphs) to multi-graphs. This process can be performed without losing any topological or semantic information. The performance of the proposed transformation algorithms (M2Simple and S2Multi) is evaluated over different synthetic multi-graph collections. Additionally, the usefulness of the patterns identified after applying our method is also shown.

As future work, we plan to develop a FAS miner allowing us to compute FASs directly from multi-graph collections, without the transformation steps.

Acknowledgment. This work was partly supported by the National Council of Science and Technology of Mexico (CONACyT) through the project grant $CB2008$-106366; and the scholarship grant 287045.

References

1. Jiang, C., Coenen, F., Zito, M.: A survey of frequent subgraph mining algorithms. Knowl. Eng. Rev. **28**(1), 75–105 (2012)
2. Holder, L., Cook, D., Bunke, H.: Fuzzy substructure discovery. In: ML92: Proceedings of the Ninth International Workshop on Machine Learning, pp. 218–223. Morgan Kaufmann Publishers Inc., San Francisco (1992)
3. Acosta-Mendoza, N., Gago-Alonso, A., Medina-Pagola, J.: Frequent approximate subgraphs as features for graph-based image classification. Knowl.-Based Syst. **27**, 381–392 (2012)
4. Li, J., Zou, Z., Gao, H.: Mining frequent subgraphs over uncertain graph databases under probabilistic semantics. VLDB J. **21**(6), 753–777 (2012)
5. Jia, Y., Zhang, J., Huan, J.: An efficient graph-mining method for complicated and noisy data with real-world applications. Knowl. Inf. Syst. **28**(2), 423–447 (2011)

6. Song, Y., Chen, S.: Item sets based graph mining algorithm and application in genetic regulatory networks. In: IEEE International Conference on Data Mining, pp. 337–340 (2006)
7. Zou, Z., Li, J., Gao, H., Zhang, S.: Finding top-k maximal cliques in an uncertain graph. In: IEEE 26th International Conference on Data Engineering (ICDE 2010), pp. 649–652 (2010)
8. Chen, C., Yan, X., Zhu, F., Han, J.: gApprox: mining frequent approximate patterns from a massive network. In: Seventh IEEE International Conference on Data Mining (ICDM 2007), pp. 445–450 (2007)
9. Flores-Garrido, M., Carrasco-Ochoa, J., Martínez-Trinidad, J.: AGraP: an algorithm for mining frequent patterns in a single graph using inexact matching. Knowl. Inf. Syst., pp. 1–22 (2014)
10. Flores-Garrido, M., Carrasco-Ochoa, J., Martínez-Trinidad, J.: Mining maximal frequent patterns in a single graph using inexact matching. Knowl.-Based Syst. **66**, 166–177 (2014)
11. Whalen, J.S., Kenney, J.: Finding maximal link disjoint paths in a multigraph. In: IEEE Global Telecommunications Conference and Exhibition. 'Communications: Connecting the Future', GLOBECOM 1990, pp. 470–474. IEEE (1990)
12. Björnsson, Y., Halldórsson, K.: Improved heuristics for optimal pathfinding on game maps. In: American Association for Artificial Intelligence (AIIDE). pp. 9–14 (2006)
13. Morales-González, A., García-Reyes, E.B.: Simple object recognition based on spatial relations and visual features represented using irregular pyramids. Multimedia Tools Appl. **63**(3), 875–897 (2013)
14. Boneva, I., Hermann, F., Kastenberg, H., Rensink, A.: Simulating multigraph transformations using simple graphs. In: Proceedings of the Sixth International Workshop on Graph Transformation and Visual Modeling Techniques, Braga, Portugal. Electronic Communications of the EASST, vol. 6, EASST (2007)
15. Finkel, R., Bentley, J.: Quad trees: a data structure for retrieval on composite keys. Acta Informatica **4**, 1–9 (1974)
16. Hall, M., Frank, E., Holmes, G., Pfahringer, B., Reutemann, P., Witten, I.: The WEKA data mining software: an update. ACM SIGKDD Explor. Newsl. **11**(1), 10–18 (2009)

Classification of Hand Movements from Non-invasive Brain Signals Using Lattice Neural Networks with Dendritic Processing

Leonardo Ojeda[1], Roberto Vega[2], Luis Eduardo Falcon[1],
Gildardo Sanchez-Ante[1], Humberto Sossa[3], and Javier M. Antelis[1,4](✉)

[1] Tecnológico de Monterrey Campus Guadalajara, Av. Gral Ramón Corona
2514,45201 Zapopan, Jalisco, Mexico
{leonardo.ojeda,luis.eduardo.falcon,gildardo,mauricio.antelis}@itesm.mx
[2] Department of Computing Science, University of Alberta,
116 St. and 85 Ave., Edmonton, AB T6G 2R3, Canada
rvega@ualberta.ca
[3] Instituto Politécnico Nacional-CIC, Av. Juan de Dios Batiz S/N,
Gustavo A. Madero, 07738 México, Distrito Federal, Mexico
hsossa@cic.ipn.mx
[4] Centro de Investigación en Mecatrónica Automotriz (CIMA),
Tecnológico de Monterrey Campus Toluca, Eduardo Monroy Cárdenas 2000,
50110 Toluca, Estado de México, Mexico

Abstract. EEG-based BCIs rely on classification methods to recognize the brain patterns that encode user's intention. However, decoding accuracies have reached a plateau and therefore novel classification techniques should be evaluated. This paper proposes the use of Lattice Neural Networks with Dendritic Processing (*LNND*) for the classification of hand movements from electroencephalographic (EEG) signals. The performance of this technique was evaluated and compared with classical classifiers using EEG signals recorded form participants performing motor tasks. The result showed that *LNND* provides: (*i*) the higher decoding accuracies in experiments using one electrode (*DA* = 80 % and *DA* = 80 % for classification of motor execution and motor imagery, respectively); (*ii*) distributions of decoding accuracies significantly different and higher than the chance level ($p < 0.05$, Wilcoxon signed-rank test) in experiments using one, two, four and six electrodes. These results shows that *LNND* could be a powerful technique for the recognition of motor tasks in BCIs.

Keywords: Lattice Neural Network · Brain-Computer Interface · Electroencephalogram · Motor imagery

1 Introduction

Brain-Computer Interfaces (BCI) have emerged as a new alternative to provide people suffering partial or complete motor impairments, with a non-muscular

© Springer International Publishing Switzerland 2015
J.A. Carrasco-Ochoa et al. (Eds.): MCPR 2015, LNCS 9116, pp. 23–32, 2015.
DOI: 10.1007/978-3-319-19264-2_3

communication channel to convey messages to the external world [1]. These systems are based on the recording and processing of the brain activity in order to obtain control signals that are used to trigger external devices [2]. Thus, the BCI design is mainly driven by the brain activity and by the mental task carried out by the user. Most of BCIs use the electroencephalogram (EEG) as this technique is non-invasive and provides high temporal resolution brain signals. The mental task is essential to induce recognizable changes or patterns in the recorded signals, and the most common mental task is the motor imagery of different parts of the body [3]. Hence, the key element in a BCI is the recognition of the changes or patterns induced in the recorded brain signals by the mental task, which is carried out by means of classification algorithms [4].

Nonetheless, the application of BCIs in real scenarios and daily live situations with final users, i.e., patients, is still limited mainly due to: (i) EEG recording systems are expensive and not fully portable, therefore, patients or health systems can not afford to obtain such technologies; (ii) the majority of research on EEG-based BCIs focuses on classical classification algorithms, but their performance has reached a plateau and therefore novel and different algorithms should be evaluated. This paper addresses these issues by studying the recognition of hand movements from EEG signals recorded with a low-cost EEG system, and by applying a novel classification technique named Lattice Neural Networks with Dendritic Processing (*LNND*) [5]. This method is commencing to be used in some applications [6], however it has not been used in BCIs.

The aim in this work was to employ and to evaluate the *LNND* classification technique in the recognition of hand movements from EEG signals (i.e. to identify whether a person is in relax or executing/imagining a hand movement), as well as to compare its performance against Linear Discriminant Analysis and Support Vector Machines, which are commonly used in EEG-based BCIs. Previous works have explored the evaluation of several classification algorithms in this context [4], however none of them have applied the *LNND* technique. To achieve this, EEG signals were obtained using a low-cost EEG system from several healthy participants performing motor execution and motor imagery tasks. These signals were used to evaluate two-class classification, concretely *relax* versus *movement*, using each electrode individually and several groups with various electrodes. The results showed that the *LNND* classification method provided the higher classification rates and thus could be used in BCIs. The paper is organized as follows: Sect. 2 introduces *LNND* classification technique; Sect. 3 describes the experiments carried out to obtain EEG signals in real settings; Sect. 4 describes the experiments and results; and Sect. 5 presents the conclusions.

2 Lattice Neural Networks with Dendritic Processing

Classical classification methods such as Fisher Linear Discriminant Analysis (*FLDA*) [7] or Support Vector Machine with Radial Basis Functions (*SVMR*) [8] compute a discriminant function or separating hyperplane (defined by $w \cdot f(x) + b = 0$, where $x \in \Re^{n \times 1}$ is a feature vector, $f(\cdot)$ is a transformation function,

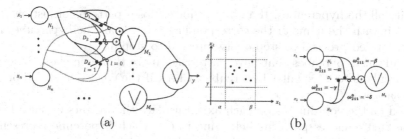

Fig. 1. (a) Diagram for the model of a *LNND*. (b) Diagram of how the weights are computed in a *LNND* using the graphical method.

$w \in \Re^{n \times 1}$ is a vector of classification weights and b is a bias term) to separate the patterns representing the classes. These methods are commonly used in EEG-based BCIs (for a review see [4]). Lattice neural networks with dendritic processing (*LNND*) is a very recent and different classification method that produces closed separation surfaces in the patterns. The technical details of this method are explained below.

Dendrites are the primary basic computational units in a biological neuron, however, they are absent in most models of artificial neural networks (ANN) [9]. Thus, the inclusion of dendrites in ANNs is essential to guarantee the creation of closed separation surfaces to discriminate the data from the different classes [10]. Recently, a new paradigm of ANN, known as Lattice Neural Networks, that considers computation in the dendritic structure as well as in the body of the neuron has been proposed [11]. This model requires no hidden layers, is capable of multiclass discrimination, present no convergence problems and produce closed separation surfaces between classes [12].

The diagram for this model of *LNND* is presented in Fig. 1a. The model consists of n input and m output neurons, where n is the number of features in the input vector, and m is the number of classes of the problem. The input and output neurons are connected via a finite number of dendrites $D_1, \ldots, D_k$. Each input neuron N_i has at most two connections on a given dendrite. The weight associated between neuron N_i and dendrite D_k of M_j is denoted by w^l_{ijk}, where the superscript $l \in \{0, 1\}$ distinguishes between excitatory ($l = 1$, marked as a black dot on Fig. 1a), and inhibitory ($l = 0$, marked as an open dot on Fig. 1a) input to the dendrite.

The training of the *LNND* can be done with the graphical training method proposed in [5], which involves four different steps. Given p classes of patterns $C^a, a = 1, 2, \ldots p$, each with n attributes:

1. Create an hypercube HC^n, that includes all the patterns in the set. In order to increase the tolerance to noise, it is possible to add a margin M on each side of the hypercube. This margin is a number greater or equal to zero, and it is a function of the length of the edges of the hyperbox.

2. Verify all the hypercubes. If an hypercube encloses patterns of just one class label it with the name of the corresponding class. If all the hypercubes have been labeled proceed to step 4, else proceed to step 3.
3. For all the hypercubes that have patterns of more than one class, divide the hypercube into 2^n smaller hypercubes. Check if the condition stated on step 2 is satisfied.
4. Based on the coordinates on each axis, calculate the weights for each hypercube that encloses patterns belonging to C^a. Each hypercube represents a dendrite. Figure 1b shows how these weights are defined.

The computation of the k-th dendrite D_k of M_j is given by:

$$\tau_k^j(x) = \bigwedge_{i \in I(k)} \bigwedge_{l \in L(i)} (-1)^{1-l}(x_i + \omega_{ijk}^l) \qquad (1)$$

where $x = (x_1, \ldots, x_n)$ denotes the input vector, $I(k) \subseteq \{1, \ldots, n\}$ corresponds to the set of all input neurons connected with the k-th dendrite D_k of M_j, $L(i) \subseteq \{0, 1\}$ corresponds to the set of excitatory and inhibitory synapses of the neuron N_i with the dendrite D_k of M_j, and the operator $\bigvee$ denotes the minimum value operator. The total output of the neuron M_j is given by the equation:

$$\tau^j(x) = \bigvee_{k=1}^{K_j} \tau_k^j(x) \qquad (2)$$

where K_j represents the total number of dendrites in the neuron M_j, and $\tau_k^j(x)$ is the output of the dendrite k of neuron M_j. The input vector x is assigned to the class whose neuron results in the biggest value [5]:

$$y = \bigvee_{j=1}^{m} \tau^j(x) \qquad (3)$$

where m is the number of classes, and $\tau^j(x)$ is the output of the neuron M_j.

In this work, we implemented the *LNND* classification technique with no margin, i.e. $M = 0$, and with a brute force search of it using 1000 values uniformly distributed between 0 and 1, *LNND0* and *LNNDB*, respectively. Therefore, four classification methods were evaluated: *FLDA*, *SVMR*, *LNND0* and *LNNDB*.

3   Methodology

3.1   Recording of EEG Signals

Participants and EEG Recording System. Twelve healthy subjects (ten males and two females) were voluntarily recruited to participate in this study (age range 20–24 years). EEG signals were recorded from 14 scalp locations according to the international 10/20 system (*AF3, F7, F3, FC5, T7, P7, O1,*

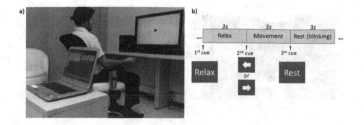

Fig. 2. (a) Snapshot of the experimental setup showing a participant seated in from of the computer screen wearing the EEG recording system. (b) Temporal sequence of a trial during the execution of the experiment.

$AF4$, $F8$, $F4$, $FC6$, $T8$, $P8$ and O2) using the low-cost commercially available Emotiv EPOC Neuroheadset system. Signals were recorded at a sampling frequency of 128 Hz with two reference electrodes CMS (on the left side) and DRL (on the right side), and no filtering was applied. The impedance for all electrodes was kept below $5k\Omega$ by using saline solution.

Experiment Design. Participants were seated in front of a computer screen with both forearms resting on their lap. The experiment consisted of many trials of hand movements in two conditions: (i) motor execution -clenching- of the left hand or right hand at a natural and effortless speed (ME); (ii) motor imagery - clenching- of the left hand or right hand at the same natural and effortless speed (MI). The experiment was controlled by visual cues presented on the screen. Figure 2a shows a snapshot of the experiment.

Each trial consisted of the time sequence depicted in Fig. 2b. The first cue instructed to relax while maintaining the initial position (relax phase). The second cue randomly displayed an arrow pointing to the left or to the right and indicated to perform or to imagine the movement of the corresponding hand (movement phase). The third cue indicated that they could rest and blink while adopting the initial position (rest phase). For each participant, the experiment was executed in four blocks (two blocks for each experimental condition) each including 48 trials, resulting in a total of 96 trials for each condition. The application that controlled the presentation of the visual cues and the recording of the EEG was developed with the BCI2000 platform [13].

3.2 Preprocessing and Features

EEG Signal Preprocessing. Recorded EEG signals were segmented in trials of 9 s using the second cue as reference. Then, each trial was trimmed from -3 to 3 s, thus the time interval $[-3, 0)$ s corresponds to the relax phase while the time interval $[0, 3)$ s corresponds to the movement phase. EEG signals were bandpass-filtered from 0.5 to 60 Hz using a zero-phase shift filter and then a common average reference (CAR) filtering was applied. Finally, two datasets were constructed by

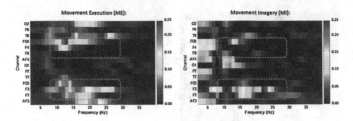

Fig. 3. r^2 analysis across all participants for (a) ME and (b) MI condition.

gathering the trials according to the experimental condition. These two datasets were used to test the four classifiers: *FLDA, SVMR, LNND0* and *LNNDB*.

Feature Extraction and Selection. The power spectral density (PSD) was used as features to discriminate between the two classes: relax and executing/imagining hand movement. The PSD was computed for the frequency range between 2 and 40 Hz at a resolution of 1 Hz. Thus, the number of the PSD features for each electrode is 39, while for all the electrodes is 546, i.e., 14 electrodes x 39 frequencies. Features computed from the relax phase $[-3, 0)s$ were labelled as *relax* while features computed in the movement phase $[0, -3)s$ were labelled as *movement*.

For the selection of features with the most discriminative power, the significant differences between the two classes were examined using an r^2 analysis [14]. The r^2 value of each feature (i.e., electrode and frequency) was computed as the squared correlation coefficient between the values of the feature and the corresponding labels of -1 for *relax* and $+1$ for *movement*. The selected PSD features at each electrode were those within the $\alpha : [8 - 12]Hz$, $\mu : [12 - 15]Hz$ and $\beta : [15 - 30]Hz$ bands presenting the highest r^2 values. This selection was performed individually for each participant and for each experimental condition. Therefore, the feature vector is $x_i \in \Re^{n \times 1}, i \in \{relax, movement\}$.

4   Experiments and Results

4.1   Classification Experiments and Evaluation Process

The classification between *relax* versus *movement* was assessed using each electrode independently and three groups of various electrodes, i.e., $Go2 \in \{FC5, FC6\}$, $Go4 \in \{FC5, FC6, F3, F4\}$ and $Go6 \in \{FC5, FC6, F3, F4, F7, F8\}$. Therefore, seventeen classification experiments were evaluated for each experimental condition. For each experiment, classification performance was assessed by a 10-fold cross-validation process, which was applied independently for each participant. To measure performance, the decoding accuracy or DA, defined as the percentage of correctly predicted trials, was computed for each fold.

The significant chance level of the decoding accuracy or DA_{sig}, was assessed using the binomial cumulative distribution at the $\alpha = 0.05$ significance level. Using 81 trials (i.e., the average of trials across all participants and experimental conditions) the significant chance level is $DA_{sig} = 60\,\%$. Finally, the Wilcoxon signed-rank test was used to test whether the distribution of DA was significantly different from DA_{sig}.

4.2 r^2 Analysis

Figure 3 shows the r^2 analysis obtained across all participants in both conditions. These results show differences between the two classes, *relax* and *movement*, mainly in the fronto-central electrodes ($FC5$, $FC6$, $F3$, $F4$, $F7$ and $F8$), and in the motor-related frequency bands (between $\sim 10\,Hz$ and $\sim 30\,Hz$). In addition, these results also reveals that the r^2 values are greater in the ME condition than in the MI condition.

4.3 Classification Results

Using Each Channel Individually. Figure 4 shows the mean DA achieved for each electrode projected in a $2D$ representation of the scalp. For the ME condition, in $FLDA$ and $SVMR$ the average DA was higher than DA_{sig} in electrodes $FC5$, $FC6$, $F3$, $F4$ and $F7$; in $LNND0$ the average DA was higher than DA_{sig} only in electrodes $FC5$ and $FC6$; while in $LNNDB$ the average DA was higher than DA_{sig} in all the electrodes. In particular, the electrode with the higher performance was $FC6$ ($DA = 75\,\%$), $FC6$ ($DA = 78\,\%$), $FC5$ ($DA = 66\,\%$) and $FC5$ ($DA = 80\,\%$) for $FLDA$, $SVMR$, $LNND0$ and $LNNDB$, respectively. For the MI condition, no electrode presented average DA greater than DA_{sig} for $FLDA$, $SVMR$ and $LNND0$, however $LNNDB$ presented average DA higher than DA_{sig} in all the electrodes. For this classifier, the best performance was achieved by electrode $FC5$ ($DA = 80\,\%$).

Using Groups of Channels. Figure 5 shows, for both experimental conditions and for each classifier, the distributions of DA across all the participants. For comparison purposes, the figure also include the distribution of DA achieved by the best electrode in the previous analysis, thus resembling a classification scenario with one electrode or $Go1$.

For the case of $Go1$, in the ME condition the median of the distributions of DA were significantly different and higher than DA_{sig} ($p < 0.05$, Wilcoxon signed-rank test) in all classifiers, however, for the MI condition solely the classifier $LNNDB$ presented a distribution of DA whose median is significantly different and higher than DA_{sig} ($p < 0.05$, Wilcoxon signed-rank test). Similarly, for $Go2$, the median of the distributions of DA in the ME condition were significantly different and higher than DA_{sig} ($p < 0.05$, Wilcoxon signed-rank test) in all classifiers, while for the MI condition the median of the distributions of

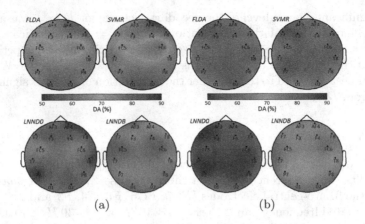

Fig. 4. Scalp topographies of average DA for each classifier in (a) ME and (b) MI condition.

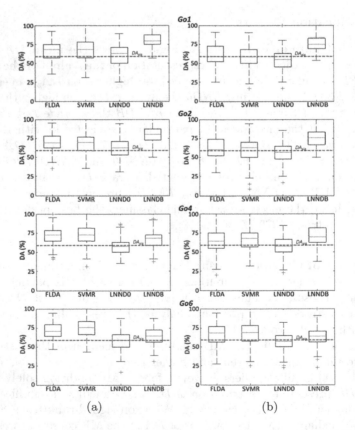

Fig. 5. Distribution DA across all participants for (a) ME and (b) MI condition. Results are presented for all classifiers when using the groups of channel $Go1$, $Go2$, $Go4$ and $Go6$.

DA is significantly different and higher than DA_{sig} ($p < 0.05$, Wilcoxon signed-rank test) with the classifiers $SVMR$ and $LNNDB$. Finally, for $Go4$ and $Go6$, the results for both conditions show that the median of the distributions of DA is significantly different and higher than DA_{sig} ($p < 0.05$, Wilcoxon signed-rank test) with the classifiers $FLDA$, $SVMR$ and $LNNDB$, however no significant differences between the median of the DA distribution and DA_{sig} is observed for the $LNND0$ classifier ($p > 0.01$, Wilcoxon signed-rank test).

5 Conclusions

This work studied the recognition of hand movements using power spectral based features extracted from EEG signals recorded with a low-cost system, and the novel Lattice Neural Networks with Dendritic Processing ($LNND$) classification technique. A battery of experiments were conducted to obtain EEG signals from several healthy participants performing motor execution (ME) and motor imagery (ME) of the hands. Then, these signals were used to evaluate two-class (*relax* versus *movement*) classification using the proposed $LNND$ technique, and to compare its performance with the classical Fisher Linear Discriminant Analysis ($FLDA$) and the Support Vector Machine with radial basis function kernel ($SVMR$), which are extensively used in BCIs.

On the one hand, the r^2 analysis showed that the features, i.e., the power spectral density in each pair electrode-frequency, with the most discriminative power are presented in electrodes surrounding the motor cortex (i.e., $FC5$, $FC6$, $F3$, $F4$, $F7$ and $F8$) and in frequencies within in motor-related frequency bands (i.e., $\alpha : [8 - 12]Hz$, $\mu : [12 - 15]Hz$ and $\beta : [15 - 30]Hz$). This revealed that the motor task performed by the participants induces recognizable changes in the recorded EEG signals, which indeed was used to select features and to perform classification. In addition, this analysis also revealed that the differences between the two classes are stronger in the ME condition than in the MI condition. This is because the PSD changes more prominent during actual movements than during imagery movements, suggesting better discriminability for ME than MI.

On the other hand, we ran seventeen experiments to evaluate classification using the proposed $LNND$ classification method, in its two versions $LNND0$ and $LNNDB$, and the classical $FLDA$ and $SVMR$. The results showed that the $LNND$ technique provides the higher decoding accuracy in experiments using a single electrode or a group of two/four electrodes (lower dimensionality of the feature space). However, the performance of the $LNND$ reduces, although maintains above the significant chance level, in experiments using a group of six electrodes (higher dimensionality of the feature space). In other words, as dimensionality of the feature space increases, the performance of the $LNND$ slightly reduces.

In summary, this work showed the possibility of using the very recent Lattice Neural Networks with Dendritic Processing classification algorithm to discriminate between motor tasks using EEG signals obtained by a commercially available low-cost system. This could be used as the basis for a low-cost and fully portable brain-computer interfaces based on motor-related mental tasks.

Acknowledgments. The authors thank Tecnológico de Monterrey, Campus Guadalajara, for their support under the Research Chair in Information Technologies and Electronics, as well as IPN-CIC under project SIP 20151187, and CONACYT under project 155014 for the economical support to carry out this research.

References

1. Wolpaw, J., Birbaumer, N., McFarland, D., Pfurtscheller, G., Vaughan, T.: Brain-computer interfaces for communication and control. Clin. Neurophysiol. **113**(6), 767–791 (2002)
2. Becedas, J.: Brain-machine interfaces: Basis and advances. IEEE Trans. Syst. Man Cybern. B Cybern. Part C: Appl. Rev. **42**(6), 825–835 (2012)
3. Graimann, B., Allison, B.Z., Pfurtscheller, G.: Brain-Computer Interfaces: Revolutionizing Human-Computer Interaction. Springer, Heidelberg (2010)
4. Lotte, F., Congedo, M., Lcuyer, A., Lamarche, F., Arnaldi, B.: A review of classification algorithms for eeg-based brain-computer interfaces. J. Neural Eng. **4**(2), R1–R13 (2007)
5. Sossa, H., Guevara, E.: Efficient training for dendrite morphological neural networks. Neurocomput. **131**, 132–142 (2014)
6. Vega, R., Guevara, E., Falcon, L.E., Sanchez-Ante, G., Sossa, H.: Blood vessel segmentation in retinal images using lattice neural networks. In: Castro, F., Gelbukh, A., González, M. (eds.) MICAI 2013, Part I. LNCS, vol. 8265, pp. 532–544. Springer, Heidelberg (2013)
7. Duda, R., Hart, P., Stork, D.: Pattern Classification. Wiley, New York (2001)
8. Cortes, C., Vapnik, V.: Support-vector networks. Mach. Learn. **20**, 273–297 (1995)
9. Ritter, G., Urcid, G.: Lattice algebra approach to single-neuron computation. IEEE Trans. Neural Networks **14**(2), 282–295 (2003)
10. Gori, M., Scarselli, F.: Are multilayer perceptrons adequate for pattern recognition and verification? IEEE Trans. Pattern Anal. Mach. Intell. **20**(11), 1121–1132 (1998)
11. Ritter, G., Schmalz, M.: Learning in lattice neural networks that employ dendritic computing. In: IEEE International Conference on Fuzzy Systems **2006**, pp. 7–13 (2006)
12. Ritter, G., Iancu, L.: Single layer feedforward neural network based on lattice algebra. In: Proceedings of the International Joint Conference on Neural Networks, vol. 4, pp. 2887–2892, July 2003
13. Schalk, G., Mcfarland, D.J., Hinterberger, T., Birbaumer, N., Wolpaw, J.R.: BCI2000: A general-purpose brain-computer interface (BCI) system. IEEE Trans. Biomed. Eng. **51**(6), 1034–1043 (2004)
14. Steel, R.G., Torrie, J.H.: Principles and Procedures of Statistics with Special Reference to the Biological Sciences. McGraw Hill, New York (1960)

A Different Approach for Pruning Micro-clusters in Data Stream Clustering

Argenis A. Aroche-Villarruel[1](✉), José Fco. Martínez-Trinidad[1],
Jesús Ariel Carrasco-Ochoa[1], and Airel Pérez-Suárez[2]

[1] Computer Science Department, Instituto Nacional de Astrofísica,
Óptica y Electrónica, Luis Enrique Erro No. 1, Sta. María Tonantzintla,
72840 Puebla, Mexico
{argenis, fmartine, ariel}@inaoep.mx
[2] Advanced Technologies Application Center, 7ma A #21406 c/214 y 216,
Rpto. Siboney, Playa, 12200 Havana, Cuba
asuarez@cenatav.co.cu

Abstract. DenStream is a data stream clustering algorithm which has been widely studied due to its ability to find clusters with arbitrary shapes and dealing with noisy objects. In this paper, we propose a different approach for pruning micro-clusters in DenStream. Our proposal unlike other previously reported pruning, introduces a different way for computing the micro-cluster radii and provides new options for the pruning stage of DenStream. From our experiments over public standard datasets we conclude that our approach improves the results obtained by DenStream.

Keywords: Clustering · Data streams · Data mining

1 Introduction

Nowadays many applications produce large volumes of information in short time intervals (data streams). A data stream is a potentially infinite sequence of objects $x_1, x_2, \ldots, x_k, \ldots$ with timestamps $T_1, T_2, \ldots, T_k, \ldots$, where each object x_i from the data stream is a vector of features containing m dimensions, denoted as $x_i = \left(x_i^1, x_i^2, \ldots, x_i^m\right)$. Processing data streams has become a current research topic in many fields of computer science, with different applications such as: intrusion detection in networks [1], observation of environments [2], medical systems [3, 4], stock exchange analysis [5], social network analysis [6], object tracking in video [7], among others.

A technique widely used for processing data streams is clustering. Due to the nature of a data stream, there are some constraints that a clustering algorithm has to consider in order to process a data stream. These constraints are: (1) the amount of data in the stream must be considered as infinite, (2) the time for processing a single object is limited, (3) memory is limited, (4) there are objects that may be noisy and (5) the data distribution may evolve. In addition, there are other problems, inherent to the clustering problem, that must be addressed: the number of clusters in the dataset is unknown, clusters can take arbitrary shapes and noise may affect the clustering process, among others.

© Springer International Publishing Switzerland 2015
J.A. Carrasco-Ochoa et al. (Eds.): MCPR 2015, LNCS 9116, pp. 33–43, 2015.
DOI: 10.1007/978-3-319-19264-2_4

STREAM [8], CluStream [9], Clus-Tree [10] and DenStream [1] are among the most relevant and cited clustering algorithms reported in the literature for processing data streams. Due to its ability to find clusters with arbitrary shapes and dealing with noisy objects, DenStream has been widely studied and some variants [2, 11–15] have been proposed. Despite the results achieved by these DenStream variants, they have some problems mainly in the pruning phase, because sometimes this phase takes long time to remove very old information and this affects the results for data streams which evolve rapidly and steadily. In addition, sometimes the pruning phase also removes information that could be useful in future cases, since there could be periods of inactivity in some kind of data.

Based on the above mentioned problems and considering the advantages of Den-Stream over other clustering algorithms, the aim of this paper is to propose solutions for some of the problems of DenStream, in order to improve the quality of its results. This paper is organized as follows: Sect. 2 describes the DenStream algorithm. Section 3 describes our proposal. Section 4 reports an experimental comparison of our proposal against DenStream. Finally, in Sect. 5 some conclusions and future research directions are presented.

2 DenStream

DenStream is a density based algorithm for clustering data streams [1], which considers a fading (Damped) window model. In this model, the weight of each object decreases exponentially through a fading function in dependence of the elapsed time t since it appears in the data stream. The exponential fading function is widely used in temporal applications where it is desirable to gradually discount the history of past behavior. The function used in DenStream is $f(t) = 2^{-\lambda t}, \lambda > 0$. The higher value for λ the lower importance of the data compared to more recent data.

DenStream uses three important concepts as the basis for its operation: core-micro-cluster (c-micro-cluster), potential core-micro-cluster (p-micro-cluster) and outlier micro-cluster (o-micro-cluster).

A c-micro-cluster at the time t for a set of nearby objects $x_{i1}, x_{i2}, \ldots, x_{in}$ with timestamps $T_{i1}, T_{i2}, \ldots, T_{in}$, is defined as a triplet $CMC(w, c, r)$, where

$w = \sum_{j=1}^{n} f(t - T_{ij})$, is its weight

$c = \frac{1}{w}\sum_{j=1}^{n} f(t - T_{ij})x_{ij}$, is the center

$r = \frac{1}{w}\sum_{j=1}^{n} f(t - T_{ij})d(x_{ij}, c)$, is the radius

where $d(x_{ij}, c)$ is the Euclidean distance between the object x_{ij} and the center c. The weight of a micro-cluster determines the minimum number of objects that a micro-cluster should have, assuming that the relevance of these objects is modified according to the fading function. With this aim, DenStream requires that the weight of each micro-cluster be less or equal than a predefined threshold μ. The radius of a micro-cluster determines which objects will be added to the micro-cluster, depending on their distance to the center of the micro-cluster. For accomplishing this, DenStream requires that the radius of a micro-cluster be less or equal than a predefined threshold ϵ. Thus, if the addition of an object to a micro-cluster makes its radius greater than ϵ, the object is not included in this micro-cluster. Due to this constraint in the radius, the number of c-micro-clusters tends to be much larger than the real number of clusters. On the other hand, the minimum weight constraint produces a significantly smaller number of c-micro-clusters than the number of objects in the data stream (see Fig. 1).

During the evolution of a data stream, the clusters and the noise are in constant change thus the c-micro-clusters are gradually built as the data stream is processed. Because of this, following a similar idea as the one proposed in [16], in DenStream potential c-micro-clusters and outlier micro-clusters for incremental computation, are introduced.

A p-micro-cluster at time t. for a set of nearby objects $x_{i1}, x_{i2}, \ldots, x_{in}$ with time-stamps $T_{i1}, T_{i2}, \ldots, T_{in}$, is defined as a triplet $\left\{ \overline{CF^1}, \overline{CF^2}, w \right\}$, where

$$w = \sum_{j=1}^{n} f(t - T_{ij}), w \geq \beta\mu,$$ is the weight of the p-micro-cluster; being $\beta \in (0, 1]$ a parameter to determine the noise threshold relative to the c-micro-clusters.

$$\overline{CF^1} = \sum_{j=1}^{n} f(t - T_{ij}) x_{ij},$$ is the weighted linear sum of the objects in the p-micro-cluster.

$$\overline{CF^2} = \sum_{j=1}^{n} f(t - T_{ij}) S(x_{ij}),$$ is the weighted squared sum of the objects in the p-micro-cluster; where $S(x_{ij}) = \left(x_{ij1}^2, x_{ij2}^2, \ldots, x_{ijm}^2 \right)$.

From $\overline{CF^1}$ and $\overline{CF^2}$, we can compute the center and radius of a p-micro-cluster, denoted as c and r, respectively, through the following expressions

$$c = \frac{\overline{CF^1}}{w} \tag{1}$$

$$r = \sqrt{\frac{\left| \overline{CF^2} \right|}{w} \left(\frac{\left| \overline{CF^1} \right|}{w} \right)^2} \tag{2}$$

An o-micro-cluster at the time t. for a set of nearby objects $x_{i1}, x_{i2}, \ldots, x_{in}$ with timestamps $T_{i1}, T_{i2}, \ldots, T_{in}$ is defined as $\left\{ \overline{CF^1}, \overline{CF^2}, w, T_o \right\}$, here $\overline{CF^1}, \overline{CF^2}$ are defined as in a p-micro-cluster. $T_o = T_{i1}$ is the creation time of the o-micro-cluster, which is used to determine its lifetime. However, in this case $w < \beta\mu$ since this parameter determines when an o-micro-cluster switches to a p-micro-cluster.

Both p-micro-clusters and o-micro-clusters can be maintained incrementally. For example, given a p-micro-cluster $c_p = \left(\overline{CF^1}, \overline{CF^2}, w \right)$. If no new objects have been added to c_p in a time interval δt, then the new representation for this micro-cluster will be $c_p = \left(2^{-\lambda\delta t} \cdot \overline{CF^1}, 2^{-\lambda\delta t} \cdot \overline{CF^2}, 2^{-\lambda\delta t} \cdot w \right)$. In the opposite case, if an object p is

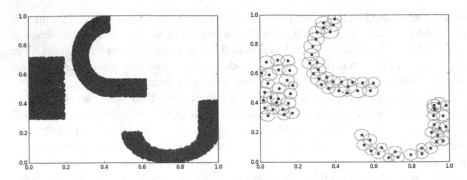

Fig. 1. Representation of a dataset using micro-clusters

added to c_p, then the micro-cluster will be represented as $c_p = \left(\overline{CF}^1 + p, \overline{CF}^2 + S(p), w + 1 \right)$ where $S(p) = (p_1^2, p_2^2, \ldots, p_m^2)$. In a similar way the o-micro-clusters are updated.

The clustering algorithm DenStream is divided in two phases: an online phase for maintaining micro-clusters, and an offline phase to generate the final clustering; the offline phase is performed when the user requests it. DenStream has an initialization stage, where k objects are taken from the data stream and DBSCAN [17] is applied to detect k p-micro-clusters.

During the online phase, when an object x_i is retrieved from the data stream, the distance between x_i and each one of the centers of the p-micro-clusters is computed, the new radius r_j of the nearest p-micro-cluster is computed, if $r_j \leq \epsilon$ then x_i is added to that p-micro-cluster. Otherwise, the same process is done but with the o-micro-clusters, first we compute the distances and then the new radius r_j of the nearest o-micro-cluster is computed, if $r_j \leq \epsilon$ then x_i is added to the nearest o-micro-cluster, in this case it should be also determined if the nearest o-micro-cluster fulfills the constraints to become a p-micro-cluster. If x_i cannot be addedo any micro-cluster a new o-micro-cluster is generated from x_i.

In order to generate the final clustering, a variant of DBSCAN is applied. In this variant the radii of the p-micro-clusters and the distance between their centers allows determining if they are density reachable. Consider two p-micro-clusters with centers c_p and c_q and radii r_p and r_q. We say that these p-micro-clusters are density reachable if $d(c_p, c_q) \leq r_p + r_q$, where $d(c_p, c_q)$ is the Euclidean distance between c_p and c_q. Those p-micro-clusters that are density reachable are merged as a cluster.

Although the expression (2), for computing the radius of a micro-cluster, is reported in all variants of DenStream, in the practice the root argument can be negative, therefore this expression cannot be used. For this reason some of the algorithms that are extensions of DenStream [14, 15] are implemented within a Framework called MOA (Massive Online Analysis),[1] and they use a variant of the expression (2) to compute the radius in the micro-clusters. The expression used in MOA is:

---

[1] http://moa.cms.waikato.ac.nz/.

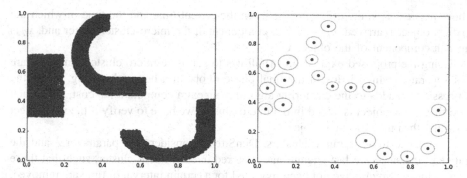

Fig. 2. p-micro-clusters obtained using the average of distances as radius.

$$r = \max_{1 \leq l \leq m} \left\{ \sqrt{\frac{CF^2_l}{w} - \left(\frac{CF^1_l}{w}\right)^2} \right\} \qquad (3)$$

In the expression (2) an approximation to the average of distances is computed, the expression (3) is similar to expression (2) but it only takes the maximum difference between the vector components instead of computing the Euclidean norm. Using the average of distances as the radius makes the radius smaller than it actually should be, which has a negative effect in the final clusters. This problem is illustrated in Fig. 2.

In the example of Fig. 2, the micro-clusters do not intersect their radii and therefore, when the DBSCAN variant is used to build the final clustering, those micro-clusters are not density reachable, and it results in many final clusters.

In MOA a constant is used to increase the size of micro-clusters radii. First the radius is computed then the product between the radius and that constant is computed. However, MOA does not explain how the value of this constant is obtained.

3   Our Proposal

In this section, we introduce some modifications of DenStream, which are depicted in Algorithm 1. In order to solve the problem of using the average of distances as radius in the micro-clusters, we propose to use as radius the distance from the center to the farthest object in the micro-cluster. In [1] this idea was previously proposed but the authors only report results over just one dataset. By using the distance from the farthest objects to the center, CF^2 can be replaced by the farthest object, the farthest object will be called x_f, also the timestamp T_f must be stored, since over time the relevance of the object also affects the radius. Considering this modification, we propose the following expression to compute the radius:

$$r = \sqrt{\left(\sum_{i=1}^{m} (c_i - x_{fi})^2\right) \times 2^{-\lambda\delta_t}} \qquad (4)$$

where $\delta t = t - T_f$ is the difference between the current time and the time in which the farthest object x_f arrived; c_i is the i-th component of the micro-cluster center and, x_{fi} is the i-th component of the object x_f.

Using the proposed expression (4) allows to get more micro-clusters and they are less separated, which helps in the final phase to obtain a better clustering. Using the expression (4) adds to the clustering algorithm an extra computational cost, since each time that a new object is added into a micro-cluster, we have to verify if the new object is farther than the current x_f object.

In order to remove micro-clusters, DenStream considers the parameter λ and the elapsed time since the last change has occurred in each micro-cluster, such that those micro-clusters which have not been modified for a certain interval of time are removed.

Algorithm 1. Improved DenStream$(DS, \epsilon, \beta, \mu, \lambda)$

---

01. $T_p = [\frac{1}{\lambda}\log(\frac{\beta\mu}{\beta\mu-1})$; (value used for the pruning phase)
02. Get the object x at current time t from data stream DS;
03. Try to Merge x into its nearest p-micro-cluster c_p;
04. if x is not merged into c_p then
05.   Try to Merge x into its nearest o-micro-cluster c_o;
06.   if x is not merged into c_o then
07.   Create a new o-micro-cluster for x
08.   else
09.   if w (the new weight of c_o) > $\beta\mu$ then
10.   Convert c_o in a new p-micro-cluster;
11. if$(t \bmod T_p) = 0$ then (pruning phase)
12. $\bar{x}_o = \frac{1}{n}\sum_{i=1}^{n} counter_i$;
13. $\bar{x}_w = \frac{1}{n}\sum_{i=1}^{n} w_i$;
14.   for each o-micro-cluster c_o do
15. $\xi = \frac{2^{-\lambda(t-to+T_p)}-1}{2^{-\lambda T_p}-1}$;
16.   if $w_o(the\,weight\,of\,c_o) < \xi$ then
17.   Try to merge c_o into its nearest p-micro-cluster c_p;
18.   if c_o is not merged into c_p then remove c_o
19.   for each p-micro-cluster cp do
20.   if $w_p(the\,weight\,of\,c_p) < \bar{x}_w$ and $counter_p < \bar{x}_o$ then
21.   Convert c_p in an o-micro-cluster;
22. $counter_p = 0$;
23.   for each p-micro-cluster cp do
24.   if $w_p(the\,weight\,of\,c_p) < \beta\mu$ then
25.   Convert c_p in an o-micro-cluster;
26. if a clustering request arrives then
27.   Generate clusters;

Another problem with DenStream is that sometimes it takes too long time to remove those p-micro-clusters which do not receive objects frequently (depending on the value of λ). The above problem is solved by using a high value for λ but another problem arises, the radii of the micro-clusters decrease too fast and therefore, when a final clustering request is done, many clusters are generated. To solve this problem, we propose to consider those micro-clusters that have received too few objects in the latest updates as noise, i.e., they become o-micro-clusters. To determine when a p-micro-cluster have received too few objects, the average number of added objects in all p-micro-clusters is computed (step 12), also the average weight from all p-micro-clusters is computed (step 13); those p-micro-clusters whose weights and number of added objects are below of both averages become o-micro-clusters (steps 19–22).

Another change included in our proposal is that before removing an o-micro-cluster, we try to merge it with its nearest p-micro-cluster (steps 16–18); in this way we may retain certain information that could be useful in the future. In order to merge two micro-clusters the distance between them must be less than the radius of the p-micro-cluster, and the radius of the new micro-cluster must not be greater than ϵ. Finally, the last change in our proposal is that before removing p-micro-clusters they become o-micro-clusters (steps 23–25), it gives to them the opportunity to grow back, since as they were considered as relevants in previous windows, maybe they are in an inactivity period and they could receive new objects in the close future.

4   Experimental Analysis

For our experimental comparison, DenStream was in the same way as in the framework MOA. The assessment of the clustering results was done through *purity* measure which is defined in (5).

$$purity = \frac{1}{k} \sum_{i=1}^{k} \frac{|c_i^d|}{|c_i|}$$

(5)

Where k is the number of clusters, $|c_i^d|$ denotes the number of objects with the most frequent label in the cluster i, and $|c_i|$ is the total number of objects in the cluster i. This evaluation is done on a predefined window, since objects are fading continuously.

Table 1 shows the characteristics of the datasets used in the experiments, KDD 99 Network Intrusion was used in [1] for evaluating DenStream, while Electricity, Forest Covertype and Poker-Hand datasets were suggested in MOA[2] to evaluate data stream clustering algorithms. In order to simulate a data stream in our experiments the objects are processed in the order that they are stored.

In our experiments all feature values were normalized in the range [0, 1] and in those datasets containing categorical features, we only use the numeric features. Different values were tested for the parameters of the algorithms and we used those in which both algorithms obtained the best results, which are: for Electricity $\epsilon = 0.12, \mu = 40$, for

---

[2] http://moa.cms.waikato.ac.nz/datasets/.

Table 1. Description of datasets used in our experiments

Name	#Objects	#Features	#Labels
Electricity	45,312	103	2
Network Intrusion (KDD 99)	494,020	41	23
Forest Covertype	581,012	54	7
Poker-Hand	829,201	11	10

KDD 99 $\epsilon = 0.15, \mu = 80$, for Forest Covertype $\epsilon = 0.16, \mu = 80$, for Poker-Hand $\epsilon = 0.17, \mu = 100$, in all datasets we use $\beta = 0.2$ *and* $\lambda = 0.25$. The window size was 10,000, and 1000 objects were used for the initialization phase. The results are shown in Fig. 3.

In the Fig. 3, we can see that our proposal performs better in most of the cases, except for Electricity, where DenStream performed better in 2 of the evaluated

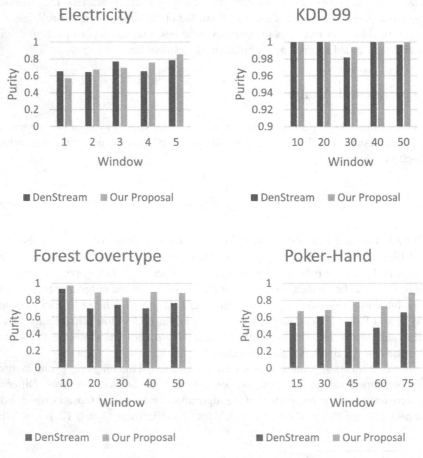

Fig. 3. Comparison of quality using purity

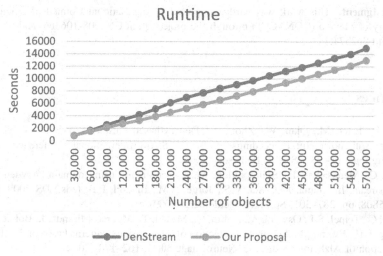

Fig. 4. Comparison of runtime in seconds

windows. For KDD 99 similar results were obtained with our proposal and DenStream, both algorithms perform well in all the windows. In Forest Covertype and Poker-Hand, our proposal preforms better in all the windows. From these experiments, we can conclude that the expression proposed for computing the radius and the changes proposed in the pruning phase of DenStream allows to improve the quality of the results provided by DenStream.

As it was previously mentioned in Sect. 3, getting the farthest object each time that a new object is added to a micro-cluster increases the computational cost of the clustering algorithm, however, in the practice there is not much difference in runtime, as it can be seen in Fig. 4, where we show a runtime comparison between our proposal and DenStream using the dataset Forest Covertype. It is important to comment that in our proposal since in the pruning phase we provide more options to change a p-micro-cluster into an o-micro-cluster, then more micro-clusters are removed and this reduces the amount of operations when we search the nearest p-micro-cluster.

5 Conclusions

In this paper a different approach to compute the radius of the micro-clusters in DenStream is proposed. Additionally, some pruning strategies are also included to remove useless micro-clusters. From our experiments we can conclude that the modifications proposed allows to improve the quality of the clustering results provided by DenStream, besides a similar runtime to that achieved by DenStream is maintained.

Since only the radius and center of micro-clusters are used to build the final clusters, and information like the weight and the lifetime of the micro-clusters are not used to build the final clustering, as future work, we propose to formulate a new strategy taking into account this additional information to build a better clustering in the final phase.

Acknowledgment. This work was partly supported by the National Council of Science and Technology of Mexico (CONACyT) through the project grant CB2008-106366; and the scholarship grant 362371.

References

1. Cao, F., Ester, M., Qian, W., Zhou, A.: Density-based clustering over an evolving data stream with noise. In: Proceedings of the 6th SIAM International Conference on Data Mining, pp. 326–337. SIAM, Bethesda (2006)
2. Ruiz, C., Menasalvas, E., Spiliopoulou, M.: C-DenStream: using domain knowledge on a data stream. In: Gama, J., Costa, V.S., Jorge, A.M., Brazdil, P.B. (eds.) DS 2009. LNCS, vol. 5808, pp. 287–301. Springer, Heidelberg (2009)
3. Plant, C., Teipel, S.J., Oswald, A., Böhm, C., Meindl, T., Mourao-Miranda, J., Bokde, A.W., Hampel, H., Ewers, M.: Automated detection of brain atrophy patterns based on MRI for the prediction of Alzheimer's disease. NeuroImage 50(1), 162–174 (2010)
4. Mete, M., Kockara, S., Aydin, K.: Fast density-based lesion detection in dermoscopy images. Comput. Med. Imaging Graph. 35(2), 128–136 (2011)
5. Yang, D., Rundensteiner, E.A., Ward, M.O.: Summarization and matching of density-based clusters in streaming environments. Proc. VLDB Endow 5(2), 121–132 (2011)
6. Lee, C.H.: Mining spatio-temporal information on microblogging streams using a density-based online clustering method. Expert Syst. Appl. 39(10), 9623–9641 (2012)
7. Yu, Y., Wang, Q., Wang, X., Wang, H., He, J.: Online clustering for trajectory data stream of moving objects. Comput. Sci. Inf. Syst. 10(3), 1293–1317 (2013)
8. Guha, S., Meyerson, A., Mishra, N., et al.: Clustering data streams: theory and practice. IEEE Trans. Knowl. Data Eng. 15(3), 515–528 (2003)
9. Aggarwal, C.C., Han, J., Wang, J., Yu, P.S.: A framework for clustering evolving data streams. In: Proceedings of VLDB (2003)
10. Kranen, P., Assent, I., Baldauf, C., Seidl, T.: The clustree: indexing micro-clusters for anytime stream mining. Knowl. Inf. Syst. 29(2), 249–272 (2011)
11. Lin, J., Lin, H.: A density-based clustering over evolving heterogeneous data stream. In: Proceeding of the 2nd International Colloquium on Computing, Communication, Control, and Management, pp. 275–277 (2009)
12. Ntoutsi, I., Zimek, A., Palpanas, T. et al.: Density-based projected clustering over high dimensional data streams. In: Proceedings of the 12th SIAM International Conference on Data Mining, pp. 987–998 (2012)
13. Forestiero, A., Pizzuti, C., Spezzano, G.: A single pass algorithm for clustering evolving data streams based on swarm intelligence. Data Min. Knowl. Disc. 26(1), 1–26 (2013)
14. Hassani, M., Spaus, P., Gaber, M.M., Seidl, T.: Density-based projected clustering of data streams. In: Hüllermeier, E., Link, S., Fober, T., Seeger, B. (eds.) SUM 2012. LNCS, vol. 7520, pp. 311–324. Springer, Heidelberg (2012)
15. Amini, A., Saboohi, H., Wah, T.Y., Herawan, T.: DMM-Stream: a density mini-micro clustering algorithm for evolving data streams. In: Proceedings of the First International Conference on Advanced Data and Information Engineering (DaEng-2013), pp. 675–682. Springer (2014)

16. Zhang, T., Ramakrishnan, R., Livny, M.: BIRCH: an efficient data clustering method for very large databases. In: Proceedings of the 1996 ACM SIGMOD International Conference on Management of Data, pp. 103–114 (1996)
17. Ester, M., Kriegel, H., Sander, J., Xu, X.: A density-based algorithm for discovering clusters in large spatial databases with noise. In: Proceeding of the 2nd International Conference on Knowledge Discovery and Data Mining, pp. 226–231 (1996)

Computing Constructs by Using Typical Testor Algorithms

Manuel S. Lazo-Cortés[1]([⊠]), Jesús Ariel Carrasco-Ochoa[1],
José Fco. Martínez-Trinidad[1], and Guillermo Sanchez-Diaz[2]

[1] Instituto Nacional de Astrofísica, Óptica Y Electrónica, Puebla, Mexico
{mlazo,ariel,fmartine}@inaoep.mx
[2] Universidad Autonóma de San Luis Potosí, San Luis Potosí, Mexico
guillermo.Sanchez@uaslp.mx

Abstract. In their classic form, reducts as well as typical testors are minimal subsets of attributes that retain the discernibility condition. Constructs are a special type of reducts and represent a kind of generalization of the reduct concept. A construct reliably provides sufficient amount of discrimination between objects belonging to different classes as well as sufficient amount of resemblance between objects belonging to the same class. Based on the relation between constructs, reducts and typical testors this paper focuses on a practical use of this relation. Specifically, we propose a method that allows applying typical testor algorithms for computing constructs. The proposed method involves modifying the classic definition of pairwise object comparison matrix adapting it to the requirements of certain algorithms originally designed to compute typical testors. The usefulness of our method is shown through some examples.

1 Introduction

All data analyses in the Rough Sets Theory [14] start with the so-called closed world assumption. According to this assumption any two objects described by two identical vectors of parameter values must be treated equal in all the subsequent analyses. Formally, the main tool that ensures this property in data analysis is the relation of indiscernibility between objects.

The partition of objects into classes is very interesting for the data analysts, and represents a key aspect in classification problems, since the classes generally represent concepts. The Rough Set Theory makes an effort to examine whether a set of descriptive attributes is sufficient to classify objects into the same classes as the original partition. In this effort, reducts play an important role.

Meanwhile, the theory of pattern recognition study the same problems and uses its own tools. Particularly in the logical combinatorial approach [16], the concept of testor [10] makes an important contribution to the problem of feature selection, reduction of the space of representation and other related problems.

Both concepts, reducts and testors have been widely studied separately. However, in the literature the study of these concepts including their points of convergence and their differences have been also studied [9].

© Springer International Publishing Switzerland 2015
J.A. Carrasco-Ochoa et al. (Eds.): MCPR 2015, LNCS 9116, pp. 44–53, 2015.
DOI: 10.1007/978-3-319-19264-2_5

The property to discern objects belonging to different classes is a common point between both concepts while the information provided by the pairs of objects belonging to the same class is sidelined. Just this point is the main contribution of the concept of construct [24]. Constructs take into account more information contained in the object pairwise comparisons, because they utilizes inter-class and intra-class information together, considering discriminating relations between objects belonging to different classes and resembling relations between objects belonging to the same class.

From the computational point of view, the most challenging problem related to testors, reducts and constructs is that of generating full sets of typical testors, reducts and constructs. This problem has been proven to be NP-hard (it is equivalent to the problem of calculate the set of all prime implicants of a disjunctive normal form) [23].

This paper addresses the problem of computing the whole set of constructs. We specifically study the relation between typical testors and constructs, and show how the algorithms for computing typical testors, designed in the logical-combinatorial pattern recognition approach, can be used to compute constructs.

Even though we do not propose a new algorithm, this research expands the set of available algorithms to calculate constructs and re-valorizes the existing typical testor algorithms, bringing new elements to the study of the relationship between the rough set theory and the testor theory.

The rest of the document is organized as follows. Section 2 provides the formal background for the study, including the definitions of reduct, testor and construct. An example is discussed. Section 3 presents the proposed method that allows using typical testor algorithms for computing constructs, illustrative examples are included. Our conclusions are summarized in Sect. 4.

2   Theoretical Foundations

2.1   Reducts

The main dataset considered in this paper is a decision table, which is a special case of an information table [8]. Formally, a decision table is defined as

Definition 1. *(decision table) A decision table is a pair $\mathcal{S}_d = (U, A_t = A_t^* \cup \{d\})$ where U is a finite non-empty set of objects, A_t is a finite non-empty set of attributes. A_t^* is a set of conditional attributes and d is a decision attribute indicating the decision class for each object in the universe. Each $a \in A_t$ corresponds to the function $I_a : U \rightarrow V_a$ called evaluation function, where V_a is called the value set of a. The decision attribute allows partitioning the universe into blocks (classes) determined by all possible decisions.*

Sometimes we will use D for denoting $\{d\}$, i.e. $(\{d\} = D)$.

A *decision table* can be implemented as a two-dimensional array (matrix), in which one usually associates rows to objects, columns to attributes and cells to values of attributes on objects.

When considering decision tables, it is important to distinguish between the so called *consistent* and the *inconsistent* ones. A decision table is said to be *consistent*, if each combination of values of descriptive attributes uniquely determines the value of the decision attribute, and *inconsistent*, otherwise. For the purpose of this paper we only consider consistent decision tables.

It is important to introduce the definition of the indiscernibility relation.

Definition 2. *(indiscernibility relation) Given a subset of conditional attributes $A \subseteq A_t^*$, the indiscernibility relation is defined as $IND(A|D) = \{(u,v) \in U \times U : \forall a \in A, [I_a(u) = I_a(v)] \vee [I_d(u) = I_d(v)]\}$*

The indiscernibility relation is an equivalence relation, so it induces a partition over the universe. Being $\mathcal{S}_d$ a consistent decision table, the partition induced by any subset of conditional attributes is finer than (or at maximum equal to) the relation determined by all possible values of the decision attribute d.

We can find several definitions of reduct (see for example, [13]), nevertheless, according to the aim of this paper, we refer to reducts assuming the classical definition of discerning decision reduct [15] as follows.

Definition 3. *(reduct for a decision table) Given a decision table S_d, an attribute set $R \subseteq A_t^*$ is called a reduct, if R satisfies the following two conditions:*

(i) $IND(R|D) = IND(A_t^*|D)$;
(ii) For any $a \in R, IND((R - \{a\})|D) \neq IND(A_t^*|D)$.

This definition ensures that a reduct has no lower ability to distinguish objects belonging to different classes than the whole set of attributes, being minimal with regard to inclusion, i.e. a reduct does not contain redundant attributes or, equivalently, a reduct does not include other reducts. The original idea of reduct is based on inter-class comparisons.

2.2 Testors

The concept of testor (initially *test*) was created by S. V. Yablonskii as a tool for analysis of problems connected with control and diagnosis of faults in circuits [26]. In publications related to this area the original term *test* is used instead of *testor* and the minimal ones (typical testors) are called *dead-end tests*.

The concept of testor (and typical testor) has had numerous generalizations and adaptations to different environments [10]. In this paper, we focus on the classical concept defined into the pattern recognition area, derived from [5], but using a notation similar to that used for reducts to make this presentation more coherent and understandable.

Definition 4. *(testor for a decision table) Let $\mathcal{S}_d = (U, A_t = A_t^* \cup \{d\})$ a decision table and let $T \subseteq A_t^*$. $I_T(u)$ denotes the partial description of u considering only attributes belonging to T. $T \subseteq A_t^*$ is a testor with respect to a decision table $\mathcal{S}_d$ if $\forall u, v \in \mathcal{U} : [I_T(u) = I_T(v)] \Rightarrow [I_d(u) = I_d(v)]$. If T is a testor such that none of its subsets is a testor, then T is a typical (irreducible) testor.*

This definition means that attributes belonging to a testor are jointly sufficient to discern between any pair of objects belonging to different classes; if a testor is typical, each attribute is individually necessary. That is exactly the same as a reduct. This means that in their classical formulations the concepts of reduct and typical testor coincide.

A more detailed study about the relation between reducts and typical testors can be found in [9].

2.3 Constructs

Both, reducts and testors are defined from an inter-class object comparison point of view. They ensure sufficient discernibility of objects belonging to different classes. The novelty of the concept of construct (introduced by R. Susmaga in 2003 [24]) is the combination of inter-class and intra-class comparisons in such a way that a resulting subset of conditional attributes would ensure not only the ability to distinguish objects belonging to different classes, but also preserves certain similarity between objects belonging to the same class.

Let us now consider the following similarity relation defined between objects belonging to the same class in a decision table $S_d = (\mathcal{U}, A_t^* \cup \{d\})$.

Definition 5. *(similarity relation) Given a subset of conditional attributes $A \subseteq A_t^*$, the similarity relation is defined as $SIM(A|D) - \{(u,v) \in U \times U : [I_d(u) = I_d(v)]$ and $\exists a \in A\ [I_a(u) = I_a(v)]\}$.*

If a pair of objects belongs to $SIM(A|D)$ then these objects belong to the same class and they are indiscernible on at least one attribute from the set A. This relation is reflexive and symmetric, but it is not transitive.

The definition of construct may be stated as follows.

Definition 6. *(cosntruct) Given a decision table S_d, an attribute set $C \subseteq A_t^*$ is called a construct, if C satisfies the following conditions:*

(i) $IND(C|D) = IND(A_t^*|D)$;
(ii) $SIM(C|D) = SIM(A_t^*|D)$;
(iii) *For any $a \in C$, $IND((C-\{a\})|D) \neq IND(A_t^*|D)$ and $SIM((C-\{a\})|D) \neq SIM(A_t^*|D)$;*

Condition (i) means that a construct retains the discernibility of objects belonging to different classes. In addition, a construct ensures similarity between objects belonging to the same class, at least at level that the whole set of condition attributes does. Condition (iii) ensures the construct's minimality regarding to inclusion relation.

Example 1. The matrix M represents a decision table, for which $U = \{u_1, u_2, u_3, u_4, u_5, u_6, u_7, u_8\}$, $A_t^* = \{a_1, a_2, a_3, a_4\}$ and $D = \{d\}$.

$$M = \begin{array}{c} \\ \\ u_1 \\ u_2 \\ u_3 \\ u_4 \\ u_5 \\ u_6 \\ u_7 \\ u_8 \end{array} \begin{pmatrix} a_1 & a_2 & a_3 & a_4 & d \\ a & 1 & 25 & red & 1 \\ a & 0 & 19 & yellow & 1 \\ b & 1 & 25 & green & 1 \\ b & 0 & 19 & red & 1 \\ a & 0 & 25 & blue & 2 \\ d & 0 & 29 & pink & 2 \\ c & 0 & 7 & blue & 3 \\ c & 1 & 5 & yellow & 3 \end{pmatrix} \qquad (1)$$

From (Eq. 1) we have that $\{a_1, a_2\}$ is not a reduct, see for example that $I_{a_1}(u_2) = I_{a_1}(u_5) = a$ and $I_{a_2}(u_2) = I_{a_2}(u_5) = 0$ being $I_d(u_2) = 1$ and $I_d(u_5) = 2$. For this table $\{a_2, a_3\}$, $\{a_1, a_4\}$ and $\{a_3, a_4\}$ are the reducts (also typical testors). Moreover, $\{a_2, a_3\}$, for example, is not a construct, see for example that $1 = I_{a_2}(u_3) \neq I_{a_2}(u_4) = 0$ and $25 = I_{a_3}(u_3) \neq I_{a_3}(u_4) = 19$ being $I_d(u_3) = I_d(u_4) = 1$ and $I_{a_1}(u_3) = I_{a_1}(u_4)$. For M the only construct is $\{a_1, a_2, a_4\}$.

3   Proposed Method

In [9] the authors study the relation between reducts and typical testors. Among the practical applications that results as consequence of this relation, they mention that algorithms for computing reducts can be used for computing typical testors, and vice versa.

In this section, we propose a method that allows using testor algorithms for computing constructs, particularly algorithms that use the binary discernibility matrix [6].

Originally, the binary discernibility matrix was defined as *the distinction table* [25] for an information table by comparing all objects regrading all attributes. If we focus on computing reducts, binary discernibility matrices only need comparisons between objects belonging to different classes. In these matrices, it is common to denote by 0 if values corresponding to the same attribute are equal and 1 otherwise.

As it is known, discernibility matrices contain redundant information, and the binary ones are not an exception, so testor algorithms usually work on basic binary discernibility matrices instead of the original discernibility ones (for more details see [12]).

Using this basic binary discernibility matrix, some testor algorithms are easily adapted for computing reducts [2, 11, 12, 19–21]. The novelty of the research reported in this paper is to provide a method that allows algorithms originally designed for computing typical testors to be used to compute constructs too. For that, we need to pre-calculate a new type of binary comparison matrix as follows.

Let $\mathcal{S}_d = (\mathcal{U}, A_t^* \cup \{d\})$ be a decision table, and let us define for each attribute a in A_t^* a dissimilarity function φ_a as follows:

$$\varphi_a : V_a \times V_a \to \{0, 1\}$$

$$\varphi_a(x,y) = \begin{cases} 0 & \text{if } x = y \\ 1 & \text{otherwise} \end{cases} \tag{2}$$

Thus, we can introduce the following definition which constitutes the basis of our proposed method.

Definition 7. *(Binary comparison matrix) Given a decision table $S_d = (U, A_t = A_t^* \cup \{d\})$, with $A_t^* = \{a_1, a_2, \ldots, a_n\}$. The binary comparison matrix $M^{\widehat{01}}$ of S_d is a matrix, in which each row is associated to a pair of objects (u, v) with $u \neq v$ and is defined by $(\zeta_{a_1}(I_{a_1}(u), I_{a_1}(v)), \zeta_{a_2}(I_{a_2}(u), I_{a_2}(v)), \ldots, \zeta_{a_n}(I_{a_n}(u), I_{a_n}(v)))$, being*

$$\zeta_{a_i}(I_{a_i}(u), I_{a_i}(v)) = \begin{cases} \varphi_{a_i}(I_{a_i}(u), I_{a_i}(v)) & \text{if } I_d(u) = I_d(v) \\ \neg\varphi_{a_i}(I_{a_i}(u), I_{a_i}(v)) & \text{otherwise} \end{cases} \tag{3}$$

Definition 7 states that pairs of objects are compared taking into account if both objects belong to different classes or to the same class. In this way, when we are comparing objects, if an attribute distinguishes two objects belonging to different classes, we put a one in the corresponding entry of the matrix, this means that this attribute should be taken into account when we are building constructs.

On the other hand, if an attribute does not distinguish between two objects (belonging to the same class), we put a one, because this attribute contributes to preserve the similarity between these objects, this means that this attribute also should be taken into account when we are building constructs. The complexity of computing the binary comparison matrix is $O(n|U|^2)$

Let us examine the following example.

Example 2. As an example, we can build the binary comparison matrix $M^{\widehat{01}}$ for the decision table (Eq. 1) shown in Example 1. For reasons of space we show $M^{\widehat{01}}$ in form of a table, the last two rows indicates the pair of objects that corresponds to the comparison in this column.

a_1	1	0	0	0	1	1	1	0	0	0	1	1	1	1	1	1	1	1	1	1	1	1	0	1	1	1	1	1
a_2	0	1	0	1	1	1	0	0	1	0	0	0	1	0	1	1	1	0	0	0	0	1	1	0	1	0	1	0
a_3	0	1	0	0	1	1	1	0	1	1	1	1	1	0	0	1	1	1	1	1	1	1	0	1	1	1	1	0
a_4	0	0	1	1	1	1	1	0	0	1	1	1	0	0	1	1	1	1	1	1	1	1	0	0	1	1	1	0
	u_1	u_1	u_1	u_1	u_1	u_1	u_1	u_2	u_2	u_2	u_2	u_2	u_2	u_3	u_3	u_3	u_3	u_3	u_4	u_4	u_4	u_4	u_5	u_5	u_5	u_6	u_6	u_7
	u_2	u_3	u_4	u_5	u_6	u_7	u_8	u_3	u_4	u_5	u_6	u_7	u_8	u_4	u_5	u_6	u_7	u_8	u_5	u_6	u_7	u_8	u_6	u_7	u_8	u_7	u_8	u_8

As usually, the comparison matrix contains redundant information, so we can apply a process of simplification that is equivalent to applying absorption laws for obtaining a matrix that just contains the information needed to compute the whole set of constructs. Obviously, rows containing only zeros are not considered. This kind of matrix is known as basic matrix [12]. For obtaining the basic binary comparison matrix from the binary comparison matrix it is necessary to compare each row of $M^{\widehat{01}}$ against the remaining rows, therefore, the complexity of this transformation is $O(n|U|^4)$.

Example 3. For the matrix $M^{\widehat{01}}$ in the previous example, we have the following basic binary comparison matrix. From this matrix is immediate that the only construct is $\{a_1, a_2, a_4\}$ as we previously said in Example 1.

$$\begin{pmatrix} 1\,0\,0\,0 \\ 0\,0\,0\,1 \\ 0\,1\,0\,0 \end{pmatrix} \tag{4}$$

So, we can enunciate the principal result of this research.

Proposition 1. *Let $\mathcal{S}_d = (U, A_t = A_t^* \cup \{d\})$ be a decision table and $M^{\widehat{01}}$ its binary comparison matrix. Then the set $CONST(\mathcal{S}_d)$ of all constructs of $\mathcal{S}_d$ is the same as the set $TT(M^{\widehat{01}})$ of all typical testors of $M^{\widehat{01}}$.*

Proof. The proposition is a straightforward consequence of the original definitions of typical testor and construct, because both may be interpreted as the prime implicants of the same disjunctive normal form. Notice that the proposition considers $M^{\widehat{01}}$ instead of the basic binary comparison matrix, nevertheless theoretically one can substitutes one matrix for the other one since the associated disjunctive normal forms are the same. Normally, the basic binary comparison matrix is used because it is simpler and usually much more smaller.

Summarizing, given a decision table, the steps that comprise the proposed method for computing constructs by using typical testor algorithms are as follows:

1. Compute the binary comparison matrix (Definition 7).
2. From matrix computed in step 1, compute the corresponding basic matrix.
3. Apply a typical testor algorithm.

3.1   Illustrative Examples

As a manner of illustrative examples, we include in this section the results obtained for three datasets from the UCI Repository of Machine Learning [3].

In Table 1, columns A, B and C contain general information of the datasets (number of conditional attributes, classes and objects respectively); columns D and E show the number of typical testors and constructs computed for each dataset, respectively. In all cases, both typical testors (reducts) and constructs were computed by using the algorithm CT-EXT [19]. Results for reducts were verified by using RSES [4,22].

These results are shown as a small example of how an area can be benefited from other one, in this case the results in the area of testors, specifically algorithms for computing typical testors, can be applied through our proposed method in the area of reducts-constructs.

Table 1. Reducts and constructs for several datasets

Data set (A)	Conditional attributes (A)	Classes (B)	Objects (C)	$|TT|$ (D)	$|CONST|$ (E)
Australian	14	2	690	44	2
German(Statlog)	20	2	1000	846	17
Shuttle	9	7	43500	19	1

However, this is not an exhausted matter, the exploration of algorithms for computing different types of typical testors [1,7,8,17,18] could be a source for further contributions in this direction.

Similarly, further study of the relation between testors and constructs can bring new benefits, not only in the area of attribute reduction but possibly also for classification.

4 Conclusions

The main purpose of the research reported in this paper has been a presentation of a novel method for computing constructs, which involves modifying the classic definition of pairwise object comparison matrix, specifically the binary comparison matrix, adapting it to the requirements of certain algorithms originally designed to compute typical testors.

As we have discussed along the paper, reducts and typical testors are concepts closely related, and in certain environments they coincide. From another point of view constructs constitute a different contribution to the attribute reduction problem.

In this paper, we show that the relation between these different concepts, which has been insufficiently studied to date, can be exploited. Specifically, we illustrate a way in which algorithms that originally were designed for computing typical testors in the framework of pattern recognition, through our proposed method can be used in a straightforward way for computing constructs, by a different implementation of the pairwise comparison matrix.

It is expected that this is not the unique benefit we can obtain of this relation therefore a deeper study is mandatory as future work.

Acknowledgements. This work was partly supported by the National Council of Science and Technology of Mexico (CONACyT) through the project grant CB2008-106366.

References

1. Alba-Cabrera, E., Lopez-Reyes, N., Ruiz-Shulcloper, J.: Generalization of the concept of testor starting from similarity function. Algorithms, Technical report. Yellow serie 134. CINVESTAV-IPN, pp. 7–28, Mexico (1994) (in Spanish)

2. Alba-Cabrera, E., Santana-Hermida, R., Ochoa-Rodriguez, A., Lazo-Cortes, M.: Finding typical testors by using an evolutionary strategy. In: Muge, F., Piedade, M., Caldas-Pinto, R. (eds.) Proceedings of the V Iberoamerican Symposium on Pattern Recognition, SIARP'2000, Universidade Tecnica de Lisboa, pp. 267–278, Lisbon, Portugal (2000)
3. Bache, K., Lichman, M.: UCI Machine Learning Repository. University of California, School of Information and Computer Science, Irvine, CA (2013). http://archive.ics.uci.edu/ml
4. Bazan, J.G., Szczuka, M.S.: The rough set exploration system. In: Peters, J.F., Skowron, A. (eds.) Transactions on Rough Sets III. LNCS, vol. 3400, pp. 37–56. Springer, Heidelberg (2005)
5. Dmitriyev, A.N., Zhuravlev, Y.I., Krendelev, F.P.: On mathematical principles for classification of objects and phenomena. Diskret. Analiz 7, 315 (1966). (in Russian)
6. Felix, R., Ushio, T.: Rough sets-based machine learning using a binary discernibility matrix. In: Proceedings of the Second International Conference on Intelligent Processing and Manufacturing of Materials, IPMM 1999, pp. 299–305. IEEE (1999)
7. Godoy-Calderon, S., Lazo-Cortes, M.S.: Δ-Testors. A generalization of the concept of testor for fuzzy environments. In: Proceedings of II Iberoamerican Workshop on Pattern Recognition. International Conference ClMAF'97. pp. 95–103, Havana, Cuba (1997) (in Spanish)
8. Goldman, R.S.: Problems on fuzzy testor theory. Automatika i Telemejanika 10, 146–153 (1980). (in Russian)
9. Lazo-Cortes, M.S., Martinez-Trinidad, J.F., Carrasco-Ochoa, J.A., Sanchez-Diaz, G.: On the relation between rough set reducts and typical testors. Inf. Sci. 294, 152–163 (2015)
10. Lazo-Cortes, M., Ruiz-Shulcloper, J., Alba-Cabrera, E.: An overview of the evolution of the concept of testor. Pattern Recogn. 34(4), 753–762 (2001)
11. Lias-Rodríguez, A., Pons-Porrata, A.: BR: a new method for computing all typical testors. In: Bayro-Corrochano, E., Eklundh, J.-O. (eds.) CIARP 2009. LNCS, vol. 5856, pp. 433–440. Springer, Heidelberg (2009)
12. Lias-Rodriguez, A., Sanchez-Diaz, G.: An algorithm for computing typical testors based on elimination of gaps and reduction of columns. Int. J. Pattern Recogn. Artif. Intell. 27(8) 1350022, 18 (2013)
13. Miao, D.Q., Zhao, Y., Yao, Y.Y., Li, H.X., Xu, F.F.: Reducts in consistent and inconsistent decision tables of the Pawlak rough set model. Inf. Sci. 179(24), 4140–4150 (2009)
14. Pawlak, Z.: Rough sets. Int. J. Comput. Inf. Sci. 11, 341–356 (1982)
15. Pawlak, Z.: Rough sets, Theoretical Aspects of Reasoning About Data. Kluwer Academic Publishers, Dordrecht (1992)
16. Ruiz-Shulcloper, J., Abidi, M.A.: Logical combinatorial pattern recognition: a review. In: Pandalai, S.G. (ed.) Recent Research Developments in Pattern Recognition, pp. 133–176. Transwold Research Network, Kerala, India (2002)
17. Ruiz-Shulcloper, J., Alba-Cabrera, E., Lopez-Reyes, N., Lazo-Cortes, M., Barreto-Fiu, E.: Topics about Testor theory CINVESTAV Technical report 134, Yellow series (1995) (in Spanish)
18. Ruiz-Shulcloper, J., Lazo-Cortes, M.: Prime k-testors. Ciencias Técnicas, Físicas y Matemáticas 9, 17–55 (1991). (in Spanish)
19. Sanchez-Diaz, G., Lazo-Cortes, M., Piza-Davila, I.: A fast implementation for the typical testor property identification based on an accumulative binary tuple. Int. J. Comput. Intell. Syst. 5(6), 1025–1039 (2012)

20. Diaz-Sanchez, G., Piza-Davila, I., Sanchez-Diaz, G., Mora-Gonzalez, M., Reyes-Cardenas, O., Cardenas-Tristan, A., Aguirre-Salado, C.: Typical testors generation based on an evolutionary algorithm. In: Yin, H., Wang, W., Rayward-Smith, V. (eds.) IDEAL 2011. LNCS, vol. 6936, pp. 58–65. Springer, Heidelberg (2011)
21. Santiesteban-Alganza, Y., Pons-Porrata, A.: LEX : a new algorithm for computing typical testors. Revista Ciencias Matematicas 21(1), 85–95 (2003). (in Spanish)
22. Skowron, A., Bazan, J., Szczuka, M., Wroblewski, J.: Rough Set Exploration System (version 2.2.1). http://logic.mimuw.edu.pl/ rses/
23. Skowron, A., Rauszer, C.: The discernibility matrices and functions in information systems. In: Słowiński, R. (ed.) Intelligent Decision Support, Handbook of Applications and Advances of the Rough Sets Theory, System Theory, Knowledge Engineering and Problem Solving. Kluwer Academic Publishers, vol. 11, pp. 331 362. Dordrecht, The Netherlands (1992)
24. Susmaga, R.: Reducts versus constructs: an experimental evaluation. Electron. Notes Theor. Comput. Sci. 82(4), 239–250 (2003). Elsevier
25. Wróblewski, J.: Genetic algorithms in decomposition and classification problem. In: Skowron, A., Polkowski, L. (eds.) Rough Sets in Knowledge Discovery 1, pp. 471–487. Physica Verlag, Heidelberg (1998)
26. Yablonskii, S.V., Cheguis, I.A.: On tests for electric circuits. Uspekhi Mat. Nauk 10(4), 182–184 (1955). (in Russian)

Improved Learning Rule for LVQ Based on Granular Computing

Israel Cruz-Vega[✉] and Hugo Jair Escalante

Instituto Nacional de Astrofísica, Óptica y Electrónica, 72840 Puebla, Mexico
isrcruz@ccc.inaoep.mx

Abstract. LVQ classifiers are particularly intuitive and simple to understand because they are based on the notion of class representatives (i.e., prototypes). Several approaches for improving the performance of LVQ in batch-learning scenarios are found in the literature. However, all of them assume a fixed number of prototypes in the learning process; we claim that the quantized approximation to the distribution of the input data using a finite number of prototypes, should not be fixed. Thus, in this paper we propose an improved learning algorithm for batch and on-line variants in LVQ. The proposed algorithm is based on a modified LVQ rule and granular computing, a simple and low cost computational process of clustering. All this, increases the dynamics in the learning process, proposing new prototypes which have a better covering of the distribution of classes, rather than using a fixed number of them. Similarly, in order to avoid an exponential growth in the number of prototypes, an automatic pruning step is implemented, respecting the desired reduction rate.

Keywords: LVQ · Granular computing · On-line learning

1 Introduction

Learning Vector Quantization (LVQ) is an effective technique used for supervised learning, mainly for classification purposes. The learning targets are called "prototypes", which can be understood as class representatives and yields class regions defined by hyperplanes between prototypes of the existing classes, this regions are known as Voronoi partitions [4,6].

LVQ induces efficient classifiers, i.e., they are simple and fast, as calculations are made over prototypes and not over all the neighbourhood of instances. This feature provides a great reduction in the computational cost and processing time (at least for homogeneous data sets). Usually, a fixed number of prototypes is considered during the learning process. However this is not necessarily a good way to achieve a better accuracy or faster convergence to desired theoretical values, as the Bayesian border. LVQ2.1, LVQ3 are improvements over the standard LVQ technique (also called LVQ1.0) with higher convergence speed and better approximation capabilities (related to the Bayesian borders) [7]. In general, these improvements over LVQ1.0 correct drawbacks such as: bad initialization, sensitivity to

© Springer International Publishing Switzerland 2015
J.A. Carrasco-Ochoa et al. (Eds.): MCPR 2015, LNCS 9116, pp. 54–63, 2015.
DOI: 10.1007/978-3-319-19264-2_6

overlapping classes distribution and, instabilities due to overly pronounced elimination of dimensions in the feature space.

In this paper, based on the rule of the LVQ1.0 algorithm, we propose an improved learning algorithm for batch and on-line learning processes, where the prototypes used per each class are not fixed. Several prototypes can be produced iteratively, increasing the accuracy power of the classifier and at the same time, respecting the desired data compression rate. Also, a pruning step is proposed, based on a simple idea of a usage-frequency variable for each one of the actual prototypes; then, at each one of the iteration steps, only prototypes with more usage are kept, avoiding the excessive computational load. The proposed algorithm has some kind of memory because the usage-frequency takes into account training samples in previous samples/iterations. This is particularly important when one tries to extend the LVQ method to work on an online setting.

Two variants are proposed, "LVQ based on Granular Computing for Batch-Learning" (LVQ-GC-BL) and, "LVQ based on Granular Computing for on-line-Learning" (LVQ-GC-OL). Both methods try to overcome deficiencies of traditional LVQ-1.0 algorithm, correcting bad initialization states by means of granular computing, a simple algorithm to find centroids; and working alongside the dynamism of incremental learning, letting the algorithm not to be fixed in the number of class representatives during the learning stage. Based on this, LVQ-GC-OL is evaluated in a simulated scenario of on-line learning. It is expected that combining by the dynamic production of new instances plus the pruning step, keeping the winner prototypes per class during the online training, the algorithm will produce an acceptable trade-off between accuracy and the reduction performances, as a typical characteristic of a vector quantization algorithm.

The remainder of this paper is organized as follows: In Sect. 2, a description of the proposed algorithms for batch learning is presented. Section 3 describes the advantages of the on-line version (LVQ-GC-OL). Section 4 presents the experimental framework and reports the results obtained. Finally, Sect. 5 outlines conclusions and future work directions.

2 Improved Rule for Batch LVQ

This section introduces the proposed extension to LVQ for batch mode.

2.1 Learning Vector Quantization

Let $X = \{(\mathbf{x}_i, y_i) \subset \mathbb{R}^D \times \{1, ..., C\} \mid i = 1, ..., N\}$ be a training data set, where $\mathbf{x} = (x_1, ..., x_D) \in \mathbb{R}^D$ are D-dimensional input samples, with cardinality $|\mathbf{X}| = N$; $y_i \in \{1, ..., C\}$ $i = 1, ..., N$ are the sample labels, and C is the number of classes. The learning structure of LVQ consists of a number of prototypes, which are characterized by vectors $\mathbf{w}_i \in \mathbb{R}^D$, for $i = 1, ..., M$, and their class labels $c(\mathbf{w}_i) \in \{1, ..., C\}$ with $Y = \{c(\mathbf{w}_j) \in \{1, ..., C\} \mid j = 1, ..., M\}$. The classification scheme is based on the best matching unit (winner-takes-all strategy), which defines the class representatives (prototypes or codebook vectors), so that the training data samples are mapped to their corresponding labels. Now, according to [8], for LVQ1: with several labeled-prototypes $\mathbf{w}_i$ of each class, the class regions in the $\mathbf{x}$ space are defined by simple nearest-neighbour comparison between $\mathbf{x}$ and the existing $\mathbf{w}_i$; the label of the closest $\mathbf{w}_i$ is the label of $\mathbf{x}$. The learning process consists in defining the optimal placement of $\mathbf{w}_i$ iteratively.

The LVQ learning tries to pull the prototypes away from the decision surfaces to demarcate the class borders more accurately. Iteratively, all of the N training instances are processed according the following rules to update the prototypes (which initially are randomly selected samples):

$$\begin{aligned}
\mathbf{w}_c\,(t+1) &= \mathbf{w}_c\,(t) + \alpha\,(t)\,[\mathbf{x}\,(t) - \mathbf{w}_c\,(t)] && \text{if } \mathbf{x}(t) \text{ is correctly classified,}\\
\mathbf{w}_c\,(t+1) &= \mathbf{w}_c\,(t) - \alpha\,(t)\,[\mathbf{x}\,(t) - \mathbf{w}_c\,(t)] && \text{if the classification of } \mathbf{x}(t) \text{ is incorrect,}\\
\mathbf{w}_i\,(t+1) &= \mathbf{w}_i\,(t) \text{ for } i \neq c && \text{for prototypes of other classes} \qquad (1)
\end{aligned}$$

where $\mathbf{w}_c(t)$ is the closest prototype to $\mathbf{x}(t)$ at iteration t in the Euclidean metric, $\alpha(t)$ is a scalar gain $(0 < \alpha < 1)$, decreasing monotonically in time.

2.2 Granular Computing

In this paper, granules play the role of prototypes. The idea of granular computing, has been developed in previous works [1,2]. In these works, the granular computing generates surrogate models in expensive fitness functions of optimization problems; then, not only the core of information is drastically reduced, but also, a basic knowledge of the structure of the model is obtained [3]. In the following we detail the granule generation process. The algorithm starts by selecting random a training sample as the center of the granule for each one of the classes. Each one of these granule-prototypes are represented by density functions. Next, for each one of the elements of the data set $(\mathbf{x}_i, y_i)$, a measure of "similarity" to the existing prototypes is computed using the following Gaussian similarity measure:

$$\mu_k\,(\mathbf{x}_j) = \exp\left(\frac{-\left(\mathbf{x}_j^i - \mathbf{w}_k^i\right)^2}{\sigma_k^2}\right) \qquad (2)$$

where $\mathbf{w}_k$ is the center of the kth granule or the prototype, for $k = 1, 2, ..., l$ number of granules, and $\mathbf{x}_j$ is the jth input of the training set, that belongs to the ith class. The radius σ_k of each granule is used to control the area of similarity degree between inputs, determining the decision boundary of inclusion into the granules. In this work, this radious σ_k is formulated as the mean value of the Euclidean distance between a granule and all the inputs,

$$\sigma_{k_i} = \frac{1}{n}\sum_{j=1}^{n}\sqrt{(\mathbf{x}_j - \mathbf{w}_k)^2} \qquad (3)$$

where n is the number of inputs of the training set.

Now, let

$$c = \arg\max\{\mu_k\,(\mathbf{x}_j)\} \qquad (4)$$

which defines the nearest $\mathbf{w}_i$ granule to $\mathbf{x}$, denoted by $\mathbf{w}_c$. Initially we have a granule for each one of the classes of the training set.

2.3   Improved Batch-Learning Rule

The proposed algorithm is designed to create new granules or prototypes and update the existing ones according to the measure of similarity mentioned above and a threshold θ. The following equations define our improved LVQ-1.0 process:

$$R_1 : \mathbf{w}_c\left(t+1\right) = \mathbf{w}_c\left(t\right) + \alpha\left[\mathbf{x}_j^i\left(t\right) - \mathbf{w}_c\left(t\right)\right]$$

$$\text{if } \mathbf{x}_j^i \text{ and } \mathbf{w}_c \text{ belongto the same class and } c \geq \theta, \tag{5}$$

$$R_2 : \mathbf{w}_c\left(t+1\right) = \mathbf{w}_c\left(t\right) - \alpha\left[\mathbf{x}_j^i\left(t\right) - \mathbf{w}_c\left(t\right)\right]$$

$$\text{if } \mathbf{x}_j^i \text{ and } \mathbf{w}_c \text{ belong to different classes} \tag{6}$$

$$R_3 : \text{Add a new } \mathbf{w}_i \text{ of the same class of } \mathbf{x}_j^i$$

$$\text{if } \mathbf{x}_j^i \text{ and } \mathbf{w}_c \text{ belong to the same class, but } c < \theta \tag{7}$$

where $\mathbf{w}_c$ represents the nearest prototype to the actual input sample $\mathbf{x}_j^i$, α represents the learning rate, and $0 < \alpha < 1$, also α will decrease monotonically with time by each iterative step. When each one of the sample input $\mathbf{x}_j^i$ of the data set arrives, it is measured by (2) to each one of the $\mathbf{w}_i$ elements, then, it is obtained the closer one, $\mathbf{w}_c$. The algorithm will perform the conventional LVQ-1.0 rule (R_1, R_2), if the measure of similarity exceeds θ, that is, the sample input $\mathbf{x}_j^i$ is not only near to any one of the prototypes and has the same class, but also, is within the receptive field of the granule. If $\mathbf{x}$ is out of this receptive field, that is, $\mu_k\left(\mathbf{x}_j^i\right) < \theta$, a new prototype will be added to the existing ones. In order to avoid increasing the number of number of prototypes without control, a pruning step will be developed in each iteration step. A better explanation of this pruning step is given in the next subsection.

Now, considering the threshold value θ, by which new prototypes are added to the pool of the existing ones W_k, we proceed as follows. First, note that all the process is performed in the hypercube $[0, 1]$. Then, since the measure of closeness between the input values $\mathbf{x}_j^i$ and the existing prototypes is given by a Gaussian measure of similarity (2), the nearest prototype $\mathbf{w}_c$ has a value close to 1 and, as the prototype is further away, the value will be near of 0. We want a dynamically process which gives the algorithm the possibility to create new prototypes every iteration, then, this value is near 1; that is, the receptive field of $\mathbf{w}_c$ is too small in the hypercube $[0, 1]$.

2.4   Pruning Step

Avoiding an exponential growth in the number of prototypes, each one of the prototypes of all classes will compete for their survival through a life index L_k. The pruning step of prototypes will be done in each iteration step. The prototypes that have the highest L_k-value for each one of the existing classes C will survive, and the others will be eliminated, according to a desired reduction rate. Hence, L_k is initially set at 1 for each one of the initial prototypes and subsequently updated as below:

$$L_k^i = \begin{cases} L_k^i + 1 & \text{If } k = K \\ L_k^i & \text{Otherwise} \end{cases} \tag{8}$$

where the constant 1 is the life reward for the winning prototype $\mathbf{w}_c$ and K is the index of this selected prototype, according to (2) and (4).

With this, the life index L_k attached to each granule that was not pruned, past information is maintained for the next iterative steps from the last ones. The flow chart of the LVQ-GC-BL algorithm is shown in Fig. 1.

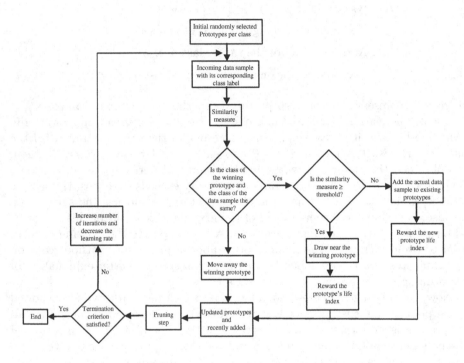

Fig. 1. LVQ-GC-BL algorithm.

3   On-line Algorithm

The online version (LVQ-GC-OL), generates a model from the training set, seeing each sample point once. That is the LVQ process the N training samples a single time. The improved LVQ not only updates granules but also allows the creation of new ones, also it is equipped with memory and punning mechanisms. Therefore, we can say that the advantages of the on-line version of our proposal are:

1. The learning process is done on-line, during every incoming sample, and the algorithm does not require of the storage of certain number of instances to perform the learning process.

2. Each prototype used and newly created, due to the attached life index to it, contains a kind of *"memory"* or *"weight"* associated to each prototype during the learning process. Once the learning process has finished (in a single iteration), the more used class representatives prototypes will be kept.
3. At every incoming sample, due to the receptive field of the prototype and the proposed threshold, not only the actual prototypes will be updated, but also the algorithm can "propose" new ones. With this, the quantized approximation to the real density distribution of classes has more "proposed" elements to cover it.

4   Experimental Framework

The experimental part of the proposed method is developed using the suite of data sets introduced in [18]. This benchmark consists of 59 data sets associated to different classification problems. The results are reported separately for small (data sets with less than $2,000$ instances) and large (data sets with at least $2,000$ instances) data sets. The evaluation methodology adopted consists of applying Prototype Generation (PG) methods to each data set in a 10-fold cross validation scheme. For the evaluation process, a 1NN rule is used by the generated prototypes. The following experimental settings, also used in [18], are defined to assess the proposed approach: 1 initial prototype per class; $Iterations = 100$; $\theta = 0.9; \alpha = 0.1$.

Now we report experimental results to show the effectiveness of the proposed method LVQ-GC-BL. The performance of LVQ-GC-BL is compared against other techniques for PG mentioned in [18], evaluated with the same benchmark conditions.

4.1   Classification Performance of Prototypes

Table 1 shows the average classification performance obtained by the LVQ-GC-BL algorithm for small and large data sets, and is compared against LVQ3 and GENN, the last one is the method that obtained the highest accuracy in [18].

Comparisons from Table 1, shows the effectiveness of our LVQ-GC-BL method in terms of classification performance. We have to remember that LVQ3 algorithm is an improvement over LVQ1.0 algorithm (by which we are basing our proposal). In this work we are improving the LVQ1.0 algorithm by making it variable in the number of class representatives. Which results in improved performance of LVQ.

Table 1. Average classification accuracy for LVQ-GC-BL and reference methods

	LVQ-GC-BL	LVQ3	GENN
Small	0.7038 ± 0.0755	0.6930 ± 0.1560	0.756 ± 0.05
Large	0.7517 ± 0.2068	0.7318 ± 0.2093	0.813 ± 0.217

GENN obtained better values in classification performance, but its reduction performance is among the worst (see below). One should note that not only a high classification performance is desired, also the data compression rate is of main concern. Next, we show results in this important parameter.

4.2   Reduction Performance of Prototypes

Table 2 shows the average instance-reduction rates comparing our algorithm LVQ-GC-BL against LVQ3 and GENN algorithms. In this paper, we consider a percentage of reduction with respect to the data set size of at least 95 %, this is according to similar reduction rates for the best algorithms in [18].

Table 2. Average reduction rates for LVQ-GC-BL and reference methods

	LVQ-GC-BL	LVQ3	GENN
Small	0.9542 ± 0.0154	0.9488 ± 0.0083	0.1862 ± 0.1206
Large	0.9688 ± 0.0180	0.9799 ± 0.0008	0.1576 ± 0.1992

We can see that our algorithm LVQ-GC-BL is highly competitive in compression of the data set. For small data sets, presents the better results and here, we can verify that, although GENN is better in classification performance, is the worst for data compression rate.

4.3   Execution Time

Next, the execution time of simulation comparing LVQ3, GENN and our algorithm is shown in Table 3. This Table shows that our algorithm has the better execution time for large data sets, and is slightly improved in execution speed by the LVQ-3 algorithm for small data sets. Our algorithm, due to the pruning step at each iteration, avoids excessive computational load in the training process.

Table 3. Execution time for LVQ-GC-BL and reference methods

	LVQ-GC-BL	LVQ3	GENN
Small	0.373	0.2316	1.4285
Large	0.541	1.7037	167.4849

4.4   On-line Learning

For the online version of the algorithm (LVQ-GC-OL), the learning process will be developed in just one iteration step (i.e., $Iterations = 1$), the learning rate is fixed and does not decrease, the rest of the parameters remained the same as above. Results of the classification performance are shown in Table 4.

Table 4. Average classification accuracy and execution time for LVQ-GC-OL

	Avg. classification accuracy	Avg. execution time
Small	0.67257 ± 0.16036	0.05017
Large	0.72264 ± 0.19731	0.06857

Although, in general the average of classification accuracy is better for batch learning rather than the online approach, in many cases, and for particular data sets, the accuracy was better in the on-line learning. This improvement in the accuracy for such particular data sets could be done due to the dynamism of the proposed algorithm. That is, every time an instance is evaluated against the existing prototypes, according to the proposed algorithm, not only the learning process can update these prototypes, but also can create new ones if the receptive field does not cover the distribution area of the corresponding class. Then, increasing the number of prototypes provides a better coverage in the class-distribution area, rather than just move a non incremental and limited number of initial prototypes according to the traditional rules. At the same time, the pruning step prevents a non-controlled increase of these class representatives.

4.5   Visualizing Learned Prototypes for a 2-D Case

The visualization of learned prototypes gives a better understanding of prototypes distribution in the training set space. For a 2-Dimensional case, the banana data set provides this possibility for visualization purposes. This data set contains 5300 instances described by 2 attributes, and for classification purposes only has 2 classes. Figure 2 shows the results of the learned prototypes for our LVQ-GC-BL (left) algorithm and the comparative LVQ-3 method (right). For comparison purposes, the training parameters are: 20 initial prototypes per class; $Iterations = 100; \theta = 0.7; \alpha = 0.1.$

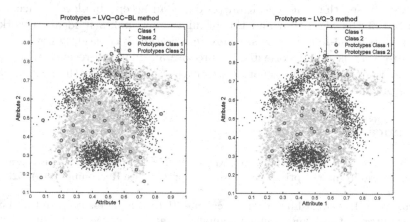

Fig. 2. Banana data set and prototypes

From Fig. 2 it can be seen that Prototypes produced by the LVQ-GC-BL algorithm, have better classes distribution than those of LVQ3, prototypes in LVQ-GC-BL, are not to close between them (i.e., they are spread through the input space), and prototypes generated by LVQ-GC-BL are covering areas that the LVQ 3 algorithm does not. These advantages are due to the dynamism in the production of new prototypes and the pruning step at every iteration. According to the proposed threshold and the life index attached to each prototype, the algorithm will only update prototypes which are more useful and will produce new ones in order to cover different zones than the existing ones by the actual prototypes. With this, a better approximation to the correct class distribution of classes is achieved.

5   Conclusion

This work presents an improved version of the classical LVQ-1.0 algorithm. The algorithm is compared against relevant classifiers such as LVQ-3 and GENN, showing acceptable values of accuracy and data compression rate. The main contributions of the algorithm can be stated as follows: (1) based on granular computing, we have a better initialization of the prototypes; which avoids the use of other algorithms like k-means clustering or SOM, (2) The proposed improvement is able to generate new prototypes at every iterative step, then, not only the actual prototypes are updated but also the new ones can have a better covering of the density distribution class, (3) the proposed life index, attached to every prototype, helps to keep the better class representatives during each iterative step and, this is a measure of "memory" persisting during the learning cycle. With all these steps, the algorithm is able to find the best location for the prototypes and, provides of a better approximation to the correct density distribution classes.

For the on-line version of the algorithm, due to the dynamism in the production of new prototypes and the update of the existing ones at every sample point, we can see that the algorithm produces an acceptable level of accuracy. The speed of the algorithm and the pruning step, makes it ideal to work in environments of on-line data sets, where information is constantly flowing.

Future work is mainly oriented to propose an adaptive receptive field and threshold value, which will generate prototypes according to the specific data features presented.

References

1. Akbarzadeh-T, M.R., Davarynejad, M., Pariz, N.: Adaptive fuzzy fitness granulation for evolutionary optimization. Int. J. Approximate Reasoning 49(3), 523–538 (2008)
2. Cruz-Vega, I., Garcia-Limon, M., Escalante, H.J.: Adaptive-surrogate based on a neuro-fuzzy network and granular computing. In: Proceedings of the 2014 Conference on Genetic and Evolutionary Computation, pp. 761–768. ACM (2014)

3. Cruz-Vega, I., Escalante, H.J., Reyes, C.A., Gonzalez, J.A., Rosales, A.: Surrogate modeling based on an adaptive network and granular computing. Soft Comput. 1–15 (2015)
4. Nova D., Estévez, P.A.: A review of learning vector quantization classifiers. Neural Comput. Appl. 1–14 (2013)
5. Bifet, A., Holmes, G., Pfahringer, B., Kirkby, R., Gavaldà, R.: New ensemble methods for evolving data streams. In: Proceedings of the 15th ACM SIGKDD International Conference on Knowledge Discovery and Data Mining, pp. 139–148. ACM (2009)
6. Kohonen, T., Maps, S.-O.: Self-organizing Maps. Springer Series in Information Sciences, vol. 30. Springer, Heidelberg (1995)
7. Kohonen, T.: Improved versions of learning vector quantization. In: 1990 International Joint Conference on Neural Networks, 1990 IJCNN, pp. 545–550. IEEE (1990)
8. Kohonen, T.: The self-organizing map. Proc. IEEE **78**(9), 1464–1480 (1990)
9. Hastie, T., Tibshirani, R., Friedman, J., Hastie, T., Friedman, J., Tibshirani, R.: The Elements of Statistical Learning, vol. 2, 1st edn. Springer, New York (2009)
10. Garcia, S., Derrac, J., Cano, J.R., Herrera, F.: Prototype selection for nearest neighbor classification: taxonomy and empirical study. IEEE Trans. Pattern Anal. Mach. Intell. **34**(3), 417–435 (2012)
11. García-Limón, M., Escalante, H.J., Morales, E., Morales-Reyes, A: Simultaneous generation of prototypes and features through genetic programming. In: Proceedings of the 2014 Conference on Genetic and Evolutionary Computation, pp. 517–524. ACM (2014)
12. Bharitkar, S., Filev, D.: An online learning vector quantization algorithm. In: ISSPA, pp. 394–397 (2001)
13. Wang, J.-H., Sun, W.-D.: Online learning vector quantization: a harmonic competition approach based on conservation network. IEEE Trans. Syst. Man Cybern. Part B Cybern. **29**(5), 642–653 (1999)
14. Poirier, F., Ferrieux, A.: {DVQ}: dynamic vector quantization-an incremental {LVQ} (1991)
15. Lughofer, E.: Evolving vector quantization for classification of on-line data streams. In: 2008 International Conference on Computational Intelligence for Modelling Control & Automation, pp. 779–784. IEEE (2008)
16. Zang, W., Zhang, P., Zhou, C., Guo, L.: Comparative study between incremental and ensemble learning on data streams: case study. J. Big Data **1**(1), 5 (2014)
17. Zadeh, L.A.: Some reflections on soft computing, granular computing and their roles in the conception, design and utilization of information/intelligent systems. Soft Comput. Fusion Found. Methodol. Appl. **2**(1), 23–25 (1998)
18. Triguero, I., Derrac, J., Garcia, S., Herrera, F.: A taxonomy and experimental study on prototype generation for nearest neighbor classification. IEEE Trans. Syst. Man Cybern. Part C Appl. Rev. **42**(1), 86–100 (2012)

Multitask Reinforcement Learning in Nondeterministic Environments: Maze Problem Case

Sajad Manteghi[1], Hamid Parvin[1], Ali Heidarzadegan[2(✉)], and Yasser Nemati[2]

[1] Department of Computer Engineering, Mamasani Branch, Islamic Azad University, Mamasani, Iran
{Manteghi,parvin}@iust.ac.ir
[2] Department of Computer Engineering, Beyza Branch, Islamic Azad University, Beyza, Iran
{heidarzadegan,Nemati}@beyzaiau.ac.ir

Abstract. In many Multi Agent Systems, under-education agents investigate their environments to discover their target(s). Any agent can also learn its strategy. In multitask learning, one agent studies a set of related problems together simultaneously, by a common model. In reinforcement learning exploration phase, it is necessary to introduce a process of trial and error to learn better rewards obtained from environment. To reach this end, anyone can typically employ the uniform pseudorandom number generator in exploration period. On the other hand, it is predictable that chaotic sources also offer a random-like series comparable to stochastic ones. It is useful in multitask reinforcement learning, to use teammate agents' experience by doing simple interactions between each other. We employ the past experiences of agents to enhance performance of multitask learning in a nondeterministic environment. Communications are created by operators of evolutionary algorithm. In this paper we have also employed the chaotic generator in the exploration phase of reinforcement learning in a nondeterministic maze problem. We obtained interesting results in the maze problem.

Keywords: Reinforcement learning · Evolutionary Q-Learning · Chaotic exploration

1 Introduction

In a multi agent system it usually is demanded to discover or observe its environment to attain its target. In a multi agent system, each agent has only a limited outlook of its own locality. In these cases each agent tries to discover a local map of its environment. It is useful for agents to share their local maps in order to aggregate a global view of the environments and to cooperatively decide about next action selection (Vidal, 2009).

Multitask learning (Caruana, 1997) is an approach in machine learning in that it is tried to learn a problem together with other related problems simultaneously, by means of a shared model. This often results to an improved model for the main task, because it permits the learner to employ the commonality among the related tasks (Wilson et al., 2007).

© Springer International Publishing Switzerland 2015
J.A. Carrasco-Ochoa et al. (Eds.): MCPR 2015, LNCS 9116, pp. 64–73, 2015.
DOI: 10.1007/978-3-319-19264-2_7

In reinforcement learning, agents learn their behaviors by interacting with an environment. An agent senses and acts in its environment in order to learn to choose optimal actions for achieving its goal. It has to discover by trial and error search how to act in a given environment. For each action the agent receives feedback (also referred to as a reward or reinforcement) to distinguish what is good and what is bad. The agent's task is to learn a policy or control strategy for choosing the best set of actions in such a long run that achieves its goal. For this purpose the agent stores a cumulative reward for each state or state-action pair. The ultimate objective of a learning agent is to maximize the cumulative reward it receives in the long run, from the current state and all subsequent next states along with goal state (Sutton et al. 1998; Sutton et al., 2007; Kaelbling et al., 1996).

If the size of problem space of such instances is huge, instead of original rein forcement learning algorithms, evolutionary computation could be more effective. Dynamic or uncertain environments are crucial issues for Evolutionary Computation and they are expected to be effective approach in such environments (Jin et al. 2005).

In this work we found that by means of communications between learning agents gets better the performance of multitask reinforcement learning. In order to apply straightforward and efficient communications between the agents in a nondeterministic environment, we have used genetic algorithm. Also, because of using genetic algorithms, it is possible to preserve the past practices of agents and to trade experiences between agents.

In the rest of the article, the literature review of the field can be presented in next section.

2 Literature Review

2.1 Reinforcement Learning

In reinforcement learning problems, an agent must learn behavior through trial-and-error interactions with an environment. There are four main elements in reinforcement learning system: policy, reward function, value function and model of the environment. The model of the environment simulates the environment's behavior and may predict the next environment state from the current state-action pair and it is usually represented as a Markov Decision Process (MDP) (Vidal, 2009). An MDP is defined as a 4-tuple $< S,A,T,R >$ characterized as follows: S is a set of states in environment, A is the set of actions available in environment, T is a state transition function in state s and action a, R is the reward function. Taking the best action available in a state is the optimal solution for an MDP. An action is best, if it collects as much reward as possible over time.

In this case, the learned action-value function, $Q : S \times A \rightarrow R$, directly approximates Q^*, the optimal action-value function, independent of the policy being followed. The current best policy is generated from Q by simply selecting the action that has the highest value from the current state.

$$Q(s,a) \leftarrow Q(s,a) + \alpha[r + \gamma \max_{a'} Q(s',a') - Q(s,a)] \tag{1}$$

In Eq. 1, s, r, α and Γ denote state, action, learning rate and discount-rate parameter respectively. In this case, the learned action-value function, Q, directly approximates Q^*, the optimal action-value function, independent of the policy being followed.

Algorithm 1. Q- Learning.

Q-Learning Algorithm:

Initialize $Q(s,a)$ arbitrarily
Repeat (for each episode):
Initialize s
Repeat (for each step of episode):
     Choose a from s using policy derived from Q
        (e.g., ε-greedy)
     Take action a, observe r, s'
     $Q(s,a) \leftarrow Q(s,a) + \alpha[r + \gamma \max_a Q(s',a') - Q(s,a)]$
$s \leftarrow s'$
Until s is terminal

The pseudo code of Q-learning algorithm is shown in Algorithm 1. One policy for choosing a proper action is ε-greedy. In this policy agent selects a random action with a chance ε, and the current best action is selected with probability $1 - \varepsilon$ (where ε is in [0,1]).

Algorithm 2. Archive based EC.

01	*Generate the initialize population of x individuals*
02	*Evaluate individuals*
03	*Each individual (parent) creates single offspring*
04	*Evaluate offsprings*
05	*Conduct pair-wise comparison over parents and offsprings*
06	*Select the x individuals,*
	which have the most wins, from parents and offsprings
07	*Stop if halting criterion is satisfied.*
	Otherwise go to Step 3

A large number of studies concerning dynamic or uncertain environments that have been performed so far have used Evolutionary Computation algorithms (Goh et al., 2009). The fitness function in EC algorithm is defined as total acquired rewards. The algorithm is presented in Algorithm 2.

2.2 Evolutionary Computations in RL

In a recent work Beigi et al. (2010; 2011) have presented a modified Evolutionary Q-learning algorithm (EQL). Also, they have shown there the superiority of EQL over QL in learning of multi task agents in a nondeterministic environment. They employ a set of possible solutions in the process of learning of agents. The algorithm is exhibited in Algorithm 3.

Algorithm 3. EQL.

01	*Initialize Q(s,a) by zero*
02	*Repeat (for each generation):*
03	*Repeat (for each episode):*
04	*Initialize s*
05	*Repeat (for each step of episode):*
06	*Choose a from s using policy derived from Q*
	(e.g., ε-greedy)
07	*Take action a, observe r, s'*
08	$s \leftarrow s'$
09	*Until s is terminal*
10	*Add visited path as a Chromosome to Population*
11	*Until population is complete*
12	*Do Crossover() by CRate*
13	*Evaluate the created Childs*
14	*Do tournament Selection()*
15	*Select the best individual for updating Q-Table as follows:*
16	$Q(s,a) \leftarrow Q(s,a) + \alpha[r + \gamma \max_a Q(s',a') - Q(s,a)]$
17	*Copy the best individual in next population*
18	*Until satisfying convergence*

2.3 Chaotic Methods in RL

Chaos theory studies the behavior of certain dynamical systems that are highly sensitive to initial conditions. Small differences in initial conditions (such as those due to rounding errors in numerical computation) result in widely diverging outcomes for chaotic systems, and consequently obtaining long-term predictions impossible to take in general. This happens even though these systems are deterministic, meaning that their future dynamics are fully determined by their initial conditions, with no random elements involved. In other words, the deterministic nature of these systems does not make them predictable if the initial condition is unknown (Kellert, 1993; Meng et al., 2008).

As it is mentioned, there are many kinds of exploration policies in the reinforcement learning, such as ε-greedy, softmax, weighted roulette. It is common to use the uniform pseudorandom number as the stochastic exploration generator in each of the mentioned policies. There is another way to deal with the problem of exploration generators which is to utilize chaotic deterministic generator as their stochastic exploration generators. As the chaotic deterministic generator, a logistic map which generates a value in the closed interval [0 1] according to Eq. 2, is used as stochastic exploration generators in this paper.

$$x_{t+1} = alpha\ x_t(1 - x_t) \tag{2}$$

Algorithm 4. CEQL.

01	*Initialize Q(s,a) by zero*
02	*Repeat (for each generation):*
03	*Repeat (for each episode):*
04	*Initialize s*
05	*Repeat (for each step of episode):*
06	*Initiate (Xcurrent) by Rnd[0,1]*
07	*Repeat*
08	*Xnext= η * Xcurrent * (1- Xcurrent)*
09	*Until (Xnext - Xcurrent <ε)*
10	*Choose a from s using Xnext*
11	*Take action a, observe r, s'*
12	*s ← s'*
13	*Until s is terminal*
14	*Add visited path as a Chromosome to Population*
15	*Until population is complete*
16	*Do Crossover() by CRate*
17	*Evaluate the created Childs*
18	*Do tournament Selection()*
19	*Select the best individual for updating Q-Table as follows:*
	$Q(s,a) \leftarrow Q(s,a) + \alpha[r + \gamma \max_a Q(s',a') - Q(s,a)]$
20	*Copy the best individual in next population*
21	*Until satisfying convergence*

In Eq. 1, x_0 is a uniform pseudorandom generated number in the [0 1] interval and alpha is a constant in the interval [0 4]. It can be showed that sequence x_i will converge to a number in the [0 1] interval provided that the coefficient alpha be a number near to and below 4. It is important to note that the sequence may be divergent for the alpha greater than 4. The closer the alpha to 4, the more different convergence points of the sequence. If alpha is selected 4, the vastest convergence points (maybe all points in the [0 1] interval) will be covered per different initializations of the sequence. So here alpha is chosen 4 to making the output of the sequence as similar as to uniform pseudo-random number (Morihiro et al., 2004).

3   Interaction Between Agents

An interaction occurs when two or more agents are brought into a dynamic relationship through a set of reciprocal actions. During interactions, agents are in contact with each other directly, through another agent and through the environment.

Multi-Agent Systems may be classified as containing (1) NI- No direct Interactions, (2) SI- Simple Interactions and (3) CI- Complex, Conditional or Collective Interactions between agents (Ngobye et al., 2010). SI is basically one-way (may be reciprocal) interaction type between agents and CI could be conditional or collective interactions. In NI, agents do not interact and just do their activities with their knowledge. In such models of MAS, inductive inference methods are used.

Some forms of learning can be modeled in NI MAS, in this case the only rule is: "move randomly until finding goal". Agents learn from evaluating their own performance and probably from those of other agents, e.g. looking at other agents' behavior and how they choose the correct actions. Therefore, agents learn either from their own knowledge or from other agents, without direct interaction with each other.

Basically, while the RL framework is designed to solve a single learning task, and such a concept as reuse of past learning experiences is not considered inside it, there are lots of cases where multiple tasks are imposed on the agents.

In this paper, we try to apply interactions between learning agents and use the best experiences have achieved by agents during multi task reinforcement learning.

4 Implementation

4.1 Maze Problem

Suppose that there are some robots in a gold mine as workers. The tasks of robots are to explore the mine to find the gold that is in an unknown location, from the start point. The mine has a group of corridors which robots can pass through them. In some of corridors there exist some obstacles which do not let robots to continue. Now, assume that because of decadent corridors, it is possible that in some places, there are some pits. If a robot enters to one of the pits, it may be unable to exit from the pit by some moves with a probability above zero. If it fails to exit by the moves, it has to try again. It makes a nondeterministic maze problem. The effort is to find the gold state as soon as possible. In such problems, the shortest path may be not the best path; because it may have some pit cells and it leads agent acts many movements until finding goal state. Thus the optimum path has less pit cells together with shorter length.

The defined time to learn a single task is called Work Time (WT). In such nondeterministic problem suppose that the maze is fixed in a WT. during a WT; agent explores its environment to find the goal as many times as possible during learning.

In mentioned dynamic version of mine problem, assume that, it is possible to repair the pits immediately even during a WT but some other pits may be generated in other locations. It means that, in dynamic version of the problem, number of pits is fixed but their locations may be changed.

It is necessary to have interactions between agents to exchange the promising experiences achieved through trial-and-error exploring of the maze. A simple interaction approach can be modeled by evolutionary computations.

4.2 Simulation

Assume that a number of robots are working in a mine and their task is to search for gold from an initial point. The mine has a group of corridors which robots can pass through them. In specific paths there exist some barriers which do not let robots to continue. Now, suppose that because of decadent corridors, it is possible that in some places, there could be some pits.

If a robot enters to such pits, it may be not to able exit from that pit with probability above zero by some moves. If it fails to exit by the moves, it has to try again. The aim of robots is finding the gold state as soon as possible. Note that the robots can use their past experiences.

In such problems, the shortest path may be not the best path; because it may have some pit cells and it leads agent acts many movement until finding goal state. Therefore the optimum path has less pit cells and short length. So applying an evolutionary version of Q-learning is more useful.

For validation of the proposed algorithm we turn to a modified version of Sutton's maze problem which is depicted in Fig. 1. The original Sutton's maze problem consists of 6 × 9 cells, 46 common states, 1 goal state and 7 collision cells (gray cells in Fig. 1). An agent can occupy each common state. It can't pass through collision cells. For each agent, there are at most four actions in each state to take: Up, Down, Left, and Right. The original Sutton's Maze problem is a deterministic problem (Sutton and Barto 1998).

In nondeterministic version of the problem one adds a number of probabilistic cells or holes (hachured cells in Fig. 1). An agent can't leave each of the holes by its taken actions certainly and it is probable for them to remain in their previous states. That is, they have to stay the same cell with above zero probability; this probability is sampled from Normal distribution with average = 0, and variance = 1. For example, if an agent take Left action in a hole with transition probability according to Fig. 2a, next state may be the same state with probability of 0.6.

Agents will gain a reward +1, if they reach goal state. All other states don't give any reward to agents. There is no punishment in the problem.

4.3 Actions

Each action of agent could be an MDP sample such as delineated in Fig. 2.

Right part shows certain case which in it agents move to next state by choosing any possible action with probability equal 1. On the other hand left part reveals that it is possible that the agents cannot move to next state by choosing any possible action and it would remain in its position. For some do actions.

These MDPs presented to learning algorithm sequentially. The presentation time of each problem instance is enough to learn. The problem is to maximize the total acquired rewards for lifespan. Agents return to start state after arriving goal state.

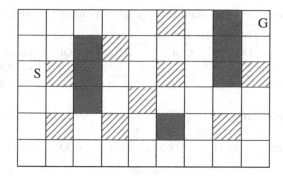

Fig. 1. Modified Sutton's maze.

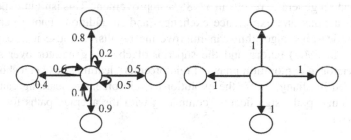

Fig. 2. Actions MDP models.

5 Experiments and Results

A set of experiments have been conducted in nondeterministic case. In the mentioned circumstance, magnitude of PCs is set to 25 throughout all the experimentations. A particular experimentation is created of 100 individual tasks. Each task represents a valid path strategy of an agent from start state to goal state. It includes a two dimensional array of the state-action pairs which the agent has tripped. A valid path is the track of agent states starting from initial state and ending to goal state, if agent has achieved the goal before the defined Max Steps. In other words, if an agent can discover the goal state before it reaches the defined Max Steps, its trajectory is accepted as an individual; and then it is inserted to initial population. Through these chromosomes, past obtained experiences of agents is maintained. In this case Max Steps is designated to 2000. Each individual has a numeric value which is calculated by Value Function.

Population size is set to 100. In every epoch, a two-point crossover method is used. Truncation selection is applied to the algorithm.

Table 1 summarizes experiment results in nondeterministic mode. As it can be inferred from the Table 1, by using interactions between agents and then applying the best achieved experiences (instead of no interaction mode), results in considerably improvement in all three terms of Best average, Worst average and Total average of path lengths.

Usage of Evolutionary Q-Learning in Original Q-Learning results in a 5.09 % improvement. It has shown that the usage of chaotic pseudorandom generator instead of

Table 1. Effect of using interactions between agents in nondeterministic environment.

Algorithms	Without interactions		With interactions		
	OQL	CQL*	EQL	CEQL*	Improvement CEQL to OQL
Best average of path length	54.26	60.49	28.34	25.11	53.7 %
worst average of path length	1600.00	1500.33	66.79	99.99	93.7 %
Total Average of path length	449.11	418.33	42.47	40.30	91 %

stochastic random generator results in a 6.85 % improvement. This isn't unexpected that usage of interactions and experience exchanges and in addition chaotic generator in reinforcement learning algorithm can improve the results. Because it is explored in (Morihiro et al., 2004) before and the superior of chaotic generator over stochastic random generator in exploration phase of reinforcement learning has been shown. Also as it is reported in (Jiang, 2007) the evolutionary reinforcement learning can improve the average found paths significantly comparing with the average paths found by the original version.

6   Conclusion

Learning during cooperating with environments and other agents is a general problem in the context of multi agent system. Multitask learning lets some related tasks to be together learned by means of a combined model. In this field, learners employ the commonality among the tasks. Also reinforcement learning is a type of learning agent concerned with how an agent should choose actions in an environment in order to get the most of agents' reward. RL is a useful approach for an agent to learn its policy in a nondeterministic environment. However it is considered such a time consuming algorithm that it can't be employed in every environment.

We found that existence of interaction between agents is a proper approach for betterment in speed and accuracy of RL methods. To improve performance of multi task reinforcement learning in a nondeterministic environment, we store past experiences of agents. By simulating agents' interactions by structures and operators of evolutionary algorithms, we explore the best exchange of agents' experiences. We also switch to a chaotic exploration instead of a random exploration. Applying Dynamic Chaotic Evolutionary Q-Learning Algorithm to an exemplary maze, we reach significantly promising results.

This can be inferred from experimental results that employing chaos as random generator in exploration phase as well as evolutionary-based computation can improve the reinforcement learning, in both rate of learning and its accuracy. It can also be concluded that using chaos in exploration phase is efficient for both evolutionary based version and non-evolutionary one.

References

Beigi, A., Parvin, H., Mozayani, N., Minaei, B.: Improving reinforcement learning agents using genetic algorithms. In: An, A., Lingras, P., Petty, S., Huang, R. (eds.) AMT 2010. LNCS, vol. 6335, pp. 330–337. Springer, Heidelberg (2010)

Beigi, A., Mozayani, N., Parvin, H.: Chaotic exploration generator for evolutionary reinforcement learning agents in nondeterministic environments. In: Dobnikar, A., Lotrič, U., Šter, B. (eds.) ICANNGA 2011, Part II. LNCS, vol. 6594, pp. 245–253. Springer, Heidelberg (2011)

Caruana, R.: Multitask learning. J. Mach. learn. **28**(1), 41–75 (1997)

Jiang, J.: A framework for aggregation of multiple reinforcement learning algorithms. Ph.D. thesis. University of Waterloo (2007)

Jin, Y., Branke, J.: Evolutionary optimization in uncertain environments - a survey. IEEE Trans. Evol. Comput. **9**(3), 303–317 (2005)

Kaelbling, L.P., Littman, M.L., Moore, A.W.: Reinforcement learning: a survey. J. Artif. Intell. Res. **4**, 237–285 (1996)

Kellert, S.H.: In the Wake of Chaos: Unpredictable Order in Dynamical Systems. University of Chicago Press, Chicago (1993). ISBN 0226429768

Meng, X.P., Meng, J., Lui, L.J.: Quantum chaotic reinforcement learning. In: Proceeding of Fourth International Conference on Natural Computation, vol. 3, pp. 662–666 (2008)

Morihiro, K., Matsui, N., Nishimura, H.: Effects of chaotic exploration on reinforcement maze learning. In: Negoita, M.G., Howlett, R.J., Jain, L.C. (eds.) KES 2004. LNCS (LNAI), vol. 3213, pp. 833–839. Springer, Heidelberg (2004)

Ngobye, M., de Groot, W.T., van der Weide, T.P.: Types and priorities of multi-agent system interactions. Interdisc. Description Complex Syst. **8**(1), 49–58 (2010)

Sutton, R., Barto, A.: Reinforcement Learning: An Introduction. MIT Press, Cambridge (1998)

Sutton, R.S., Koop, A., Silver, D.: On the role of tracking in stationary environments. In: Proceedings of the 24th International Conference on Machine Learning, pp. 871–878 (2007)

Wilson, A., Fern, A., Ray, S.: Tadepalli, P.: Multi-task reinforcement learning: a hierarchical Bayesian approach. In: International Conference on Machine Learning (ICML), Corvallis, OR, USA, pp. 1015–1022 (2007)

Vidal, J.M.: Fundamentals of Multi Agent Systems (2009). http://multiagent.com/2008/12/fundamentals-of-multiagent-systems.html

A Modification of the TPVD Algorithm for Data Embedding

J.A. Hernández-Servin$^{(\boxtimes)}$, J. Raymundo Marcial-Romero,
Vianney Muñoz Jiménez, and H.A. Montes-Venegas

Facultad de Ingeniería, Universidad Autónoma Del Estado de México, CU,
Cerro Coatepec, Toluca, Edo México, México
{xoseahernandez,jrmarcialr,vmunozj,hamontesv}@uaemex.mx

Abstract. Pixel-Value-Differencing (PVD) methods for data hiding have the advantage of a high payload. These algorithms have however the problem of overflow/underflow pixels, thus a location map for those pixels usually ignored when embedding the message is necessary. In this paper, we modified the Tri-way Pixel-Value Differencing method that removes the need of the location map and fix the problem. Our proposal replaces the table of ranges to estimate the amount of information to be embedded by a function based on the floor and ceil functions. As for the problem of overflow/underflow pixels we tackle it by means of a linear transformation. The linear transformation is based on the floor function, so information is lost therefore a location map to compensate for this data lost is necessary to recover the embedded message. The inclusion of the map in the algorithm is also discussed. The technique uses two steganographic methods, namely, the tri-way method to store the message and a reversible steganographic method to store the map needed to invert the linear function in order to recover the encoded message.

Keywords: Steganography · Tri-way pixel-value differencing · TPVD · Reversible · Data hiding

1 Introduction

The technique for embedding useful information into digital covers such as images is called data hiding or steganography. The medium to be used as a cover is not restricted to images, it can be any form of digital signal. In this paper, 8-bits digital images are taken as digital covers. There is a characteristic that distinguishes steganography as opposed to cryptography in that not necessarily the message has to be encrypted but must be an innocuous-looking object. In general terms, a good steganographic algorithm for data hiding must have: (1) high embedding rate measured by bits per pixel (bpp), (2) distortion of the host must be low; the distortion is normally measured by the peak signal to noise ratio (PSNR) in dB units and (3) the method has to be immune to steganalysis, as much as possible, in the sense that a hacker must not even suspect there is a hidden message in an image. The third characteristic is specially difficult to comply. One could

© Springer International Publishing Switzerland 2015
J.A. Carrasco-Ochoa et al. (Eds.): MCPR 2015, LNCS 9116, pp. 74–83, 2015.
DOI: 10.1007/978-3-319-19264-2_8

say, that there are three categories of methods for doing steganography that can loosely be classified as pixel-value-differencing (PVD) [1,2], least-significant-bit (LSB) [3] and methods based on linear transformations such as Fourier [4] and Wavelet transforms [5,6]. Even though there is a growing interest among *Coding Theorist* for developing new steganographic techniques from a code theory point of view [7]. Many methods had been subject to steganalysis and LSB methods, as originally published, are the weakest to such analysis [8]. The PVD methods have been also analysed by steganalysis, some authors [9] claim that PVD methods and their derivations generate abnormal high fluctuations in PVD histograms making this particular method prone to detection. Despite these drawbacks there is still high interests in PVD methods and their improvement for the high payload they provide.

Many steganographic methods have the problem that when embedding a message some pixels of the stego image exceed the 8-bit range [0, 255]. This problem constrain the algorithm to ignore those pixels that fall off boundaries for message insertion. Additionally, the decoder has to know the exact location of pixels that are used for insertion and the location of the pixels being ignored.

In this paper, we propose a simple modification to the tri-way pixel-value differencing (TPVD) method [2] to avoid building a location map of the pixels ignored for embedding. We are also dealing with the problem of overflow/ underflow pixels by proposing a simple linear transformation that potentially increase the payload while keeping a reasonable peak signal to noise ratio. In fixing overflow/underflow pixels we ran into additional difficulties as we also need a map to revert the process in order to recover the embedded message. To deal with this problem, we propose to use the resultant stego image, that is the image after message insertion, to embed the map using a reversible data hiding method. This simple idea has been explored elsewhere for different purposes [10]. Reversible data hiding is the subject of intense research where the purpose is not only to recover the embedded message but also to recover the host image. This problem has been tackled with a wavelet approach as reported in [5,6].

The paper is organised as follows. In Sect. 2, we briefly review the TPVD method, present our proposal and introduce some necessary notation for the rest of the paper. Section 3 presents the method to fix overflow/underflow pixels of the TPVD method. Finally, in Sect. 4 we show that our technique really works by analysing some images.

2   A Modified TPVD Method for Data Hiding

This section explains in detail the *Tri-way Pixel-Value Differencig* (TPVD) [2] method for steganographic data embedding. Let us assume that M is an image with pixel value range in $J = [0, 2^8 - 1] \cap \mathbb{N}$. It will be understood that M is a matrix in $J^{m \times n}$ which can be partioned in blocks of size 2×2. That is, $M = \{[B_{uv}]\}_{u,v}$ with $[B_{uv}] \in J^{2 \times 2}$. For every block $[B]$ in M we define the distance block matrix as $[d] \in J^{2 \times 2}$ where $[d]_{ij} = [B]_{ij} - [B]_{11}$. In the TPVD method a set of ranges are defined in order to decide the amount of information to be embedded in every block. The set of ranges $R = \{[l, u]|l \leq u; l, u \in J\}$ are supposed to be fix and shall be shared to the decoder. Analogously, to the definition of block distance, we can define the lower and upper block for every block $[B]$ in M. Thus, $[[l]_{ij}, [u]_{ij}]$ are those intervals in R such that $[l]_{ij} \leq |[d]_{ij}| \leq [u]_{ij}$. Likewise, it is defined the block $[w] = [u] - [l] + 1$ and $[t] = \lfloor \log_2([w]) \rfloor$ where the $\log_2$ is performed on each entry of the block $[w]$. Each entry in $[t]$ is the amount of bits to be embedded on each pixel of block $[B]$ from the binary message. The message is normally plain text where each character is converted into its decimal equivalent following the ascii table standard then into its binary counterpart.

2.1   Our Proposal

Instead of considering a set of ranges we propose to embed $[t]$ bits from message where each entry is defined as

$$[t]_{ij} = \begin{cases} 0 & i, j = 1 \\ \phi\left(|[d]_{ij}|\right) & \text{otherwise} \end{cases} \tag{1}$$

where $\phi(z) = \lceil \log_2(z + 2H(1 - z)) \rceil$ and H is the heaviside function defined as $H(z) = 1$ if $z \geq 0$ and 0 otherwise. If we assume the message to embed is given as a string b of $0's$ and $1's$, for instance $b = 001010101010 \cdots$, then we can take blocks of size $\sum_{i,j} [t]_{ij}$ then we build the message block

$$[b]_{ij} = \sum_{s=0}^{[t]_{ij}-1} b_{ijs} 2^{[t]_{ij}-s-1} \tag{2}$$

where the b_{ijs} are the bits that compose the message b. For example, taken an arbitrary b we have that

$$b = \cdots 0 \cdots \overbrace{1}^{[t]_{12}} 0 \cdots 0 \overbrace{1}^{[t]_{21}} \cdots \overbrace{1}^{[t]_{22}} \cdots . \tag{3}$$

To embed the message let us consider the block matrix $2^{[t]}$ where each entry is defined as $2^{[t]_{ij}}$ for evey i, j. Define a new block difference $[d']$ on each entry i, j as

$$[d']_{ij} = (2 \cdot H([d]_{ij}) - 1)(2^{[t]_{ij}} + [b]_{ij}). \tag{4}$$

Three more blocks will be defined in order to obtain the stego image. The difference between $[d]$ and $[d']$ is denoted by $[m]$ that is $[m] = [d] - [d']$ and based on $[m]$ we define the auxiliary blocks $[\alpha]$, $[\beta]$ as

$$\alpha_{ij} = \left\lceil \frac{m_{ij}}{2} \right\rceil \cdot H(m_{ij}) + \left\lfloor \frac{m_{ij}}{2} \right\rfloor H(-m_{ij}) \tag{5}$$

and

$$\beta_{ij} = \left\lceil \frac{m_{ij}}{2} \right\rceil \cdot H(-m_{ij}) + \left\lfloor \frac{m_{ij}}{2} \right\rfloor H(m_{ij}) \tag{6}$$

and finally define the block matrix Φ

$$[\Phi(p,q)]_{ij} = \begin{cases} -[\alpha]_{pq} & i,j = 1 \\ [\alpha]_{ij} + [\beta]_{ij} - [\alpha]_{pq} & \text{otherwise} \end{cases} \tag{7}$$

Therefore the stego block $[B'(p,q)]$ is given by $[B'(p,q)] = [B] - [\Phi(p,q)]$

Optimal Stego-block. The mean square error for any pair of matrices $M, N \in J^{s \times t}$ is defined as

$$|M - N| = \frac{1}{st} \sum_i^s \sum_j^t (M_{ij} - N_{ij})^2. \tag{8}$$

Thus, the optimal stego-block can be choosen as

$$[B'] = \min_{\substack{p \neq 1 \\ q \neq 1}} \{|[\Phi(p,q)]|\} \tag{9}$$

2.2 Decoding

To decode, that is to obtain the embedded message from the stego-block $[B']$ we define the distance block matrix as $[d^*] = [B'] - \mathbf{I} \cdot [B']_{11}$ where $\mathbf{I}$ is the matrix with $\mathbf{I}_{ij} = 1$ for all i, j. The block $[t*]$ is calculated replacing $[d]$ by $[d^*]$ in Eq. (1). Therefore the message is recovered by

$$[b] = \left(\frac{[d^*]}{2H([d^*]) - 1} \right) \mathrm{mod} 2^{[t^*]} \tag{10}$$

where the function $m \mod n$ denotes the residual part obtained after dividing the integer m over n. One of the features of the PVD methods is that the differences are kept unmodified which can readily be verified.

Proposition 1. *The differences between neighbour pixels before and after embedding are the same. That is, $[d'] = [d^*]$ for every block in which the host image is being partioned.*

Proof. Straightforward.

Claim. By defining the amount of bits to be embedded through the function ϕ as in Eq. 1 we avoid the problem of having a map of exact locations of the ignored pixels in order to fully recover the message.

Proof. By Proposition 1 above we have that for all i, j except when $i = j = 1$

$$\lfloor \log_2 |[d^*]_{ij}| \rfloor = \left\lfloor \log_2 \left(2^{[t]_{ij}} + [b]_{ij} \right) \right\rfloor = [t]_{ij} \tag{11}$$

Thus, from Eq. 11, the block $[t]$ gives a clear indication of which blocks are being ignored. When $[t] = 0$ the coder ignores that particular block.

3   Correcting Stego–blocks

In general any steganographic PVD method has the problem that when inserting a message some pixel might fall off the valid boundary. That is, the stego-block $[B']$ might not be in $J^{2\times 2}$. To correct it, we are proposing a simple linear transformation that results into a valid stego-block. For this, let us assume that $[B']$ is in $I^{2\times 2}$ where $I = [c^-, c^+]\cap\mathbb{N}$ is a finite interval of integers such that $c^- = \min I \leq 0$ and $255 \leq c^+ = \max I$. Let us define $C = \frac{J_1-J_0}{c^+-c^-}$, $D = \frac{J_0 c^+ - J_1 c^-}{c^+-c^-}$ then the transformation proposed is given by $\psi : I \longrightarrow J$, $\psi(x) = Cx+D$ and obviously its inverse is given by $\psi^{-1} : J \longrightarrow I$, $\psi^{-1}(y) = \frac{y}{C} - \frac{D}{C}$.

The simplicity of the function does give an equally simple scheme to produce a valid stego image, we can simply define $S = \lfloor\psi(B')\rfloor \in J^{2\times 2}$. However, to recover the message is not that straighforward. The function ψ is clearly bijective, and it can be checked that $J = \psi^{-1}(\psi(J))$ but $J \neq \psi(J)$ since $\psi(J)$ are not integers but real numbers. However, $\psi(J) \in [0, 255] \subset \mathbb{R}$ that is $J_0 = \min\psi(J)$ and $J_1 = \min\psi(J)$. In other words, the function ψ is evenly embedding J into J. We can naively say that $\psi^{-1}(S)$ recovers B', however ψ^{-1} is not a function formally speaking but an inverse relation.

In order to fully recover B', we must be able to build the inverse relation both accurately and efficiently. As S is the image that the decoder must have, thus we must be able to recover B' from it. Let us consider $S = \lfloor\psi(I)\rfloor$, and denote the inverse relation by $\Xi = (\xi, \tau)$ such that $\Xi(s) = x$ for every $x \in I$ and $s \in S$. We will show that Ξ in fact has the following form $x = \Xi(s) = \min\xi_s + \tau(s)$ with $\tau \in \{0,1\}$ and ξ_s is a set of integer in I that depends on s. Let us define the set of indices that depend on C as $K = \{0, ..., 2\lceil\frac{1}{C}\rceil + 1\}$.

Claim. For every $s \in S$ such that $s = \lfloor\psi(x)\rfloor$ for some $x \in I$; let us consider the sequence $s_k = \psi\lceil\psi^{-1}(s)\rceil + Ck$ for all $k \in K$ then if $A_s = \{k \in K | s \leq s_k \leq s+1\}$ and $\lceil\frac{1}{C}\rceil \leq 2$

(i) the set ξ_s is given by $\xi_s = \left\{\lceil\psi^{-1}(s)\rceil + k\right\}_{k\in A_s}$,
(ii) and $\tau(s) = x - \min\xi_s$ defines the map τ.

Proof. Straightforward.

The map τ is a binary matrix of same dimension as the stego image, that is all entries in τ are 0's and 1's. This only happens when $|I| \leq 2|J|$ or equivalently when $c^- \leq 126$ and $c^+ \leq 255 + 126$. Beyond that point the inverse relation $\Xi(s) \geq 2$ and the map needs more than two integers to encode the inverse relation. This obviously, although possible, will make the stego image file bigger as it requires more space.

3.1   Encoding the Map

Once the stego image S is obtained from the modified TPVD scheme, the image S is, in general, out of the proper pixel value range. By applying a correction, as explained in Sect. 3, we end up with a valid image M and a map τ. The decoder must know the map τ somehow. We propose to use a reversible steganographic method to embed the map τ. Here, the image to be used as host is again the stego image M. This simple idea was first explored by [10]. They, however, used a different method for embedding the message.

A reversible steganographic scheme is a method where the embedded message as well as the host image are fully recovered. The first method with such characteristics was proposed by Tian [6]; later improvements and generalisations were carried out by [5]. There are some other methodologies proposed for reversible data hiding, for a brief review see for instance [11]. The problem with reversible techniques is that they usually need a map to reverse the process. However, the authors do not discuss the problem of where to embed the map to let the decoder extract the message.

In this paper, we are proposing to use the steganographic scheme by [5] or simply use the proposed by [6]. The scheme in [6] is based on the well known Haar wavelet. This technique has also the problem that some pixels may fall off the valid boundaries when embedding data. Those pixels must be ignored since they cannot be used for embedding. However, the decoder must know the exact location of the ignored pixels. The important part of the algorithm by [5,6] is that they efficiently managed to embed the map of the ignored pixels as part of the message.

4 Experimental Work

The experimental work was carried out on 512×512 standard images as shown in Fig. 1. The level of distortion is measured using the standard peak signal to noise ratio (PSNR) and embedding capacity is measured in bits per pixel (bpp). We choose those images in order to compare our results with [5]. The authors in [5] do compare their findings with some other proposals, so we think is an interesting comparison.

The algorithm proposed in this paper has the flexibility to ignore overflow/underflow pixels, as other PVD methods, but without the need of marking off the ignored pixels. We run two sets of experiments on five images, the first one ignores overflow/underflow pixels and the second one allows overflow/underflow pixels and then fixes them according to Sect. 3.

4.1 Ignoring Overflow/Underflow Pixels

Table 1(a) shows the results of embedding data at its maximum capacity for the five images shown in Fig. 1. We are reproducing the values reported in [5] for the same set of images for the comparison.

We also carried out experiments by restricting the maximum value for $[t]$ to be less than four. That is, we allow the algorithm to embed each pixel difference at most four bits of message.

At first glance, by comparing results from Table 1(a) and (b) we observe an unexpected and counter-intuitive behaviour. That is, less data embedded should result in less distortion of the host image which is not observed by comparing the PSNR values obtained in the two experiments. It seems that this is normal

(a) Lena   (b) Airplane   (c) Barbara   (d) Boat   (e) Goldhill

Fig. 1. Original images (first row), and the images obtained after data embedding (second row) using our modified TPVD.

behaviour in the TPVD method. Restricting the maximum value for the $[t]$ blocks to less than seven, the algorithm behaves in such a way that introduces some pepper noise in the edge regions of the stego image; as a result, the PSNR goes below the optimum (Fig. 2).

The amount of bpp for each image in Table 1(a), is calculated according to the following formula, bpp $= \frac{1}{mn} \sum_{[t] \in M} \sum_{\substack{i=1 \\ j=1}}^{2} [t]_{ij}$ where M denotes the host image seen as a matrix of dimension $m \times n$. In the implementation, the entry $[t]_{11} = 0$ since the pivot pixel is not used for embedding, thus the entry does not contribute to the sum, even so it is always modified as it can be seen from Eq. 4.

Figure 3 shows the effect on fixing overflow/underflow pixels in the Boat image.

Table 1. (a) Comparison of the results obtained by ignoring overflow/underflow pixels against results published by [5] using the Haar wavelet technique. This experiment is run allowing the maximun in t to be ≤ 7 for every block (b) Comparison of the results obtained by ignoring pixels out of range against results published by [5]. The maximum bits permitted for embedding at each pixel difference is four, that is max $t \leq 4$.

	Our proposal		F. Peng [5]	
	PSNR	bpp	PSNR	bpp
Lena	36.04	1.38	30.75	1.2
Airplane	36.09	1.3	33.45	1.2
Barbara	34.56	1.8	26.66	1.2
Boat	37.03	1.66	30.7	1.2
Goldhill	37.55	1.6	26.89	1.2

(a)

	Our proposal		F. Peng [5]	
	PSNR	bpp	PSNR	bpp
Lena	32.53	1.36	30.75	1.2
Airplane	28.72	1.25	33.45	1.2
Barbara	25.22	1.68	26.66	1.2
Boat	30.29	1.61	30.7	1.2
Goldhill	31.68	1.57	26.89	1.2

(b)

(a) Original image (b) Stego image   (c)   (d)

Fig. 2. The boat image shows the method explained in Sect. 3 to fix the over-flow/underflow pixels common in PVD methods. Images shown in (c), (d) are the difference between host image and stego image with max[t] $\leq$ 4 and max[t] $\leq$ 7 respectively.

4.2 Comparison Against TPVD

The function ϕ in Eq. 1, proposed as a modification to the range table of the original TPVD scheme is twofold. We can either define ϕ using $\lfloor \cdot \rfloor$ or $\lceil \cdot \rceil$, floor and ceiling functions respectively. The differences, although subtle, have some impact on the message payload.

Figure 3 shows the empirical probability distribution for both the Goldhill and stego image for comparison. It can be seen some similarities, however, there are some shifts in the histogram that have some impact on the PSNR value.

By comparing the values betwen Tables 2 and 3b, it can clearly be seen that the choice of the ceiling rather than the floor function makes a difference on the message payload. It does increase the payload but the PSNR, in some cases, gets not desirable values. We can see that if we keep max[t] $\leq$ 4 we can substantially increase the payload without compromising the distortion of the host image. However, in the original paper [2], the authors reported that for the Lena image they managed to embed 75,836 bytes which roughly correspond to a $\frac{(75836)(8)}{512^2} \approx 2.3$ bpp in an image of size 512 $\times$ 512 and PSNR of 38.8. This is much higher than expected. By comparing Table 3a and b we can clearly see that by choosing the ceiling function and ignoring the overflow/underflow pixels, we can improve the PSNR values and get higher payload, even so, they are not that

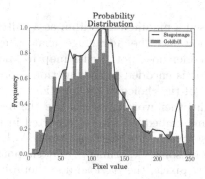

Fig. 3. Emprirical dsitribution of the stegoimage and Goldhill image.

Table 2. Experimental results where pixels are allowed to fall off boundaries, that is $c^- = -126$ and $c^+ = 126$. In this instance the function ϕ is defined by means of the floor function $\lfloor \cdot \rfloor$.

	max[t] ≤ 4		max[t] ≤ 7	
	PSNR	bpp	PSNR	bpp
Lena	22.12	1.36	23.58	1.38
Airplane	24.7	1.25	21.03	1.3
Barbara	21.42	1.69	22.17	1.81
Boat	20.21	1.62	22.06	1.67
Goldhill	31.68	1.57	37.05	1.6

Table 3. (a) The values shown in this table were performed with $c^- = c^+ = 0$ and the function ϕ is defined by means of the ceiling function $\lceil \cdot \rceil$. (b) The values shown in this table were performed with $c^- = -126$ and $c^+ = 126$ and the function ϕ is defined by means of the ceiling function $\lceil \cdot \rceil$.

	max[t] ≤ 4		max[t] ≤ 7	
	PSNR	bpp	PSNR	bpp
Lena	30.31	1.77	27.4	1.79
Airplane	28.02	1.64	27.34	1.63
Barbara	24.85	2.04	24.19	1.98
Boat	28.48	2.04	25.65	2.05
Goldhill	29.83	2.0	25.78	2.05

(a)

	max[t] ≤ 4		max[t] ≤ 7	
	PSNR	bpp	PSNR	bpp
Lena	21.95	1.77	20.43	1.85
Airplane	24.6	1.64	17.24	1.75
Barbara	21.35	2.05	18.91	2.33
Boat	20.18	2.04	20.49	2.18
Goldhill	28.6	2.0	19.76	2.12

(b)

higher as those reported by [2]. We must, however, emphasize that the choice of ϕ instead of the range table removes the need of a map for the exact location of the ignored pixels. Also, it is worth noting that the authors in [2] do not discuss how the decoder knows which pixels or blocks are being ignored. We think is an important part of the problem of steganographic methods.

5   Conclusions

In this paper, we proposed a modified algorithm as an alternative to the TPVD scheme. The modification has the advantage that when ignoring overflow/underflow pixels the decoder does not need a map to know the exact location of the ignored pixels as it is encoded in the function ϕ proposed. The proposal of Sect. 3 is independent of the choice of the function ϕ for the amount of information to embedded. In other words, we can use the original table of ranges in order to keep the same embedded rate as reported in [2], while keeping the flexibility of fixing those overflow/underflow pixels. Although not reported in this paper, we carried out some experiments that supports the thesis that, by fixing overflow/underflow pixels, the payload does notably increase while keeping a reasonable PSNR value. These still remains to be reported and further investigated.

Although our results did not improve the original TPVD scheme, it still performs much better compared to others methods reported in the literature. Also, we have highlighted some fundamental problems that are not studied or discussed in depth in most PVD methods. We are developing a methodology to tackle those problems but this is still an ongoing research.

References

1. Wu, D.-C., Tsai, W.-H.: A steganographic method for images by pixel-value differencing. Pattern Recogn. Lett. **24**(9–10), 1613–1626 (2003). doi:10.1016/ S0167-8655(02)00402-6
2. Changa, K.-C., Changa, C. P., Huangb, P.S., Tu, T.-M.: A novel image steganographic method using tri-way pixel-value differencing. J. Multimed. **3**(2), 37–44 (2008)
3. Qazanfari, K., Safabakhsh, R.: A new steganography method which preserves histogram: generalization of lsb++. Inf. Sci. **277**, 90–101 (2014). doi:10.1016/j.ins. 2014.02.007
4. Rekik, S., Guerchi, D., Selouani, S.-A., Hamam, H.: Speech steganography using wavelet and fourier transforms. EURASIP J. Audio, Speech, Music Process. **2012**, 1123–1125 (2012)
5. Peng, F., Li, X., Yang, B.: Adaptive reversible data hiding scheme based on integer transform. Signal Process. **92**, 54–62 (2012)
6. Tian, J.: Reversible data embedding using a difference expansion. IEEE Trans. Circuits Syst. Video Technol. **13**(8), 890–896 (2003)
7. Zhang, W., Li, S.: Steganographic codes-a new problem of coding theory. J. Latex Class Files **1**(11), 1–7 (2002)
8. Fridrich, J., Goljan, M., Hogea, D.: Steganalysis of jpeg images: breaking the f5 algorithm. In: Petitcolas, F.A.P. (ed.) IH 2002. LNCS, vol. 2578, pp. 310–323. Springer, Heidelberg (2003)
9. Bui, C.-N., Lee, H.-Y., Joo, J.-C., Lee, H.-K.: Steganalysis method defeating the modified pixel-value differencig steganography. Int. J. Innovative Comput. Inf. Control **6**(7), 3193–3203 (2010)
10. Liu, G., Dai, Y., Wang, Z., Shiguo, L.: Adaptive image steganography by reversible data hiding. In: Li, T., Xu, Y., Ruan, D. (Eds.) International Conference on Intelligent Systems and Knowledge Engineering (ISKE 2007) (2007)
11. Sarkar, T., Sanyal, S.: Reversible and irreversible data hiding technique, CoRR abs/1405.2684. http://arxiv.org/abs/1405.2684

Prototype Selection for Graph Embedding Using Instance Selection

Magdiel Jiménez-Guarneros[(⊠)], Jesús Ariel Carrasco-Ochoa,
and José Fco. Martínez-Trinidad

Instituto Nacional de Astrofísica, Óptica y Electrónica (INAOE), Luis Enrique Erro
No. 1, Sta. María Tonantzintla, 72840 San Andrés Cholula, Puebla, Mexico
{magdiel.jg,ariel,fmartine}@inaoep.mx

Abstract. Currently, graph embedding has taken a great interest in
the area of structural pattern recognition, especially techniques based
on representation via dissimilarity. However, one of the main problems
of this technique is the selection of a suitable set of prototype graphs that
better describes the whole set of graphs. In this paper, we evaluate the
use of an instance selection method based on clustering for graph embedding, which selects border prototypes and some non-border prototypes.
An experimental evaluation shows that the selected method gets competitive accuracy and better runtimes than other state of the art methods.

Keywords: Structural pattern recognition · Graph embedding · Prototype selection · Graph classification

1 Introduction

Selecting a suitable representation for different types of objects is very important
in pattern recognition [1]. There are two approaches that have been studied:
the statistical and the structural. In the statistical approach [2], objects are
represented as feature vectors, being its main advantage the wide collection of
algorithms developed for its treatment. However, this type of representation
involves constraints such as the use of a certain number of characteristics or the
lack of usefulness in the representation of the components of an object and the
relationships that exist between them. Moreover into the structural approach,
especially with representations based on graphs [3], it is possible to overcome
these major statistical approach's drawbacks. As a result, there is a growing
interest in the use of this type of representation for different applications, such
as: document analysis [1], image segmentation [4], medical diagnosis [5], face
recognition [6], fingerprint classification [7], among others. Nevertheless, the main
disadvantage of the graph-based representation is the high complexity of the
algorithms for their treatment.

In the literature [8–12], graph embedding emerges as a solution to deal with
the major problems of the structural approach, especially for graph-based representations. The key idea of this technique is transforming a graph into an

© Springer International Publishing Switzerland 2015
J.A. Carrasco-Ochoa et al. (Eds.): MCPR 2015, LNCS 9116, pp. 84–92, 2015.
DOI: 10.1007/978-3-319-19264-2_9

n-dimensional feature vector, making possible the use of those methods and techniques developed for the statistical approach [10]. Although different solutions for graph embedding have been proposed in the literature, this paper follows the approach initially proposed by Riesen and Bunke in [8–11], called graph embedding via dissimilarity. This technique transforms a input graph g into an n-dimensional feature vector by means of a set of prototype graphs P and a dissimilarity measure. This feature vector is built by computing the dissimilarities between g and the n prototypes in P. However, one of the major drawbacks of using this technique is the search for a suitable set of prototypes, since the number of prototypes in P determines the dimensionality of the feature vectors. A small number of prototypes is desirable because it will lead to lower computational cost.

In [8–12], several approaches for prototype selection for graph embedding are proposed. In works as [8,10] for example, the authors propose different prototype selection methods by means of two selection strategies, the first one [8] is carried out on the whole graph dataset regardless of the class each selected prototype belongs to; while in the second [10], the prototypes are selected separately by class. The second strategy provides better classification results than the first one. Moreover in [11], the authors propose extensions of the prototype selection methods proposed in [10]. The key idea of their proposal involves the discrimination between classes through a weight assigned to the intra-class compactness and the inter-class separation in a prototype selection. The methods presented in [11] show better classification results in contrast to the methods reported in [10]. On the other hand, the method proposed in [12] tries to optimize the dimensionality of the feature vectors; such as some reduction methods studied in [9]. The method proposed in [12] uses a genetic algorithm to perform prototype selection by means of two operations called compression and expansion. In this paper, we propose the use of an instances selection method based on clustering for prototype selection for graph embedding via dissimilarity.

The rest of the paper is organized as follows: Sect. 2 presents a brief description of the technique of graph embedding via dissimilarity. Section 3 describes the instance selection method used in our proposal. Experimental results are presented in Sect. 4. Finally, Sect. 5 presents some conclusions and future work.

2   Graph Embedding via Dissimilarity

Graph embedding consists in transforming a graph into a feature vector of suitable dimensionality [10]. The main motivation of this approach is to overcome the lack of algorithmic tools in the graph domain and taking advantage the existing statistical pattern recognition techniques, preserving the representational power the graphs provide.

Given a labeled set of sample graphs $T = \{g_1, \dots, g_m\}$ from some graph domain G, and a dissimilarity measure $d(g_i, g_j)$. Graph embedding consists in selecting a set $P = \{p_1, \dots, p_n\} \subseteq T$ with $n < m$ prototype graphs (see Sect. 3), and by computing the dissimilarity measure of a input graph g with each of the

n prototypes in P an n-dimensional feature vector $(d(g, p_1), ..., d(g, p_n))$ that represents the graph g is built. Based on this technique a set of graphs can be transformed into n-dimensional feature vectors allowing the use of the different techniques of the statistical approach to pattern recognition.

Definition 1 (Graph Embedding). Given a graph domain G, a dissimilarity function $d : G \times G \longrightarrow \mathbb{R}$, and $P = \{p_1, ..., p_n\} \subseteq G$ a prototype graph set. The mapping $\varphi_n^P : G \to \mathbb{R}^n$ is defined as:

$$\varphi_n^P(g) = (d(g, p_1), ..., d(g, p_n))$$

In this work, a dissimilarity measure based on the Graph Edit Distance (GED) was used, which defines the dissimilarity, or distance between two graphs as the minimum number of edit operations (insertion, deletion and substitution of nodes or edges) necessary to transform a graph into another [13]. A variety of algorithms have been proposed in the literature to calculate the edit distance between two graphs [13–18], in this work we use the approach based on bipartite graphs; initially presented in [15], improved in [16] and proposed as a toolkit in [17]. Although other improvements have been introduced for this approach [18], not all improvements have good performance for some application domains.

3   Prototype Selection

The graph embedding method described in Sect. 2 strongly depends on the selected prototypes, since they determine the quality of the embedding, and the number of prototypes determines the dimensionality of the feature vectors. A good prototype selection is very important because it should include as much descriptive information of the original graph dataset as possible [10]; furthermore, a small number of prototypes is desirable. The selected prototypes should avoid redundancy or even noise from the set of graphs.

In the literature, different prototype selection methods have been proposed. Strategies as those presented in [8,10,11] show good results, the most remarkable are those that consider those discrimination between classes; however, its main characteristic is that the number of prototypes is determined by the user. Other strategies optimize the size of the prototype set over a validation set [9,12]. Although these strategies infer automatically the number of prototypes to select, their selection cost is high in comparison with those proposed in [8,10,11]. On the other hand, to the best of our knowledge, strategies based on instance selection have not been previously used for graph embedding. Therefore, in this work, we propose and empirically evaluate the use of an instance selection method for prototype selection for graph embedding.

The following section describes the instance selection method that we used in this paper. This method, based on clustering, aims to retain prototypes that are located in the decision border and some others in the interior of each class (non-border prototypes).

3.1 Prototype Selection by Clustering (PSC)

The Prototype Selection by Clustering (PSC) method [19] was proposed for n-dimensional vector spaces, and among the different selection methods shown in the literature, PSC is one of the most recent; which has a fast runtime compared with other instances selection methods [19]. Thus, we adapted PSC to be used for prototype selection by using the dissimilarity measure $d(g_i, g_j)$ proposed in [17].

The main idea of PSC is to retain border prototypes and some non-border prototypes. For this, PSC initially divides the graph set T in small regions to find border and non-border prototypes into these regions, instead of finding them over the whole set T. In [19], originally *K-Means* was used to create these regions, however, in this work, we use the *K-Medoids* algorithm [20]. This clustering algorithm works similar to *K-Means*, but the major difference is on how the cluster centers are computed in each iteration. In this case, the median determines the center of each cluster A_j, and it is computed as follows:

$$center(A_j) = arg \min_{g_1 \in A_j} \sum_{g_2 \in A_j} d(g_1, g_2) \qquad (1)$$

PSC creates C clusters from a graph set T using *K-Medoids*. Subsequently, for each cluster A_j, it is necessary to determine if it is homogeneous or not. A pseudo-code description of this algorithm can be seen in Algorithm 1.

Algorithm 1. $PSC(T, C)$

Input: $T = (g_1, \ldots, g_m)$: A graph set; C: Number of clusters
Output: $P = (p_1, \ldots, p_n)$: Prototype set

1 $P \leftarrow \emptyset$;
2 $Clusters \leftarrow K\text{-}Medoids(T, C)$;
3 **foreach** *cluster A_j in Clusters* **do**
4   **if** A_j *is homogeneous* **then**
5          $c \leftarrow$ the center of A_j;
6          $P \leftarrow P \cup \{c\}$;
7   **else**
8          $C_M \leftarrow$ the majority class in A_j;
9   **foreach** *class C_K in A_j, where $C_K \neq C_M$* **do**
10   **foreach** *prototype p_i belonging to class C_K* **do**
11   Let $p_c \in C_M$ be the nearest prototype p_i;
12                 $P \leftarrow P \cup \{p_c\}$;
13   Let $p_M \in C_K$ be the nearest prototype to p_c;
14                 $P \leftarrow P \cup \{p_M\}$;

A cluster A_j is considered homogeneous if all the prototypes in the cluster belong to the same class. Therefore, these prototypes are non-border and they can be represented by the center of the cluster, discarding all other prototypes in the cluster (steps 4–6).

A cluster A_j is non-homogeneous if it has prototypes from two or more classes. PSC analyses the non-homogeneous clusters in order to find prototypes located in the border of the classes. For this purpose, PSC finds the most frequent class C_M (majority class) in each non-homogeneous cluster A_j. Then, the border prototypes in C_M are the most similar prototypes in A_j that belong to different class C_K (see Algorithm 1, steps 10–12). The border prototypes for C_K are obtained in the same way (steps 13–14).

As result, PSC obtains the most representative prototype for each homogeneous cluster (centers) and the border prototypes from non-homogeneous clusters. The number of clusters C to build in PSC, will be discussed in the next section.

4   Experimental Results

This section shows an experimental evaluation on 10 databases taken from IAM repository [21]. Table 1 describes the main features of each database divided into training, validation, and test, sets. The main purpose of these experimental tests is to evaluate the prototype selection obtained through PSC for graph embedding. In our experiments, we include an assessment in terms of accuracy, number of selected prototypes and runtime.

For each database, the training set is used to build and train the classifier, while the validation set was used to determine the appropriate number of prototypes, and the test set was used for evaluation. In order to carry out the classification task we use Support Vector Machines (SVM) with a Radial Basis

Table 1. Summary of characteristics of the graphs datasets used in our experiments. The size of training (Tr), validation (Va), testing (Te), number of classes (#Cls), average (Avg.) and maximum (Max.) of nodes and edges, as well as whether or not the classes are balanced (Bal.), are shown.

Dataset	Size (Tr,Va,Te)	#Cls.	Nodes (Avg.)	Edges (Avg.)	Nodes (Max.)	Edges (Max.)	Bal.
Letter low	750,750,750	15	4.7	3.1	8	6	Yes
Letter medium	750,750,750	15	4.7	3.2	9	7	Yes
Letter high	750,750,750	15	4.7	4.5	9	9	Yes
GREC	286,286,528	22	11.5	12.2	25	30	Yes
Fingerprints	500,300,2000	4	5.4	4.4	26	24	No
Protein	200,200,200	6	32.6	62.1	126	149	Yes
AIDS	250,250,1500	2	15.7	16.2	95	103	No
Mutagenicity	1500,500,1500	2	30.3	30.8	417	112	Yes
Coil-RAG	2400,500,1000	100	3	3	11	13	Yes
Coil-DEL	2400,500,1000	100	21.5	54.2	77	222	Yes

Function (RBF) kernel, provided by the WEKA Toolkit [22]. This classifier was chosen since it has been used in recent works as [11], and it has had a remarkable performance in n-dimensional vector spaces.

4.1   Baseline Methods

The graph embedding strategy proposed by Riesen and Bunke [8–11] can work without making a previous selection of prototypes. However, as it was mentioned previously, the dimensionality of the vectors in the embedding depends on the number of prototypes. So a small set of prototypes is desirable since it will lead to lower computational cost. For us, it is important to make a comparison between the prototypes selected by PSC and using all prototypes in the dataset, being this our first baseline method.

Our second baseline method is the Spanning Prototype Selection Discriminative (SPS-D) proposed in [11]. This method was chosen because it shows very good results, and is one of the most recent methods reported in the literature. The main idea of SPS-D consists in selecting those prototypes that are located in the center of each class. Subsequently, SPS-D selects the farthest prototype to the prototypes already selected, trying to cover the whole training set. This method, considers discrimination between classes by assigning a weight of compactness intra-class and separation inter-class. For our experiments, the weights assigned to each dataset were fixed as in [11].

4.2   Experimental Comparison

In order to compare PSC and SPS-D, we carry out an experiment by using different percentages of retention. A number of prototypes per class was designed for SPS-D, according to retention levels from 5 to 95 % (see Fig. 1). The retention percentage in PSC is of interest, because this method uses the number of clusters C as its unique parameter. In [19], the authors fixed for C an even number of clusters per class for instance selection. However, in our experiment C was set trying to get a number of prototypes similar to the number of prototypes used by SPS-D for each retention level.

Table 2 shows the classification results after generating the graph embedding using all the prototypes of the training set, as well as the prototypes selected by SPS-D and PSC. This table shows the accuracy on the test set, the number of selected prototypes, and the retention percentage for each method. The best accuracy results appear boldfaced.

The experimental results show that PSC gets better average accuracy than SPS-D, we can observe that PSC's average accuracy was not better than the one obtained by using all graphs of the training set. However, a Friedman test [23] on the classification results with a significance level $\alpha = 0.5$ and 2 degrees of freedom distributed according to chi-square, shows that there is not a statistical significant difference between PSC's results and the two baseline methods. It is important to comment that these results were performed with the same parameter values for the SVM classifier, we did not perform a custom configuration for each database as it was the made in previous works [10,11].

Table 2. Accuracy of SVM applied over the graph embedding using all graphs of the training set, and the prototypes selected by SPS-D and PSC. Acc.: Accuracy, Prots.: number of prototypes, Ret.: retention percentage.

Dataset	All Graphs			SPS-D			PSC		
	Acc.	Prots.	Ret.	Acc.	Prots.	Ret.	Acc.	Prots.	Ret.
Letter low	**99.20**	750	100.00	99.07	226	30.13	**99.20**	272	36.27
Letter Medium	94.67	750	100.00	**94.80**	706	94.13	94.40	152	20.27
Letter High	**92.80**	750	100.00	92.40	301	40.13	91.60	302	40.27
GREC	**98.30**	286	100.00	98.11	67	23.43	**98.30**	156	54.55
Fingerprint	80.45	500	100.00	80.60	299	59.80	**81.05**	286	57.20
Protein	65.00	200	100.00	61.50	189	94.50	**66.00**	194	97.00
AIDS	99.27	250	100.00	**99.40**	26	10.40	**99.40**	50	20.00
Mutagenicity	**66.54**	1500	100.00	66.37	1125	75.00	65.17	484	32.27
Coil-RAG	**94.80**	2400	100.00	94.70	2201	91.71	94.70	2202	91.75
Coil-DEL	95.00	2400	100.00	**95.20**	1401	58.38	**95.20**	1802	75.08
Average	**88.60**		100.00	88.21		57.76	88.50		52.46

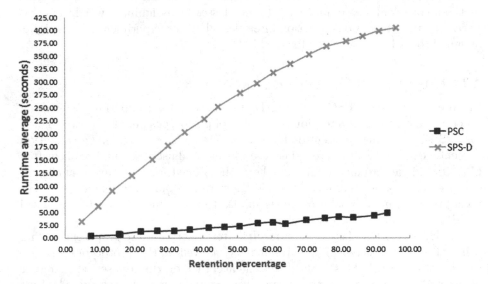

Fig. 1. Average runtime spent by the PSC and SPS-D for prototype selection on the 10 graphs datasets.

The optimal number of prototypes used to perform the graph embedding is also of interest. The results show that PSC get competitive classification results with a lower retention. In the last row of Table 2, we can see that average retention obtained by PSC is lower compared with the one of SPS-D. These findings are important because the use of a small number of prototypes implies a lower computational cost in the classification stage in many real applications.

A second experiment is related to the runtime of SPS-D and PSC, the results are shown in Fig. 1. This experiment shows the average runtime on the ten datasets regarding the percentage of selected prototypes. The runtimes that appear in Fig. 1 do not include the time for computing the distance matrix involved in the graph embedding, since both methods use this matrix.

In Fig. 1, we can observe that PSC is faster than SPS-D. We can also notice that the percentage of retention of PSC does not coincide with that of SPS-D because the number of clusters for PSC was assigned trying to select the same number of prototypes as SPS-D, but it was not always possible. Finally, we can see that PSC obtains very good runtimes with competitive classification results.

5   Conclusions

In graph embedding, especially in techniques based on dissimilarities, the search for a suitable set of prototypes proves to be an important task. A small number of prototypes will reduce the dimensionality of the feature vectors and therefore leads to lower computing costs in real world applications.

The main contribution of our work is to show that instance selection methods are useful for selecting prototypes for graph embedding. In this paper, we have evaluated the PSC instance selection method for the selection of prototypes for graph embedding via dissimilarity. PSC selects graphs located on the border of the class, with the aim of preserving the discrimination, and some prototypes that are non-borders. The experimental results show that the use of PSC for selecting prototypes for graph embedding allows obtaining competitive classification results compared to SPS-D, a state of the art prototype selection method for graph embedding, but requiring less runtime.

As it was described in this paper, an instance selection method was evaluated for selecting prototypes for the graph embedding via dissimilarity, however, given the promising results, as future, we will conduct a full study including other instance selection methods.

Acknowledgment. This work was partly supported by the National Council of Science and Technology of Mexico (CONACyT) through the project grant $CB2008$-106366; and the scholarship grant 298513.

References

1. Bunke, H., Riesen, K.: Recent advances in graph-based pattern recognition with applications in document analysis. Pattern Recogn. 44(5), 1057–1067 (2011)
2. Bishop, C.M.: Pattern Recognition and Machine Learning. Information Science and Statistics. Springer, New York (2006)
3. Conte, D., Foggia, P., Sansone, C., Vento, M.: Thirty years of graph matching in pattern recognition. Int. J. Pattern Recognit. Artif. Intell. 18(3), 265–298 (2004)
4. Harchaoui, Z., Bach, F.: Image classification with segmentation graph kernels. IEEE Computer Society (2007)

5. Sharma, H., Alekseychuk, A., Leskovsky, P., Hellwich, O., Anand, R., Zerbe, N., Hufnagl, P.: Determining similarity in histological images using graph-theoretic description and matching methods for content-based image retrieval in medical diagnostics. Diagn. Pathol. **7**(1), 134 (2012)

6. Han, P.Y., San, H.F., Yin, O.S.: Face recognition using a kernelization of graph embedding. World Acad. Sci. Eng. Technol. **6**(2), 460–464 (2012)

7. Marcialis, G., Roli, F., Serrau, A.: Graph-based and structural methods for fingerprint classification. In: Kandel, A., Bunke, H., Last, M. (eds.) Applied Graph Theory in Computer Vision and Pattern Recognition. SCI, vol. 52, pp. 205–226. Springer, Heidelberg (2007)

8. Riesen, K., Neuhaus, M., Bunke, H.: Graph embedding in vector spaces by means of prototype selection. In: Escolano, F., Vento, M. (eds.) GbRPR. LNCS, vol. 4538, pp. 383–393. Springer, Heidelberg (2007)

9. Bunke, H., Riesen, K.: Graph classification based on dissimilarity space embedding. In: da Vitoria Lobo, N., Kasparis, T., Roli, F., Kwok, J.T., Georgiopoulos, M., Loog, M. (eds.) SSPR & SPR 2008. LNCS, vol. 5342, pp. 996–1007. Springer, Heidelberg (2008)

10. Riesen, K., Bunke, H.: Graph classification based on vector space embedding. Int. J. Pattern Recognit. Artif. Intell. **23**(06), 1053–1081 (2009)

11. Borzeshi, E.Z., Piccardi, M., Riesen, K., Bunke, H.: Discriminative prototype selection methods for graph embedding. Pattern Recogn. **46**(6), 1648–1657 (2013)

12. Livi, L., Rizzi, A., Sadeghian, A.: Optimized dissimilarity space embedding for labeled graphs. Inf. Sci. **266**, 47–64 (2014)

13. Gao, X., Xiao, B., Tao, D., Li, X.: A survey of graph edit distance. Pattern Anal. Appl. **13**(1), 113–129 (2010)

14. Vento, M.: A one hour trip in the world of graphs, looking at the papers of the last ten years. In: Kropatsch, W.G., Artner, N.M., Haxhimusa, Y., Jiang, X. (eds.) GbRPR 2013. LNCS, vol. 7877, pp. 1–10. Springer, Heidelberg (2013)

15. Riesen, K., Bunke, H.: Approximate graph edit distance computation by means of bipartite graph matching. Image Vis. Comput. **27**(7), 950–959 (2009)

16. Fankhauser, S., Riesen, K., Bunke, H.: Speeding up graph edit distance computation through fast bipartite matching. In: Jiang, X., Ferrer, M., Torsello, A. (eds.) GbRPR 2011. LNCS, vol. 6658, pp. 102–111. Springer, Heidelberg (2011)

17. Riesen, K., Emmenegger, S., Bunke, H.: A novel software toolkit for graph edit distance computation. In: Kropatsch, W.G., Artner, N.M., Haxhimusa, Y., Jiang, X. (eds.) GbRPR 2013. LNCS, vol. 7877, pp. 142–151. Springer, Heidelberg (2013)

18. Riesen, K., Bunke, H.: Improving bipartite graph edit distance approximation using various search strategies. Pattern Recogn. **48**(4), 1349–1363 (2015)

19. Olvera-Lpez, J., Carrasco-Ochoa, J., Martnez-Trinidad, J.: A new fast prototype selection method based on clustering. Pattern Anal. Appl. **13**(2), 131–141 (2010)

20. Kaufman, L., Rousseeuw, P.: Clustering by Means of Medoids. Reports of the Faculty of Mathematics and Informatics, Faculty of Mathematics and Informatics (1987)

21. Riesen, K., Bunke, H.: Iam graph database repository for graph based pattern recognition and machine learning. In: da Vitoria Lobo, N., Kasparis, T., Roli, F., Georgiopoulos, M., Anagnostopoulos, G.C., Kwok, J.T., Loog, M. (eds.) SSPR & SPR 2008. LNCS, vol. 5342, pp. 287–297. Springer, Heidelberg (2008)

22. Hall, M., Frank, E., Holmes, G., Pfahringer, B., Reutemann, P., Witten, I.H.: The weka data mining software: an update. SIGKDD Explor. Newsl. **11**(1), 10–18 (2009)

23. Demšar, J.: Statistical comparisons of classifiers over multiple data sets. J. Mach. Learn. Res. **7**, 1–30 (2006)

Correlation of Resampling Methods for Contrast Pattern Based Classifiers

Octavio Loyola-González[1,2(✉)], José Fco. Martínez-Trinidad[2],
Jesús Ariel Carrasco-Ochoa[2], and Milton García-Borroto[3]

[1] Centro de Bioplantas, Universidad de Ciego de Ávila, Carretera a Morón km 9,
69450 Ciego de Ávila, Cuba
octavioloyola@bioplantas.cu
[2] Instituto Nacional de Astrofísica, Óptica y Electrónica, Luis Enrique Erro No. 1,
Sta. María Tonanzintla, 72840 San Andrés Cholula, Puebla, México
{octavioloyola,fmartine,ariel}@inaoep.mx
[3] Instituto Superior Politécnico José Antonio Echeverría, Calle 114 No. 11901,
Marianao, 19390 La Habana, Cuba
mgarciab@ceis.cujae.edu.cu

Abstract. Applying resampling methods is an important approach for working with class imbalance problems. The main reason is that many classifiers are sensitive to class distribution, biasing their prediction towards the majority class. Contrast pattern based classifiers are sensitive to imbalanced databases because these classifiers commonly find several patterns of the majority class and only a few patterns (or none) of the minority class. In this paper, we present a correlation study among resampling methods for contrast pattern based classifiers. Our experiments performed over several imbalanced databases show that there is a high correlation among different resampling methods. Correlation results show that there are nine different groups with very high inner correlation and very low outer correlation. We show that most resampling methods allow improving the accuracy of the contrast pattern based classifiers.

Keywords: Supervised classification · Contrast patterns · Resampling methods · Imbalanced databases

1 Introduction

The main aim of a supervised classifier is to classify a query object using a model based on a representative sample of the problem classes. Sometimes, this model can be used to gain understanding of the problem domain or to make the problem easier to understand by experts in the application domain [13]. An important family of understandable classifiers is based on contrast patterns. Nevertheless, contrast pattern classifiers are sensitive to the class imbalance problems [18].

This work was partly supported by National Council of Science and Technology of Mexico under the project CB2008-106366 and scholarship grant 370272.

© Springer International Publishing Switzerland 2015
J.A. Carrasco-Ochoa et al. (Eds.): MCPR 2015, LNCS 9116, pp. 93–102, 2015.
DOI: 10.1007/978-3-319-19264-2_10

In some imbalanced real-world problems, the objects in a class can be under-represented regarding the remaining problem classes. Oftentimes, the most important class contains significantly less objects because it could be associated to rare cases or because the data acquisition of these objects is costly [26]. This type of problems is known as the class imbalance problems.

Some contrast pattern based classifiers, which show good performance in problems with balanced classes, are degraded in class imbalance problems [16]. A common way to deal with the class imbalance problem is applying resampling methods. Resampling methods modify the dataset in order to produce a balanced class distribution. Resampling methods are more versatile than other approaches to deal with class imbalance problems because they do not depend on the learning algorithm [2].

Many comparative studies have been published about the application of resampling methods to improve the accuracy of several contrast pattern based classifiers [17–19, 24, 27]. Although, up to our knowledge, there is no correlation study among different resampling methods for contrast pattern classifiers.

In this paper, we present a correlation study about the effects of the most used resampling methods for improving the accuracy of a contrast pattern based classifier over several imbalanced databases. Our main goal is to offer an insight about which resampling methods have similar behavior for improving contrast pattern based classifiers. This knowledge would be helpful to simplify future research regarding resampling methods for contrast pattern based classifiers.

The rest of the paper has the following structure. Section 2 provides a brief introduction to contrast patterns. Section 3 reviews the most popular resampling methods. Section 4 presents our correlation study about the methods presented in Sect. 3, the experimental setup, and a discussion of the results. Finally, Sect. 5 provides conclusions and future work.

2   Contrast Patterns

A *pattern* is an expression defined in a certain language that describes a collection of objects. For example, a pattern that describes a set of sick plants can be expressed as: $[Necrosis = "Yes"] \land [StemHigh \in [0.6, 1.5]] \land [Leaves \leq 2]$. Then, a *contrast pattern* is a pattern appearing frequently in a class and infrequently in the remaining problem classes [30].

In some domains, contrast pattern based classifiers have shown to make consistently more accurate predictions than popular classification models like Naive Bayes, Nearest Neighbor, Bagging, Boosting, and even Support Vector Machines (SVM) [12, 30].

Many algorithms have been proposed for mining contrast patterns but those based on decision trees gain special attention because they obtain a small collection of high quality patterns [11]. In this paper, we used Logical Complex Miner (LCMine) [12], a contrast pattern miner that extracts contrast patterns from a collection of diverse decision trees. Moreover, we used Classification by Aggregating Emerging Patterns (CAEP) [9] as a contrast pattern based classifier. LCMine jointly CAEP attains higher accuracies than other contrast pattern based classifiers (like SJEP [10]) and comparable accuracies to some state-of-the-art classifiers like SVM [12].

Contrast pattern based classifiers are sensitive to class imbalance problems [16,18]. The main reasons are the following: first, contrast pattern miners are based on patterns' frequency, therefore they are prone to generate more patterns for the majority class than for the minority class. Second, contrast patterns that predict the minority class are often highly specific and thus their support is very low, hence they are prone to be discarded in favor of more general contrast patterns that predict the majority class.

3   Resampling Methods

There are three approaches to deal with the class imbalance problem: *data level*, *algorithm level*, and *cost-sensitive* [16,27]. Resampling methods, belonging to the data level approach, are more versatile than the other two approaches since resampling methods can be applied independently of the supervised classifier, therefore most of the research has been done in this direction [2,17].

We can group resampling methods into three types: *oversampling* methods, which create new objects in the minority class, *undersampling* methods, which remove objects from the majority class, and *hybrid* methods that combine both oversampling and undersampling methods [5,16–18,21,23–25,28,29].

In this paper, we selected the most popular state-of-the-art resampling methods (see Table 1) including nine oversampling methods, three hybrid methods, and eight undersampling methods. All resampling methods with their default parameter values were executed using the KEEL Data-Mining software tool [4]. The main goal of our work is to offer researchers information regarding which resampling methods have similar behavior in order to simplify future research on resampling methods for contrast pattern based classifiers.

4   Correlation Study

This section presents the correlation study developed in this research. First, in Sect. 4.1, we describe the experimental setup. Then, in Sect. 4.2 we analyze the correlation obtained among the resampling methods and the base classifier selected in our study. Finally, in Sect. 4.3, we provide some discussion about the results.

4.1   Experimental Setup

For our experiments, we used 95 databases taken from the KEEL dataset repository[1] [3]. The databases have different characteristics regarding to the number of objects, number of features, and class imbalance ratio (see Table 2).

---

[1] http://www.keel.es/datasets.php.

Table 1. Summary of resamplig methods used in our study. No: the index associated to each resampling method in this paper; Abbreviation: the abbreviation name used in the literature and in this paper; Name and Reference: full name and reference; Type: the main approach used, Hybrid sampling (Hybrid), Oversampling (Over) or Undersampling (Under).

No	Abbreviation	Name and Reference	Type
1	SPIDER	Selective Preprocessing of Imbalanced Data [21]	Over
2	TL	Tomek's modification of Condensed Nearest Neighbor [5]	Under
3	ROS	Random oversampling [5]	Over
4	SPIDER2	Selective Preprocessing of Imbalanced Data 2 [21]	Over
5	NCL	Neighborhood Cleaning Rule [5]	Under
6	Borderline-SMOTE	Borderline Synthetic Minority Oversampling TEchnique [17]	Over
7	AHC	Aglomerative Hierarchical Clustering [7]	Over
8	SMOTE	Synthetic Minority Oversampling Technique [5]	Over
9	SMOTE-ENN	SMOTE + Edited Nearest Neighbor [5]	Hybrid
10	SMOTE-TL	SMOTE + Tomek's modification of Condensed Nearest Neighbor [5]	Hybrid
11	OSS	One Sided Selection [5]	Under
12	ADASYN	ADAptive SYNthetic Sampling [14]	Over
13	ADOMS	Adjusting the Direction Of the synthetic Minority classS examples [25]	Over
14	Safe Level SMOTE	Safe Level Synthetic Minority Oversampling TEchnique [17]	Over
15	CNN	Condensed Nearest Neighbor [5]	Under
16	CNNTL	CNN + Tomek's modification of Condensed Nearest Neighbor [5]	Under
17	RUS	Random undersampling [5]	Under
18	CPM	Class Purity Maximization [29]	Under
19	SMOTE-RSB	Hybrid Preprocessing using SMOTE and Rough Sets Theory [23]	Hybrid
20	SBC	Undersampling Based on Clustering [28]	Under

There are several measures to evaluate the performance of a classifier. Nevertheless the most used measure for class imbalance problems is the Area Under the Receiver Operating Characteristic curve (AUC) [15–17]. All our results are based on the AUC measure, which are averaged over 5-fold-cross-validation. Although the standard stratified cross-validation (SCV) is the most commonly employed method in the literature, we performed a Distribution optimally balanced-SCV (DOB-SCV) in order to avoid problems due to data distribution, especially for highly imbalanced databases [20]. All original dataset partitions with 5-fold-cross-validation used in this paper are available for downloading at the KEEL dataset repository.

We used Kendall's τ correlation, which is more closely related to the ranking task than correlations like Pearson's or Spearman's ρ [6]. Kendall's τ values range from -1 (perfect negative correlation) to 1 (perfect positive correlation).

We also used the Friedman test and the Bergmann-Hommel dynamic post-hoc procedure to compare all the results [8]. Post-hoc results will be shown using CD (*critical distance*) diagrams. In a CD diagram, the rightmost classifier is the best classifier, the position of the classifier within the segment represents its rank value, and if two or more classifiers share a thick line it means they have statistically similar behavior.

Table 2. Summary of the imbalanced databases used in our study. Name: the related name in the KEEL dataset repository; #Obj: number of objects; #Feat.: number of features; IR: class imbalance ratio [22].

Name	#Objects	#Feat.	IR	Name	#Objects	#Feat.	IR
glass1	214	9	1.82	ecoli0146vs5	280	6	13.00
ecoli0vs1	220	7	1.86	shuttlec0vsc4	1829	9	13.87
wisconsin	683	9	1.86	yeast1vs7	459	7	14.30
pima	768	8	1.87	glass4	214	9	15.46
iris0	150	4	2.00	ecoli4	336	7	15.80
glass0	214	9	2.06	pageblocks13vs4	472	10	15.86
yeast1	1484	8	2.46	abalone9vs18	731	8	16.40
haberman	306	3	2.78	dermatology6	358	34	16.90
vehicle2	846	18	2.88	zoo3	101	16	19.20
vehicle1	846	18	2.90	glass016vs5	184	9	19.44
vehicle3	846	18	2.99	shuttle2vsc4	129	9	20.50
glass0123vs456	214	9	3.20	shuttle6vs23	230	9	22.00
vehicle0	846	18	3.25	yeast1458vs7	693	8	22.10
ecoli1	336	7	3.36	glass5	214	9	22.78
newthyroid1	215	5	5.14	yeast2vs8	482	8	23.10
newthyroid2	215	5	5.14	lymphography normalfibrosis	148	18	23.67
ecoli2	336	7	5.46	flareF	1066	11	23.79
segment0	2308	19	6.02	cargood	1728	6	24.04
glass6	214	9	6.38	carvgood	1728	6	25.58
yeast3	1484	8	8.10	krvskzeroonevsdraw	2901	6	26.63
ecoli3	336	7	8.60	krvskonevsfifteen	2244	6	27.77
pageblocks0	5472	10	8.79	yeast4	1484	8	28.10
ecoli034vs5	200	7	9.00	winequalityred4	1599	11	29.17
yeast2vs4	514	8	9.08	poker9vs7	244	10	29.50
ecoli067vs35	222	7	9.09	yeast1289vs7	947	8	30.57
ecoli0234vs5	202	7	9.10	abalone3vs11	502	8	32.47
glass015vs2	172	9	9.12	winequalitywhite9vs4	168	11	32.60
yeast0359vs78	506	8	9.12	yeast5	1484	8	32.73
yeast0256vs3789	1004	8	9.14	krvskthreevseleven	2935	6	35.23
yeast02579vs368	1004	8	9.14	winequalityred8vs6	656	11	35.44
ecoli046vs5	203	6	9.15	ecoli0137vs26	281	7	39.14
ecoli01vs235	244	7	9.17	abalone17vs78910	2338	8	39.31
ecoli0267vs35	224	7	9.18	abalone21vs8	581	8	40.50
glass04vs5	92	9	9.22	yeast6	1484	8	41.40
ecoli0346vs5	205	7	9.25	winequalitywhite3vs7	900	11	44.00
ecoli0347vs56	257	7	9.28	winequalityred8vs67	855	11	46.50
yeast05679vs4	528	8	9.35	abalone19vs10111213	1622	8	49.69
vowel0	988	13	9.98	krvskzerovseight	1460	6	53.07
ecoli067vs5	220	6	10.00	winequalitywhite39vs5	1482	11	58.28
glass016vs2	192	9	10.29	poker89vs6	1485	10	58.40
ecoli0147vs2356	336	7	10.59	shuttle2vs5	3316	9	66.67
led7digit02456789vs1	443	7	10.97	winequalityred3vs5	691	11	68.10
ecoli01vs5	240	6	11.00	abalone20vs8910	1916	8	72.69
glass06vs5	108	9	11.00	krvskzerovsfifteen	2193	6	80.22
glass0146vs2	205	9	11.06	poker89vs5	2075	10	82.00
glass2	214	9	11.59	poker8vs6	1477	10	85.88
ecoli0147vs56	332	6	12.28	abalone19	4174	8	129.44
cleveland0vs4	177	13	12.62				

4.2   Correlation Analysis

In this section, we analyze different levels of correlation over the AUC results obtained from LCMine+CAEP before and after applying resampling methods. We include, as *base* classifier, to LCMine+CAEP without applying resampling methods.

For the correlation analysis, we performed a Kendall's τ correlation based on the AUC results of the contrast pattern based classifier before and after applying resampling methods. Figure 1 shows, in grayscale, the correlation results regarding to the values obtained in the Kendall's τ correlation. Darker values are associated to correlations closer to one, while lighter values are associated to values closer to zero.

Then, using an agglomerative clustering [1], the resampling methods were clustered in nine different groups with very high inner correlation and very low outer correlation. In Fig. 1, squares with a thick line group those methods belonging to the same cluster. The groups are the following:

Group 1. {AHC, Base, Boderline-SMOTE, ROS, SPIDER, SPIDER2, TL, NCL}
Group 2. {SMOTE, SMOTE-ENN, SMOTE-TL}
Group 3. {ADASYN, ADOMS, Safe Level SMOTE}
Group 4. {CNN, CNNTL}
Group 5. {OSS}
Group 6. {RUS}
Group 7. {CPM}
Group 8. {SMOTE-RSB}
Group 9. {SBC}

Our analysis shows that resampling methods into Group 1 have high correlation with the base classifier. Group 2 contains three resampling methods that have a similar behavior, that can be explained because SMOTE-ENN and SMOTE-TL are extensions of SMOTE. Results in Group 3 have high correlation because ADOMS and Safe Level SMOTE are modifications of SMOTE; and ADASYN produces similar results than SMOTE [14]. Group 4 has two undersampling methods based on Condensed Nearest Neighbor (CNN) which presents a high correlation among them. The rest of the groups have only one resampling method. Group 9 has negative correlation (close to zero) regarding to the remaining groups.

Figure 2 shows a CD diagram with a statistical comparison of the AUC results obtained from LCMine before and after applying resampling methods. Note that Group 1 does not have statistical difference among the resampling methods into this group, with the exception of TL and NCL. Nevertheless, TL and NCL have high correlation with the base classifier, they always improved the AUC results regarding to the base classifier. Group 2 achieved the best AUC results regarding all resampling methods selected and the base classifier. Groups 3 and 5 have no statistical difference with the base classifier and they have a similar position into the Friedman ranking. Groups 4, 7, and 9 shown statistical difference with the

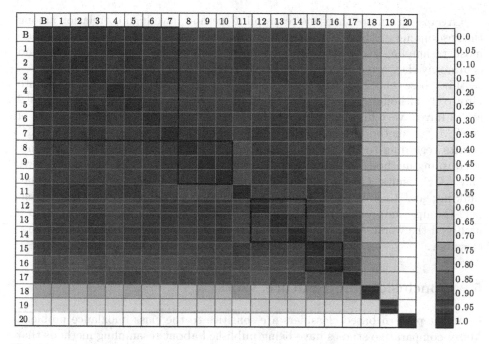

Fig. 1. Table of correlation among resampling methods and the base classifier ("B") using grayscale. The intensity of gray color is proportional to the positive correlation values.

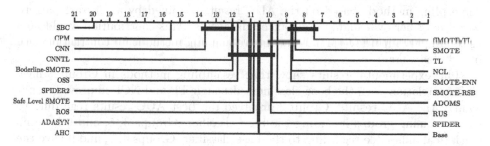

Fig. 2. CD diagram with a statistical comparison of the AUC results for the base classifier before and after using resampling methods over all the tested databases.

base classifier and they have the worst AUC results. Groups 6 and 8 have no statistical difference between them. These groups have a good position into the Friedman ranking and they have statistical difference with the base classifier.

4.3   General Concluding Remarks

The results shown in the previous section lead us to conclude that there are five resampling methods not correlated with any of the remaining 15 resampling methods or the base classifier.

Groups with more than one resampling method have high correlation among the resampling methods within each group. These groups are significant because most resampling methods contained in a group exhibit similar behavior and commonly they are extensions of the same resampling methods.

The base classifier has high correlation with resampling methods into Group 1, although only TL and NCL improved the AUC results. Groups 2 and 3 have a very high inner correlation because they contain only extensions of the SMOTE method. Resampling methods into Group 2 archived the best AUC results regarding to the remaining resampling methods. Group 4 contains only resampling methods based on Condensed Nearest Neighbor (CNN) which have bad AUC results. Groups 3, 5, and 6 have similar position into the Friedman ranking, and they have no statistical difference regarding to the base classifier. Group 8 improved the AUC results regarding the base classifier. Group 9 archived the worst AUC results regarding all resampling methods and the base classifier.

5   Conclusions and Future Work

Contrast pattern based classifiers are sensitive to the class imbalance problem. Many comparative studies have being published about resampling methods that aim to improve the accuracy in contrast pattern based classifiers. Nevertheless, no study have being published about correlations among resampling methods.

The main contribution of this paper is a correlation study among several resampling methods based on the AUC results obtained by a contrast pattern based classifier over highly imbalanced databases. This contribution would help us to simplify future research regarding resampling methods for contrast pattern based classifiers.

The experimental results show that resampling methods in Group 1 have high correlation with the base classifier, although TL and NCL improved significantly the AUC results. Group 2 archived the best AUC results regarding to the remaining groups including the base classifier. Groups 3, 5, and 6 have no statistical difference regarding to the base classifier. Groups 4, 7, and 9 have the worst AUC results. Groups 6 and 8 improved the AUC results regarding the base classifier. Finally, although the base classifier has a high correlation with some resampling methods, most of resampling methods improve the AUC results for the contrast pattern based classifier.

As future work, we plan to investigate about the influence of the imbalance ratio on these results. This way, we could suggest what resampling method would perform better for a given imbalanced dataset.

Acknowledgment. The authors want to thank Rebekah Hosse Clark for her valuable contributions improving the grammar and style of this paper.

References

1. Aggarwal, C.C., Reddy, C.K.: Data Clustering: Algorithms and Applications, 1st edn. Chapman & Hall/CRC, Boca Raton (2013)
2. Albisua, I., Arbelaitz, O., Gurrutxaga, I., Lasarguren, A., Muguerza, J., Pérez, J.: The quest for the optimal class distribution: an approach for enhancing the effectiveness of learning via resampling methods for imbalanced data sets. Prog. Artif. Intell. **2**(1), 45–63 (2013)
3. Alcalá-Fdez, J., Fernández, A., Luengo, J., Derrac, J., García, S.: KEEL data-mining software tool: data set repository, integration of algorithms and experimental analysis framework. J. Multiple-Valued Logic Soft Comput. **17**(2–3), 255–287 (2011)
4. Alcalá-Fdez, J., Sánchez, L., García, S., del Jesús, M.J., Ventura, S., Garrell, J.M., Otero, J., Romero, C., Bacardit, J., Rivas, V.M., Fernández, J.C., Herrera, F.: KEEL: a software tool to assess evolutionary algorithms for data mining problems. Soft Comput. **13**(3), 307–318 (2009)
5. Batista, G.E.A.P.A., Prati, R.C., Monard, M.C.: A study of the behavior of several methods for balancing machine learning training data. SIGKDD Explor. Newsl. **6**(1), 20–29 (2004)
6. Bruning, J.L., Kintz, B.L.: Computational Handbook of Statistics, 4th edn. Longman, New York (1997)
7. Cohen, G., Hilario, M., Sax, H., Hugonnet, S., Geissbuhler, A.: Learning from imbalanced data in surveillance of nosocomial infection. Artif. Intell. Med. **37**, 7–18 (2006)
8. Demšar, J.: Statistical comparisons of classifiers over multiple data sets. J. Mach. Learn. Res. **7**, 1–30 (2006)
9. Dong, G., Zhang, X., Wong, L., Li, J.: CAEP: classification by aggregating emerging patterns. In: Arikawa, S., Furukawa, K. (eds.) DS 1999. LNCS (LNAI), vol. 1721, pp. 30–42. Springer, Heidelberg (1999)
10. Fan, H., Ramamohanarao, K.: Fast discovery and the generalization of strong jumping emerging patterns for building compact and accurate classifiers. IEEE Trans. Knowl. Data Eng. **18**(6), 721–737 (2006)
11. García-Borroto, M., Martínez-Trinidad, J.F., Carrasco-Ochoa, J.A.: Finding the best diversity generation procedures for mining contrast patterns. Expert Syst. Appl. **42**(11), 4859–4866 (2015)
12. García-Borroto, M., Martínez-Trinidad, J.F., Carrasco-Ochoa, J.A., Medina-Pérez, M.A., Ruiz-Shulcloper, J.: LCMine: an efficient algorithm for mining discriminative regularities and its application in supervised classification. Pattern Recogn. **43**(9), 3025–3034 (2010)
13. García-Borroto, M., Martínez-Trinidad, J., Carrasco-Ochoa, J.: A survey of emerging patterns for supervised classification. Artif. Intell. Rev. **42**(4), 705–721 (2014)
14. He, H., Bai, Y., Garcia, E.A., Li, S.: ADASYN: adaptive synthetic sampling approach for imbalanced learning. In: 2008 International Joint Conference on Neural Networks (IJCNN 2008), pp. 1322–1328 (2008)
15. Huang, J., Ling, C.X.: Using AUC and accuracy in evaluating learning algorithms. IEEE Trans. Knowl. Data Eng. **17**(3), 299–310 (2005)
16. López, V., Fernández, A., García, S., Palade, V., Herrera, F.: An insight into classification with imbalanced data: empirical results and current trends on using data intrinsic characteristics. Inf. Sci. **250**, 113–141 (2013)

17. López, V., Triguero, I., Carmona, C.J., García, S., Herrera, F.: Addressing imbalanced classification with instance generation techniques: IPADE-ID. Neurocomputing **126**, 15–28 (2014)
18. Loyola-González, O., García-Borroto, M., Medina-Pérez, M.A., Martínez-Trinidad, J.F., Carrasco-Ochoa, J.A., De Ita, G.: An empirical study of oversampling and undersampling methods for LCMine an emerging pattern based classifier. In: Carrasco-Ochoa, J.A., Martínez-Trinidad, J.F., Rodríguez, J.S., di Baja, G.S. (eds.) MCPR 2012. LNCS, vol. 7914, pp. 264–273. Springer, Heidelberg (2013)
19. Menardi, G., Torelli, N.: Training and assessing classification rules with imbalanced data. Data Min. Knowl. Disc. **28**(1), 92–122 (2014)
20. Moreno-Torres, J.G., Saez, J.A., Herrera, F.: Study on the impact of partition-induced dataset shift on k-Fold cross-validation. IEEE Trans. Neural Netw. Learn. Syst. **23**(8), 1304–1312 (2012)
21. Napierała, K., Stefanowski, J., Wilk, S.: Learning from imbalanced data in presence of noisy and borderline examples. In: Szczuka, M., Kryszkiewicz, M., Ramanna, S., Jensen, R., Hu, Q. (eds.) RSCTC 2010. LNCS, vol. 6086, pp. 158–167. Springer, Heidelberg (2010)
22. Orriols-Puig, A., Bernadó-Mansilla, E.: Evolutionary rule-based systems for imbalanced data sets. Soft. Comput. **13**(3), 213–225 (2009)
23. Ramentol, E., Caballero, Y., Bello, R., Herrera, F.: SMOTE-RSB*: a hybrid preprocessing approach based on oversampling and undersampling for high imbalanced data-sets using SMOTE and rough sets theory. Knowl. Inf. Syst. **33**(2), 245–265 (2011)
24. Sáez, J.A., Luengo, J., Stefanowski, J., Herrera, F.: Managing borderline and noisy examples in imbalanced classification by combining SMOTE with ensemble filtering. In: Corchado, E., Lozano, J.A., Quintián, H., Yin, H. (eds.) IDEAL 2014. LNCS, vol. 8669, pp. 61–68. Springer, Heidelberg (2014)
25. Tang, S., Chen, S.: The Generation mechanism of synthetic minority class examples. In: 5th International Conference on Information Technology and Applications in Biomedicine (ITAB 2008), pp. 444–447 (2008)
26. Weiss, G., Tian, Y.: Maximizing classifier utility when there are data acquisition and modeling costs. Data Min. Knowl. Disc. **17**(2), 253–282 (2008)
27. Yap, B., Rani, K., Rahman, H., Fong, S., Khairudin, Z., Abdullah, N.: An application of oversampling, undersampling, bagging and boosting in handling imbalanced datasets. In: Herawan, T., Deris, M.M., Abawajy, J. (eds.) Proceedings of the First International Conference on Advanced Data and Information Engineering (DaEng 2013). LNEE, vol. 285, pp. 13–22. Springer, Heidelberg (2014)
28. Yen, S.-J., Lee, Y.-S.: Under-sampling approaches for improving prediction of the minority class in an imbalanced dataset. In: Huang, D.-S., Li, K., Irwin, K. (eds.) ICIC 2006. LNCIS, vol. 344, pp. 731–740. Springer, Heidelberg (2006)
29. Yoon, K., Kwek, S.: An unsupervised learning approach to resolving the data imbalanced issue in supervised learning problems in functional genomics. In: 5th International Conference on Hybrid Intelligent Systems (HIS 2005), pp. 303–308 (2005)
30. Zhang, X., Dong, G.: Overview and analysis of contrast pattern based classification. In: Dong, G., Bailey, J. (eds.) Contrast Data Mining: Concepts, Algorithms, and Applications. Data Mining and Knowledge Discovery Series, vol. 11, pp. 151–170. Chapman & Hall/CRC, Boca Raton (2012)

Boosting the Permutation Based Index
for Proximity Searching

Karina Figueroa[1]([✉]) and Rodrigo Paredes[2]

[1] Facultad de Ciencias Físico-Matemáticas, Universidad Michoacana, Morelia, Mexico
karina@fismat.umich.mx
[2] Departamento de Ciencias de la Computación, Universidad de Talca, Talca, Chile
raparede@utalca.cl

Abstract. Proximity searching consists in retrieving objects out of a database *similar* to a given query. Nowadays, when multimedia databases are growing up, this is an elementary task. The permutation based index (PBI) and its variants are excellent techniques to solve proximity searching in high dimensional spaces, however they have been surmountable in low dimensional ones. Another PBI's drawback is that the distance between permutations cannot allow to discard elements safely when solving similarity queries.

In the following, we introduce an improvement on the PBI that allows to produce a better promissory order using less space than the basic permutation technique and also gives us information to discard some elements. To do so, besides the permutations, we quantize distance information by defining distance rings around each permutant, and we also keep this data. The experimental evaluation shows we can dramatically improve upon specialized techniques in low dimensional spaces. For instance, in the real world dataset of NASA images, our boosted PBI uses up to 90 % less distances evaluations than AESA's, the state-of-the-art searching algorithm with the best performance in this particular space.

Keywords: Permutation based index · Distance quantization · Proximity searching

1 Introduction

Nowadays, similarity searching has become an important task for retrieving objects in a multimedia database; with applications in pattern recognition, data mining and computational biology, to name a few. This task can be mapped into a metric space problem. A Metric Space is a pair $(\mathbb{X}, d)$, where $\mathbb{X}$ is a universe of objects, and d is a distance function $d : \mathbb{X} \times \mathbb{X} \to \mathbb{R}^+ \cup \{0\}$. The distance function is a *metric* if it satisfies, for all $x, y, z \in \mathbb{X}$, the following properties: reflexivity $d(x, y) = 0$ iff $x = y$, symmetry $d(x, y) = d(y, x)$, and triangle

This work is partially funded by National Council of Science and Technology (CONACyT) of México, Universidad Michoacana de San Nicolás de Hidalgo, México, and Fondecyt grant 1131044, Chile.

© Springer International Publishing Switzerland 2015
J.A. Carrasco-Ochoa et al. (Eds.): MCPR 2015, LNCS 9116, pp. 103–112, 2015.
DOI: 10.1007/978-3-319-19264-2_11

inequality $d(x, y) \leq d(x, z) + d(z, y)$. The last one being useful to discard objects when solving similarity queries.

In practical applications, we have a subset $\mathbb{U}$ of n objects taken from the universe $\mathbb{X}$. So, the similarity searching problem can be defined as the problem of finding a small subset $\mathbb{S} \subset \mathbb{U}$ of the objects that are close to a given query q with respect to a particular metric function.

Basically, there are two types of queries: range query $(q, r)_d$ and k-nearest neighbor query $kNN(q)_d$. The first one retrieves all the objects within a given radius r measured from q, that is, $(q, r)_d = \{u \in \mathbb{U}, d(q, u) \leq r\}$. The second retrieves the k objects in $\mathbb{U}$ that are the closest to q. Formally, $|kNN(q)_d| = k$, and $\forall\, u \in kNN(q)_d, v \in \mathbb{U} \setminus kNN(q)_d, d(u, q) \leq d(v, q)$.

There are some indices to speed up similarity queries. One of them, the permutation-based index (PBI) [3] has a competitive performance in the hard case of high dimensional spaces. Oddly, in low dimensional spaces, this technique has poor performance when compared with, for instance, the pivot-based index (which is particularly well suited for low dimensional spaces, as reported in [4]). Another important drawback of the PBI is that it does not allow to discard objects when solving similarity queries.

Our contribution consists in granting the basic PBI the capability of safely discard objects. For this sake, we enhance the PBI with distance information in a convenient way, so that the increment of the space requirement is very small. Our technique allows to improve the retrieval of the PBI in low and medium dimensional metric spaces in a dramatic way. As a matter of fact, in the real world metric space of NASA images, we obtain the true answer of the $1NN(q)_d$ using 90 % less distances evaluations than AESA, the best pivot index.

2   Related Work

2.1   Metric Space Indices

There are three kinds of indices for proximity searching in metric spaces, namely, pivot-based indices, partition-based indices and permutation-based indices. In [4], the reader can find a complete survey of the first two kinds.

Pivot-based indices consider a subset of objects $A = \{a_1, a_2, \ldots, a_{|A|}\} \subseteq \mathbb{U}$, called the pivots. These indices keep all the $n|A|$ distances between every object of the dataset $\mathbb{U}$ to all $a_i \in A$. Later, to solve a range query $(q, r)_d$, the pivot-based searching algorithms measure $d(q, a_1)$ and, by virtue of the triangle inequality, for every $u \in \mathbb{U}$ they lower bound $d(q, u) \geq |d(q, a_1) - d(u, a_1)|$. So, if $|d(q, a_1) - d(u, a_1)| > r$ then $d(q, u) > r$ and they discard u avoiding the computation of $d(q, u)$. Once they are done with a_1, they use a_2 to try to discard elements from the remaining set, and so on, until using all the pivots in A. The elements that still cannot be discarded at this point are directly compared with q.

There are many techniques based on this idea. Some of them try to reduce the memory requirements in order to get a small index, and others to reduce the number of distance evaluations. These kinds of techniques have a competitive performance in low dimensional spaces. One can imagine that a low dimensional space is one that can be embedded in a uniformly distributed vector space whose dimension is lower than 16, *preserving* the relative distances among objects.

Among the pivoting techniques, AESA [10] excels in the searching performance. To do this, AESA considers that every object could be a potential pivot. So, it stores the whole matrix of distances between every object pair. The matrix requires $\frac{n(n-1)}{2}$ memory. Since every object can operate as a pivot, the authors also define a scheme to sequencing the pivot selection. For this sake, they use an array $SumLB$ which accumulates the lower bounds of the distances between the query and every non-discarded object, with respect to all the previous objects in $\mathbb{U}$ used as pivots. Formaly, if it has previously selected i pivots, $SumLB(u) = \sum_{j=1}^{i} |d(q, a_j) - d(u, a_j)|$. So, the first pivot is an object chosen at random, and from the second pivot, AESA uses the non-discarded object that minimize $SumLB(u)$. AESA is the bottom line of exact algorithms. However, in several real-world scenarios, AESA is impractical to use, as its memory requirement is only plausible for reduced size dataset (up to tens of thousands).

Partition-based indices split the space into zones as compact as possible and assign the objects to these zones. To do this, a set of centers $\{c_1, \ldots, c_m\} \in \mathbb{U}$ is selected, so that each other object is placed in the zone of its closest center. These indices have a competitive performance in high dimensional spaces.

2.2   The Permutation Based Index (PBI)

Let $\mathbb{P} \subset \mathbb{U}$ be a subset of permutants of size m. Each element $u \in \mathbb{U}$ induces a preorder $\leq_u$ given by the distance from u towards each permutant, defined as $y \leq_u z \Leftrightarrow d(u, y) \leq d(u, z)$, for any pair $y, z \in \mathbb{P}$.

Let $\Pi_u = i_1, i_2, \ldots, i_m$ be the permutation of u, where permutant $p_{i_j} <_u p_{i_{j+1}}$. Permutants at the same distance take an arbitrary but consistent order. Every object in $\mathbb{U}$ computes its preorder of $\mathbb{P}$ and associates it to a permutation which is stored in the index (PBI does not store distances). Thus, a simple implementation needs nm space. Nevertheless, as only the permutant identifiers are necessary, several identifiers can be compacted in a single machine word.

The hypothesis of the PBI is that two equal objects are associated to the same permutation, while similar objects are, hopefully, related to similar permutations. So, if Π_u is similar to Π_v one expects that u is close to v.

Later, given the query $q \in \mathbb{X} \setminus \mathbb{U}$, the PBI search algorithm computes its permutation Π_q and compares it with all the permutations stored in the index. Then, the dataset $\mathbb{U}$ is traversed in increasing permutation dissimilarity, comparing the objects in $\mathbb{U}$ with the query using the distance d of the particular metric space. Regrettably, PBI cannot discard objects at query time. Instead, a premature cut off in the reviewing process produces a probabilistic search algorithm (as it reports the right answer to the query with some probability).

There are many similarity measures between permutations. One of them is the L_p family of distances [5], that obeys Eq. (1), where $\Pi^{-1}(i_j)$ denotes the position of permutant i_j within the permutation Π.

$$L_p(\Pi_u, \Pi_q) = \sum_{j=[1,|\mathbb{P}|]} |\Pi_u^{-1}(i_j) - \Pi_q^{-1}(i_j)|^p \tag{1}$$

With $p = 1$, we obtain *Spearman Footrule* (S_F) and for $p = 2$, *Spearman Rho* (S_ρ). For example, let $\Pi_u = (42153)$ and $\Pi_q = (32154)$ be the object u and query q permutations, respectively. So, $S_F(\Pi_u, \Pi_q) = 8$, $S_\rho(\Pi_u, \Pi_q) = 32$. As reported in [3], S_F has a good performance, requiring less computational effort than S_ρ.

There have been several works trying to improve the PBI's performance. There are some techniques to select good permutants obtaining little improvement [1,7]. Other variants have been proposed with the aim of reducing the PBI's space requirement [2,8,9] by discarding some portions of the permutations; however, these techniques lose precision in the query retrieval.

In general terms, all of the PBI's variants are designed for high dimensional spaces and none of them can prune the dataset when solving similarity queries.

2.3 Distance Quantization

A simple way to reduce the memory requirement when representing the distances among objects is to quantize them. Of course, there is a tradeoff between memory requirement and precision. However, there are some cases where the quantization is effective. For instance, in BAESA [6], the authors quantize the whole distance matrix of AESA [10]. Using only four bits per distance, eight times less space than AESA, BAESA needs just 2.5 times the distance computations of AESA.

3 Our Contribution

Essentially, we introduce a novel modification into the PBI. It consists in enhancing it with quantized distance information. Since we now have distance data, we can safely discard objects, and this dramatically improves the performance of the PBI search algorithm in low and medium dimensionality metric spaces.

In the following, we describe the modification of the PBI and also how to adapt its searching algorithm in order to benefit from the extra data.

3.1 Enhancing PBI with Distance Quantization

For each object $u \in \mathbb{U}$, the index stores not only its permutation, but also quantized distance information. To do that, for each permutant $p \in \mathbb{P}$, we define ζ concentric zones $z_1, \ldots, z_\zeta$ limited by $\zeta - 1$ radii $r_1, \ldots, r_{\zeta-1}$ and two extra values $r_0 = 0$ and $r_\zeta = \infty$ (for limiting the first and last zone). So, each object $u \in \mathbb{U}$ is located in the zone z_i where it satisfies the relation $r_{i-1} < d(u, p) \leq r_i$.

To compute the zones, we do not need extra distance computations. In fact, we compute them during the calculation of the permutation. We call Π_u the permutation for u and Z_u the zones information for object u. So, for the permutant in the j-th cell of the permutation, $Z_u(j)$ indicates in which of its zones the object u lies. Hence, every object has its permutation and new extra information, the zone where it belongs according to every permutant.

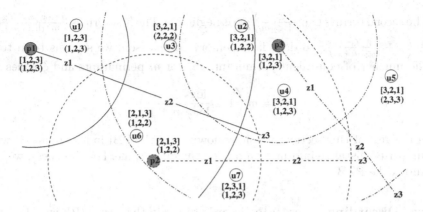

Fig. 1. Example of our technique, permutations and zones for every object.

Geometrically, for each $p_i \in \mathbb{P}$, this can be seen as partitioning the space with ζ distance rings centered in p_i. Figure 1 illustrates the idea, where the dataset is $\mathbb{U} = \{p_1, p_2, p_3, u_1, \ldots, u_7\}$ and the first three objects are the permutants ($\mathbb{P} = \{p_1, p_2, p_3\}$). Each one has 3 zones (which are denoted by $z1, z2, z3$). For each object in $\mathbb{U}$, we show its permutation Π_u in the sequence closed by [ ] and its zone information Z_u in the one closed by ( ). For example, for u_6, its permutation is $[2, 1, 3]$ and for permutants p_2, p_1 and p_3, u_6 belongs to zones 1, 2 and 2. Notice that with our proposal, now we can figure out whether two objects with equal permutations are close or not.

Our proposal has two significant advantages, namely, (*i*) now we can discard some elements without compute their distances directly, and (*ii*) we improve the prediction power of the PBI.

In terms of space, besides the table of permutations, we need to store the distance information. Each object needs $m\lceil \log_2 m \rceil$ bits for the permutation (remember that $m = |\mathbb{P}|$) and also needs to store the zones. In order to represent the m zones (one for each permutant), we need $m\lceil \log_2 \zeta \rceil$ bits (where ζ is the number of zones). Finally, for each zone of a permutant, we need a float number to store the radius. This counts for $m\zeta 32$ bits. Adding up for the n objects we obtain $nm(\lceil \log_2 m \rceil + \lceil \log_2 \zeta \rceil) + \zeta m 32$ bits.

For the sake of determining an equivalence in space requirement between permutants with or without zones, we follow this rationale. The standard PBI uses $nm\lceil \log_2 m \rceil$ bits. The extra memory used by the zones allows to add more permutants to the plain PBI, but is not enough to double the number of permutants. Assuming that m is a power of 2, any extra permutant forces to use another bit. We also assume that ζ is a power of 2. So, in the numerator of Eq. (2), we have the space used by the PBI with zones, and in its denominator, the space used by a plain PBI with m^* permutants, where $m^* \in (m, 2m)$.

$$\frac{nm(\log_2 m + \log_2 \zeta) + \zeta m 32}{nm^*(\log_2 m + 1)} = \frac{m(\log_2 m + \log_2 \zeta)}{m^*(\log_2 m + 1)} + \frac{\zeta m 32}{nm^*(\log_2 m + 1)} \qquad (2)$$

The second term is $O\left(\frac{\zeta}{n\log_2 m}\right)$, so is negligible. The first term is $\frac{m(\log_2 m+\log_2 \zeta)}{m^*(\log_2 m+1)}$ $= \frac{m}{m^*}\left(1+\frac{\log_2 \zeta-1}{\log_2 m+1}\right)$. To do a fair memory comparison, we set this term to 1. So, the number of equivalent permutants m^* for m permutants and ζ zones is

$$m^* = m\left(1+\frac{\log_2 \zeta - 1}{\log_2 m + 1}\right) \tag{3}$$

In the rest of this section, we show how to use the PBI index enhanced with the quantized distance information during the query time. For shortness, we call this index the PZBI.

Object Discarding. In order to discard objects in the new PZBI, we adapt the pivot excluding criterion. To do so, after computing the distance from q to all the permutants in order to produce Π_q, we compute the zones where q belongs (Z_q) and the zones where the query ball $(q,r)_d$ intersects. For this sake, we manage two arrays FZ and LZ. For each permutant, in FZ and LZ we store the first and last zone, respectively, where the query ball intersects. Therefore, given the query radius r, for each permutant $p \in \mathbb{P}$, FZ_p and LZ_p store the number of the zone that contains $d(p,q) - r$ and $d(p,q) + r$, respectively.

With this, when we are reviewing the objects, for each permutant $p \in \mathbb{P}$ we discard, by virtue of the triangle inequality, every element in $\mathbb{U}$ that belongs to any zone that is not in the range $[FZ_p, LZ_p]$. This allows discarding some elements without performing the direct comparison with the query.

Note that we can simulate $kNN(q)_d$ queries with range queries whose initial radius is ∞, and that its radius reduces to the final one as long as the search process progresses.

Improving the Distance Permutation. Since we have more information per object (its permutation and corresponding zones), we can use the zone information so as to improve the reviewing order when solving a similarity query. We propose several strategies as follow. Let us define $Z_D(Z_u, Z_v) = \sum_{j=[1,|\mathbb{P}|]} |Z_u(j) - Z_v(j)|$. Z_D accumulates the sum of absolute values of differences in the zone identifiers between the objects u and $v \in \mathbb{X}$ for all the permutants. With this, we define the following variants:

- **PsF:** Just computing Spearman Footrule S_F as the plain PBI, see Eq. (1).
- **SPZ:** It is the sum of $S_F(\Pi_u, \Pi_q)$ and $Z_D(Z_u, Z_q)$.
- **PZ:** We compute $S_F(\Pi_u, \Pi_q)$ and $Z_D(Z_u, Z_q)$, separately. Next, we sort by S_F in increasing order and use Z_D to break ties.
- **ZP:** Analogous to **PZ**, but we sort by $Z_D(Z_u, Z_q)$ and break ties with $S_F(\Pi_u, \Pi_q)$.

Computing Radius. Another important issue to define is how to establish the radii of the concentric zones. Every radius can be asigned as follow:

- **Uniformly distributed by distances.** (REQ). We obtain a sample of objects and compute their distance towards the permutant. This gives us a good approximation of the distribution of distances with respect to that permutant. Let us call r_{max} and r_{min} the maximum and minimum distance computed in the sample for that permutant. Then, we basically use $(r_{max} - r_{min})/\zeta$ as the increment between one radius to the next.
- **Uniformly distributed by elements** (EEQ). Once again we obtain a sample of objects and compare them with the permutant. Later, we sort the distances and select points taken at regular intervals (a fraction $\frac{1}{\zeta}$ of the sample size).

4 Experimental Evaluation

We tested our contribution with two kinds of databases: uniformly distributed vectors in the unitary cube of dimension in [8, 32] and NASA images. The first one allows us to know the performance of the search algorithm and how the parameters can change the results. The second one gives us a good idea of how our technique works in real life.

4.1 Unitary Cube

We use a dataset with 80,000 vectors uniformly distributed in the unitary cube using the Euclidean distance. We tested $1NN(q)_d$ queries with a query set of size 100 in dimension 8 to 32.

In Fig. 2, we show the performance of the PZBI using 8 permutants in dimension 8, with REQ parameter. In Fig. 2(a), the y-axe shows the average of the percentage of elements retrieved by the $1NN(q)_d$, as we review an increasing portion of the database (x-axe). Figure 2(b) shows the impact of the number of zones in the performance of the PZBI. Notice that we can process $1NN(q)_d$ queries faster than pivot-based algorithm. For instance, with 8 zones and EEQ, PZBI requires up to 81 % less distance evaluations. Pivots are represented with a horizontal line just to allow a visual comparison of the results. In this plot, we can notice that the strategy SPZ is better than PZ or ZP.

In Fig. 3, we use dimensions 8, 16 and 32. We show that the PBI is beated by the PZBI using 8 zones and SPZ in low dimensional spaces. Notice that the PBI becomes better in high dimension. Pivot-based algorithm in dimension 16 is out the plot with almost 54,000 distances. In this figure we use m^* permutants and we show that in high dimension is better to use more permutants, however, in low dimension, is better to split the space with zones.

4.2 NASA Images

We use a dataset of 40,150 20-dimensional feature vectors, generated from images downloaded from NASA[1], where duplicated vectors were eliminated. We also use Euclidean distance.

---

[1] At http://www.dimacs.rutgers.edu/Challenges/Sixth/software.html.

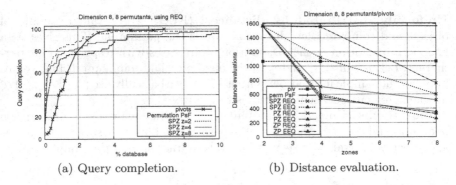

(a) Query completion.   (b) Distance evaluation.

Fig. 2. Searching performance in the space of vectors of dimension 8, using 8 permutants/pivots.

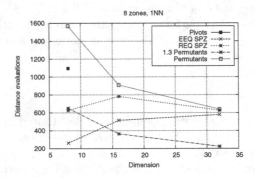

Fig. 3. Searching performance in the space of vectors, with 8 zones, using SPZ for dimensions 8, 16 and 32.

In Fig. 4, we use 8 and 16 permutants with 8 zones (Fig. 4 (a) and (b), respectively). Notice that using the strategy SPZ we have an excellent improvement in the distance evaluations. In this figure, we are comparing our results with AESA (see Sect. 2.1) and its quantized version BAESA (see Sect. 2.3). Notice that our PZBI needs $\frac{1}{3}$ of the distance computations used by AESA, as can be seen when using 8 zones, EEQ space partitioning and SPZ. AESA and pivots are represented with an horizontal line in order to simplify the visual comparison.

In Fig. 5, we test $kNN(q)_d$ queries, varying $k \in [1, 10]$ (that is, increasing the size of the query answer), the number of zones $\zeta \in [8, 32]$, with two different size of permutants, $m = |\mathbb{P}| = 8$ and 64. In Fig. 5(a), we notice that using $m = 8$ permutants, we compute less than 50 % of distance evaluations of AESA for $1NN(q)_d$ and just 3 times or $10NN(q)_d$ using significantly less space in the index. We also notice that with just 8 permutants and 8 zones, we can retrieve up to $4NN(q)_d$ faster than AESA. On the other hand, using $m = 64$ permutants with 32 zones (5 bits per element) we are computing just 25 % of the pivot technique even in the hardest case.

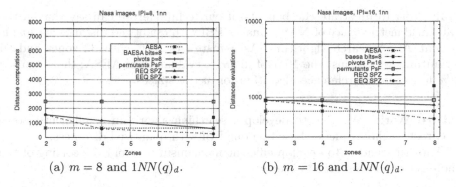

(a) $m = 8$ and $1NN(q)_d$. (b) $m = 16$ and $1NN(q)_d$.

Fig. 4. Nasa image dataset. Changing the number of zones and the impact on the selection of REQ or EEQ.

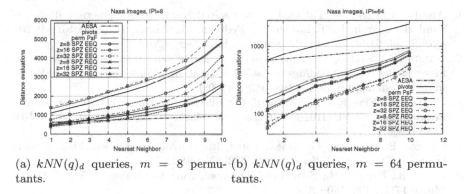

(a) $kNN(q)_d$ queries, $m = 8$ permutants. (b) $kNN(q)_d$ queries, $m = 64$ permutants.

Fig. 5. NASA image dataset. $kNN(q)_d$ queries, varying k.

In Fig. 5(b), we use 64 permutants and we have a better performance. For example, with 32 zones, 64 permutants and REQ we use just the 10 % of distance of AESA, that is to say 90 % less than the reference-algorithm for metric spaces.

In both plots (Figs. 4 and 5), our technique beats AESA. This is really surprising, as AESA used to be the lower bound of searching in metric spaces.

5   Conclusions and Future Work

The basic Permutant Based Index (PBI) has shown an excellent performance in high dimensional metric spaces. Its main drawback is that it does not allow to safely discard objects when solving similarity queries. Our contribution consists in granting the basic PBI the capability of safely discard objects. For this sake, we enhance the PBI with distance information in a quantized way.

Our technique allows to improve the retrieval of the PBI in low and medium dimensional metric spaces in a dramatic way.

In order to illustrate the benefits of our technique, we can say that in the real world metric space of NASA images PBI with quantized distance is capable to beat AESA search algorithm. As a matter of fact, with 32 zones and 64 permutants we use just the 10 % of distance evaluation of AESA, that is to say 90 % less than the reference-algorithm for metric spaces.

Future Work. We plan to develop a searching algorithm to efficiently solve k-nearest neighbor queries based on our PBI with quantized distances.

Another trend is to develop efficient mechanism to avoid the sorting of the whole set of non-discarded objects.

References

1. Amato, G., Esuli, A., Falchi, F.: Pivot selection strategies for permutation-based similarity search. In: Brisaboa, N., Pedreira, O., Zezula, P. (eds.) SISAP 2013. LNCS, vol. 8199, pp. 91–102. Springer, Heidelberg (2013)
2. Amato, G., Savino, P.: Approximate similarity search in metric spaces using inverted files. In: Proceedings of the 3rd International Conference on Scalable Information Systems, pp. 28:1–28:10. ICST (2008)
3. Chávez, E., Figueroa, K., Navarro, G.: Effective proximity retrieval by ordering permutations. IEEE Trans. Pattern Anal. Mach. Intell. (TPAMI) **30**(9), 1647–1658 (2009)
4. Chávez, E., Navarro, G., Baeza-Yates, R., Marroquín, J.: Searching in metric spaces. ACM Comput. Surv. **33**(3), 273–321 (2001)
5. Fagin, R., Kumar, R., Sivakumar, D.: Comparing top k lists. SIAM J. Discrete Math. **17**(1), 134–160 (2003)
6. Figueroa, K., Fredriksson, K.: Simple space-time trade-offs for AESA. In: Demetrescu, C. (ed.) WEA 2007. LNCS, vol. 4525, pp. 229–241. Springer, Heidelberg (2007)
7. Figueroa, K., Paredes, R.: An effective permutant selection heuristic for proximity searching in metric spaces. In: Martínez-Trinidad, J.F., Carrasco-Ochoa, J.A., Olvera-Lopez, J.A., Salas-Rodríguez, J., Suen, C.Y. (eds.) MCPR 2014. LNCS, vol. 8495, pp. 102–111. Springer, Heidelberg (2014)
8. Figueroa Mora, K., Paredes, R.: Compact and efficient permutations for proximity searching. In: Carrasco-Ochoa, J.A., Martínez-Trinidad, J.F., Olvera López, J.A., Boyer, K.L. (eds.) MCPR 2012. LNCS, vol. 7329, pp. 207–215. Springer, Heidelberg (2012)
9. Mohamed, H., Marchand-Maillet, S.: Quantized ranking for permutation-based indexing. In: Brisaboa, N., Pedreira, O., Zezula, P. (eds.) SISAP 2013. LNCS, vol. 8199, pp. 103–114. Springer, Heidelberg (2013)
10. Vidal, E.: An algorithm for finding nearest neighbors in (approximately) constant average time. Pattern Recogn. Lett. **4**, 145–157 (1986)

Image Processing and Analysis

Rotation Invariant Tracking Algorithm Based on Circular HOGs

Daniel Miramontes-Jaramillo[1]([⊠]), Vitaly Kober[1,2],
and Víctor Hugo Díaz-Ramírez[3]

[1] Department of Computer Science, CICESE, 22860 Ensenada, BC, Mexico
dmiramon@cicese.edu.mx
[2] Department of Mathematics, Chelyabinsk State University,
Chelyabinsk, Russian Federation
vkober@cicese.mx
[3] CITEDI-IPN, 22510 Tijuana, BC, Mexico
vhdiaz@citedi.mx

Abstract. Tracking of 3D objects based on video information has become an important problem in the last years. We present a tracking algorithm based on rotation-invariant HOGs over a data structure of circular windows. The algorithm is also robust to geometrical distortions. The performance of the algorithm achieves a perfect score in terms of objective metrics for real-time tracking using multicore processing with GPUs.

Keywords: Tracking · Histogram of oriented gradients · Circular window

1   Introduction

Among image processing problems, tracking has become popular, and it is expected to be further upraised in the next decade owing to revolutionary technology such as intelligent TVs, Oculus Rift[1], Google Project Glass[2], and Microsoft HoloLens[3]. These vision technologies need reliable tracking algorithms to automatically control various objects such as eyes for positioning or hands for different gestures.

Tracking systems can be classified by the type of descriptor [1] as follows: template trackers [2–4] use histograms and other data structures to describe objects; silhouette trackers [5–7] use shapes and edges of objects; feature trackers [8–10] extract interest points of targets.

In this paper we deal with the tracking problem based on iterative matching algorithm and position prediction. First, we extract a data structure containing circular windows (CWMA) [11] with histograms of both oriented gradients

---

[1] Oculus VR, Oculus Rift, http://www.oculusvr.com/.
[2] Google, Project Glass, http://www.google.com/glass/.
[3] Microsoft Corporation, Microsoft HoloLens, http://www.microsoft.com/microsoft-hololens/.

© Springer International Publishing Switzerland 2015
J.A. Carrasco-Ochoa et al. (Eds.): MCPR 2015, LNCS 9116, pp. 115–124, 2015.
DOI: 10.1007/978-3-319-19264-2_12

(HOG) [12] and radial gradients [13] to provide rotation invariance. After finding the object in the first frame, kinematic prediction model [14] is utilized to follow the target across the video sequence. In order to achieve real-time processing, the proposed algorithm is implemented with GPUs. Such implementation takes advantage of multicore and SIMD technologies to process multiple histograms and to achieve 30 FPS rate of tracking processing.

The performance of the proposed algorithm in a test database is compared with that of other trackers based on matching algorithms such as SIFT [15], SURF [16] and ORB [17] in terms of success rate percentage and processing time.

The paper is organized as follows. Section 2 describes noise removal preprocessing. Computation of radial HOGs in circular windows is presented in Sect. 3. Section 4 introduces object descriptors. Sections 5 and 6 present the proposed tracking algorithm. Finally, computer simulations and discussion are given in Sect. 8.

2   Noise Removal

The first step in common tracking algorithms is suppression of additive sensors noise by a Gaussian filter. In order to adaptively remove additive noise first it is necessary to estimate its parameters in each frame. We assume that an input frame contains additive white noise. The autocorrelation function of the noise is the Kronecker delta function. Suppose the additive noise model given as

$$f'(x,y) = f(x,y) + n(x,y),$$ (1)

where f' is the observed noisy image, f is an original image, and n is white noise. Since the image signal and noise are statistically independent, the autocorrelation function of the observed image consists of the autocorrelation function of the original image and the noise variance added only to the origin. Assuming that the autocorrelation function of the original image is a monotonic function in the vicinity of the origin, the signal variance can be estimated as the difference between the value at the origin of the autocorrelation function of the noisy image and the value obtained by linearly extrapolating to the origin with other values of the autocorrelation function of noisy image. It is illustrated in Fig. 1.

The estimated noise standard deviation (σ_n) is used as parameter for the Gaussian filter. Note that white noise affects the orientation of pixel gradients. The number of quantized directions Q for the histograms can be chosen as a function of the noise standard deviation as follows:

$$Q = \left\lceil \frac{360}{\sigma_n} \right\rceil.$$ (2)

3   Radial HOGs on Circular Windows

Let R be a closed disk with radius r and center point $c = (x,y)$, and let $p_i = (x_i, y_i)$ be a point within the disk. At each point p_i we define two orthogonal vectors relative to the origin [13]. If $(u_i, v_i) = p_i - c$ and $\psi_i = atan(u_i, v_i)$, then

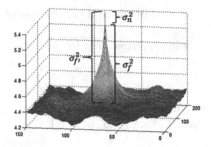

Fig. 1. Estimation of white noise variance.

$$rad_i = (cos\psi, sin\psi), \ tan_i = (-sin\psi, cos\psi),\qquad(3)$$

where rad_i and tan_i are the radial and tangential directions at $p_i - c$. Figure 2 shows the angles $\psi_i \in R$.

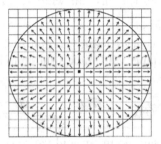

Fig. 2. Mask of radial angles R.

The size of R is the same as that of the sliding window running across the frame image. Let be a matrix that transforms the Cartesian gradients to radial gradients, that is,

$$T = \begin{bmatrix} rad \\ tan \end{bmatrix} = \begin{bmatrix} cos\psi & sin\psi \\ -sin\psi & cos\psi \end{bmatrix}.\qquad(4)$$

First, let us compute the gradient $g = (gx, gy)$ at each point in the window with the Sobel operator [18], and then the radial gradient can be computed as follows:

$$g_r = Tg^T = \begin{bmatrix} cos\psi & sin\psi \\ -sin\psi & cos\psi \end{bmatrix} \begin{bmatrix} gx \\ gy \end{bmatrix}.\qquad(5)$$

It can be seen that this transform yields rotation-invariant gradients if the mask and the window have the same size and rotated pixels are at the same distance from the origin as non-rotated pixels.

Next, let us compute the magnitude and orientation of pixels in the window,

$$mag\left(x,y\right) = \sqrt{gx_r^2 + gy_r^2}, \tag{6}$$

$$ori\left(x,y\right) = \arctan\left(gy_r/gx_r\right). \tag{7}$$

Using the gradient magnitudes $\{mag(x,y) : (x,y) \in R\}$ and orientation values quantized for Q levels $\{\varphi(x,y) : (x,y) \in R\}$,

$$\varphi\left(x,y\right) = \left\lfloor \frac{Q}{360} ori\left(x,y\right) + \frac{1}{2} \right\rfloor, \tag{8}$$

the histogram of oriented gradients can be computed as follows,

$$HOG\left(\alpha\right) = \begin{cases} \sum_{(x,y) \in R} \delta\left(\alpha - \varphi\left(x,y\right)\right), & mag\left(x,y\right) \geq Med \\ 0, & otherwise \end{cases}, \tag{9}$$

where $\alpha = (0,,Q-1)$ are histogram values (bins), Med is the median of the gradient magnitudes inside of the circular window, and $\delta\left(z\right) = \begin{cases} 1 & z = 0 \\ 0 & otherwise \end{cases}$ is the Kronecker delta function.

Finally, we compute a centered and normalized histogram, which further improves the rotation invariance and makes the matching robust to slight scale change,

$$\overline{HOG}\left(\alpha\right) = \frac{HOG\left(\alpha\right) - Mean}{\sqrt{Var}}, \tag{10}$$

where $Mean$ and Var are sample mean and variance of the histogram, respectively.

4   Object Descriptor

Let W_i be a set of closed disks, with distances between disks D_{ij} and angles between every three adjacent centers of the closed disks γ_i [11],

$$W_i = \left\{(x,y) \in \mathbb{R} : (x - x_i)^2 + (y - y_i)^2 \leq r, \, i = 1, ... M\right\}, \tag{11}$$

$$D_{ij} = \left\{ \sqrt{(x_i - x_j)^2 + (y_i - y_j)^2}, \, i = 1, ..., M; \, j = i + 1, ... M\right\}, \tag{12}$$

$$\gamma_i = \left\{ \cos^{-1}\left[\frac{D_{i,j+1}^2 + D_{i,j+2}^2 + D_{i,j+3}^2}{2D_{i,j+1}D_{i,j+2}} \right], \, i = 1, ..., M - 2\right\}, \tag{13}$$

where (x,y) are coordinates at the center of the disks, r is the radius of the disks, M is the number of circular disks inside the borders of the target object.

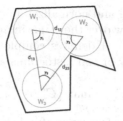

Fig. 3. Circular disk structure computed from the target object.

The circular disks form a geometric structure that fills inside the target object as shown in Fig. 3

Histograms of oriented gradient for the target object (HOG_i^O) are computed from each disk W_i in the structure using the radial mask R as shown in Sect. 3, and further used for matching. It is interesting to note that at any position of the structure, each disk contains image area that is unchangeable during rotation.

5   Iterative Matching

In order to locate the target object, we perform a recursive calculation of the HOG over the frame image by means of a sliding window the size of R as depicted in Sect. 3. Since the sliding window is symmetric, it can easily move in horizontal or vertical direction one pixel at the time, as shown in Fig. 4,

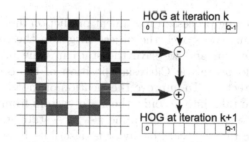

Fig. 4. Recursive update of a histogram along columns.

The iterative computation of histograms allows for fast computation of the frame image as only $2\pi r$ pixels in the circular window are processed instead of πr^2 pixels of the circular window area at each iteration.

At any step k of the iterative process we compute a scene histogram HOG_k^S. With both histograms, we correlate the ith circular window at position k in the frame image by means of the Inverse Fast Fourier Transform,

$$C_i^k(\alpha) = IFT\left[HS_{ik}(\omega)\, HO_i^*(\omega)\right], \tag{14}$$

where $HS_{ik}(\omega)$ is the centered and normalized Fourier Transform of $HOG^S_k(\alpha)$ inside of the ith circular window over the frame image, and $HO_i(\omega)$ is the Fourier Transform of $HOG^O_i(\alpha)$, the $(*)$ denotes complex conjugate. The correlation peak is a measure of similarity of two histograms and its calculated as follows,

$$P_i^k = \max_{\alpha} \left\{ C_i^k(\alpha) \right\}. \tag{15}$$

The correlation peaks are in the range of $[-1, 1]$. We suggest a M-pass procedure: first perform a matching of the first circular window in the structure; the objective is to reject as much as possible points in the frame image by applying a threshold Th to the correlation peaks to conserve the higher valued points and keep a low probability of miss errors. Second, only accepted points are considered to carry out the matching with the second circular window of the structure, taking into account the threshold value and the center to center distance D_{ij} to the first window, rejecting another set of points; and so on, evaluating the M windows in the structure. The final decision about the presence of the target object is taken considering the joint distribution of the correlation peaks for all windows. In this way, a trade-off between the probabilities of miss and false alarm errors is achieved.

Matching algorithms in sliding windows can be computationally exhaustive and time consuming, then, to improve the processing time we use a GPU to compute a huge set of histograms in the frame and compare them to the target object HOG in parallel with the multicore approach that GPUs provide [19].

6 Prediction

To speed up the tracking process, a movement prediction model is implemented [14]. In the first frame, we search the object target in the complete frame, as the position is unknown, and we save the state in time in form of a vector $[x, y, \theta]$, where x and y are the Cartesian position of the object and θ is the object movement direction. To improve the location of the object in subsequent frames at time τ, we take into account the state vectors from past and current frames to predict the state vector at frame $\tau + 1$. A state-space motion model is implemented. The target object behavior is described by a coordinated turn model as follows:

$$x_{\tau+1} = x_\tau + \frac{\sin(\theta_\tau \Delta)}{\theta_\tau} \hat{x}_\tau - \frac{1 - \cos(\theta_\tau \Delta)}{\theta_\tau} \hat{y}_\tau + a_{x,\tau} \frac{\Delta^2}{2}, \tag{16}$$

$$y_{\tau+1} = y_\tau + \frac{1 - \cos(\theta_\tau \Delta)}{\theta_\tau} \hat{x}_\tau + \frac{\sin(\theta_\tau \Delta)}{\theta_\tau} \hat{y}_\tau + a_{y,\tau} \frac{\Delta^2}{2}, \tag{17}$$

$$\hat{x}_{\tau+1} = \cos(\theta_\tau \Delta) \hat{x}_\tau - \sin(\theta_\tau \Delta) \hat{y}_\tau + a_{x,\tau} \Delta, \tag{18}$$

$$\hat{y}_{\tau+1} = \sin\left(\theta_\tau \Delta\right) \hat{x}_\tau - \cos\left(\theta_\tau \Delta\right) \hat{y}_\tau + a_{y,\tau} \Delta, \tag{19}$$

$$\theta_{\tau+1} = \theta + a_{\theta,\tau}, \tag{20}$$

where, x_τ and y_τ are the position of the target at frame τ in Cartesian coordinates, $\hat{x}_\tau$ and $\hat{y}_\tau$ are velocity components in x and y directions, θ_τ is the target's angular rate, $a_{x,\tau}$ and $a_{y,\tau}$ are random variables representing acceleration in x and y directions, and $a_{\theta,\tau}$ is the angular acceleration. In case that the target object has an unexpected turn and the predicted position does not coincide with the actual target position, we take a frame fragment of size equal to $1.5r$ around the predicted coordinates for further precise matching. This improves the target location and makes the algorithm tolerant to a sudden direction change. If the object is lost, then we search within the frame fragment with increased size of $0.1r$ for five consecutive frames while the target object remains lost; in case it is not found, we resume the search in the entire frame.

7   Computer Simulations

This section provides results over a set of computer simulations. The experiments are carried out over 10 synthetic sequences of 240 frames each one. The objects are taken from the Amsterdam Library of Object Images (ALOI) [20], each one contains a structure of two circular windows ($M = 2$) and variable radius in the range of $r = [28, 32]$. The frames are of standard size of 640×480 pixels each one, and are free source landscape photos from the Internet.

Each sequence features the target object with varied deformations such as rotation in-plane from 0 to $360°$, rotation out-of plane or perspective changes as high as $35°$, a slight scaling in the interval of $[0.8, 1.2]$ and additive white noise up to $\sigma_n = 15$. The parameters of the proposed algorithm referred to as CWMA are as follows: $M = 2$, $Q = \begin{cases} \lceil 360/\sigma_n \rceil & , 1.5 < \sigma_n < 40 \\ 64 & , otherwise \end{cases}$ and $Th = 0.8$.

The algorithm was implemented in a standard PC with Intel Core i7 processor with 3.2 GHz and 8 GB of RAM and a GPU ATI RADEON 6450, using OpenCV for basic image processing algorithms and OpenCL for parallelization. Trackers based on matching algorithms SIFT, SURF and ORB were implemented with the same technologies taking the core implementation from OpenCV.

We test our algorithm in percentage of the success rate of locating the target object and processing time in terms of Frames Per Second (FPS).

Figure 5 shows the hit rate, as can be seen, the proposed algorithm is the most stable of the algorithms achieving almost perfect score across all the sequences, while the other algorithms based on features are more chaotic; we attribute this behavior to the structure of circular windows and the truly rotation invariant descriptors that keep the object information and modify gradient directions with the radial mask.

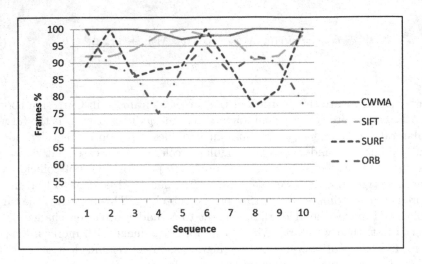

Fig. 5. Success rate in percentage of frames by sequence.

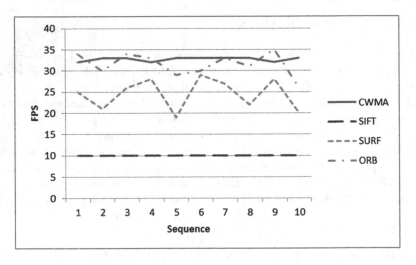

Fig. 6. Frames per second achieved at each sequence.

Figure 6 shows the frames per second of the algorithms. Our proposed algorithm and the one based on ORB run at similar peace; however, as the graphic shows, our implementation is more stable across the sequences. This is caused by the structure stability, contrary to the feature based algorithms where the processing time can change drastically depending on the quantity of keypoints found within the image.

Overall, our algorithm shows a better performance when compared with other matching based algorithms based on popular descriptors such as SIFT, SURF and ORB.

8 Conclusion

In this work we presented a robust tracking algorithm based on rotation-invariant HOGs over a data structure of circular windows. The proposed algorithm utilizes a prediction stage based on modeling the kinematic behavior of a target in two-dimensional space. According to computer simulation results the proposed algorithm showed a superior performance in terms of tracking accuracy and speed of processing comparing with similar tracking techniques based on features matching. Implementation of the proposed algorithm with GPU devices achieves real-time processing.

Acknowledgments. This work was supported by the Ministry of Education and Science of Russian Federation (grant 2.1766.2014K).

References

1. Yilmaz, A., Javed, O., Shah, M.: Object tracking: a survey. ACM Computing Surveys **38**(4), 1–45 (2006)
2. Schweitzer, H., Bell, J.W., Wu, F.: Very fast template matching. In: Heyden, A., Sparr, G., Nielsen, M., Johansen, P. (eds.) ECCV 2002, Part IV. LNCS, vol. 2353, pp. 358–372. Springer, Heidelberg (2002)
3. Nejhum, S., Ho, J., Yang, M.H.: Online visual tracking with histograms and articulating blocks. Computer Vision and Image Understanding **114**(8), 901–914 (2010)
4. Fieguth, P., Terzopoulos, D.: Color-based tracking of heads and other mobile objects at video frame rates. In: Proceedings of the IEEE Conference on the Computer Vision and Pattern Recognition, pp. 21–27 (1997)
5. Haritaoglu, I., Harwood, D., Davis, L.: W4: real-time surveillance of people and their activities. IEEE Transactions Pattern Analysis and Machine Intelligence **22**(8), 809–830 (2000)
6. Sato, K., Aggarwal, J.: Temporal spatio-velocity transform and its application to tracking and interaction. Computer Vision and Image Understanding **96**(2), 100–128 (2004)
7. Talu, F., Turkoglu, I., Cebeci, M.: A hybrid tracking method for scaled and oriented objects in crowded scenes. Exp. Syst. App. **38**, 13682–13687 (2011)
8. Buchanan, A., Fitzgibbon, A.: Combining local and global motion models for feature point tracking. In: IEEE Conference on Computer Vision and Pattern Recognition, pp. 1–8 (2007)
9. Intille, S., Davis, J., Bobick, A.: Real-time closed-world tracking. In: Proceedings of IEEE Conference on Computer Vision and Pattern Recognition, pp. 697–703 (1997)
10. Sbalzarini, I.F., Koumoutsakos, P.: Feature point tracking and trajectory analysis for video imaging in cell biology. J. Struct. Bio. **151**(2), 182–195 (2005)
11. Miramontes-Jaramillo, D., Kober, V., Díaz-Ramírez, V.H.: CWMA: circular window matching algorithm. In: Ruiz-Shulcloper, J., Sanniti di Baja, G. (eds.) CIARP 2013, Part I. LNCS, vol. 8258, pp. 439–446. Springer, Heidelberg (2013)
12. Dalal, N., Triggs, B.: Histograms of oriented gradients for human detection. In: Proceedings of the IEEE Computer Society Conference on Computer Vision and Pattern Recognition, vol. 1, pp. 886–893 (2005)

13. Takacs, G., Chandrasekhar, V., Tsai, S.S., Chen, D., Grzeszczuk, R., Girod, B.: Fast computation of rotation-invariant image features by an approximate radial gradient transform. IEEE Trans. Image Proc. **22**(8), 2970–2982 (2013)
14. Hu, W., Tan, T., Wang, L., Maybank, S.: A survey on visual surveillance of object motion and behavior. IEEE Trans. Syst. Man Cybern. C Appl. Rev. **34**(3), 334–352 (2004)
15. Lowe, D.G.: Object recognition from local scale-invariant features. In: Proceedings of International Conference on Computer Vision, vol. 2, pp. 1150–1157 (1999)
16. Bay, H., Ess, A., Tuytelaars, T., Van Gool, L.: SURF: speeded up robust features. Comput. Vis. Image Underst. **110**(3), 346–359 (2008)
17. Rublee, E., Rabaud, V., Konolige, K., Bradski, G.: ORB: an efficient alternative to SIFT or SURF. In: Proceedings of the IEEE International Conference on Computer Vision, pp. 2564–2571 (2011)
18. Pratt, W.K.: Digital Image Processing. Wiley, New York (2007)
19. Miramontes-Jaramillo, D., Kober, V., Daz-Ramrez, V.H.: Multiple objects tracking with HOGs matching in circular windows. In: Proceedings of SPIE Application of Digital Image Processing XXXVII, vol. 9217, 92171N-8 (2014)
20. Geusebroek, J.M., Burghouts, G.J., Smeulders, A.W.M.: The Amsterdam library of object images. Int. J. Comp. Vis. **61**(1), 103–112 (2005). http://staff. science.uva.nl/aloi/

Similarity Analysis of Archaeological Potsherds Using 3D Surfaces

Edgar Roman-Rangel[1](✉) and Diego Jimenez-Badillo[2]

[1] CVMLab - University of Geneva, Geneva, Switzerland
edgar.romanrangel@unige.ch
[2] National Institute of Anthropology and History of Mexico (INAH),
Mexico city, Mexico
diego_jimenez@inah.gob.mx

Abstract. This work presents a new methodology for efficient scanning and analysis of 3D shapes representing archaeological potsherds, which is based on single-view 3D scanning. More precisely, this work presents an analysis of visual diagnostic features that help classifying potsherds, and that can be automatically identified by the appropriate combination of the 3D models resulting from the proposed scanning technique, with robust local descriptors and a bag representation approach, thus providing with enough information to achieve competitive performance in the task of automatic classification of potsherd. Also, this technique allows to perform other types of analysis such as similarity analysis.

Keywords: 3D shapes · Potsherds · Classification · Similarity analysis

1 Introduction

The analysis of archaeological potsherds is one of the most demanding tasks in Archaeology, as it helps understanding both cultural evolution and relations (economic and military) between different cultural groups from the past [1].

One of the trending methods for studying potsherds in the last decade consists on generating 3D models by scanning surviving pieces of ceramics found at excavation sites [2]. This approach enables new capabilities with respect to analyzing the original pieces, e.g., automatic and semi-automatic content analysis [2], virtual reconstructions [3], more efficient archiving and sharing of documents [4,5], etc. However, it results in a very time-consuming task, as it requires scanning each piece several times from different viewpoints and a later, often manual, registration of the different shots to build the full 3D model [2].

In this work, we present a new approach based on fast scanning of single-view 3D surfaces, which is a much faster process to obtain 3D information. We complemented the proposed approach by evaluating the descriptive performance of the 3D SIFT descriptor [6], which is a relatively new extension of the well known SIFT descriptor [7] used for description of 2D images. This evaluation consisted in computing 3D SIFT descriptors for 3D surfaces of potsherds, create

© Springer International Publishing Switzerland 2015
J.A. Carrasco-Ochoa et al. (Eds.): MCPR 2015, LNCS 9116, pp. 125–134, 2015.
DOI: 10.1007/978-3-319-19264-2_13

bag representation, and perform classification experiments. Our results show that the 3D surfaces contain enough information to achieve competitive classification performance. Also, we conducted a similarity analysis to evaluate the potential risk for confusing potsherd coming from different ceramic pieces.

The remaining of this paper is organized as follows. Section 2 gives an introduction to the type of data used in this work. Section 3 explains the method used for description of 3D surfaces, namely the 3D SIFT descriptor [6]. Section 4 explains the two types of experiments we performed, and Sect. 5 presents our results. Finally, Sect. 6 presents our conclusions.

2   3D Potsherds

This section describes the dataset of 282 potsherds used in this work. Namely, it explains the origin of the potsherds, the type of digital data generated from them, and basic statistics of the resulting dataset.

2.1   The Data Set

Origin. The ceramic style of the potsherds used in this work is known as "Aztec III Black on Orange", and it constitutes one of the most important wares in Mesoamerican archaeology [1]. Geographically speaking, this ceramic style corresponds to the entire Basin of Mexico, i.e., the geographical extension of the former Aztec empire, and it used to be part of the common utilitarian assemblage in households during the late Aztec period (1350–1520 C.E.).

During the 1970's, a large amount of these ceramics was collected on the surface by the Valley of Mexico Survey Projects [8,9]. Later, stylistic analyses of their materials was performed during the 1990's, along with an important comparison of their geographic distribution with historically-documented polities [10,11], which helped understand the economic relations between dependent communities and the Aztec capital.

Single View Scanning. As part of our proposed approach, we decided to rely on using 3D meshes generated from scanning the potsherds from a single viewpoint. This is, the type of data we analyzed consists on 3D surfaces rather than the traditional approach of using volumetric models [2]. Figure 1 shows a few examples of these single view 3D surfaces.

Scanning these single view 3D surfaces is a faster data acquisition process, in comparison to the traditional approach of using full 3D models, which can take much longer time as it requires the scanning from different viewpoints and a later, often manual, registration step for generating the full model [2].

As shown in the examples of Fig. 1, there are many section of the potsherds that are common across classes, e.g., central sections of the potsherd, which intuitively suggests that the risk for obtaining poor classification performance is high. However, note that there are also some sections of the potsherd containing visual information that is diagnostic for specific classes, thus increasing the potential for achieving a good classification performance.

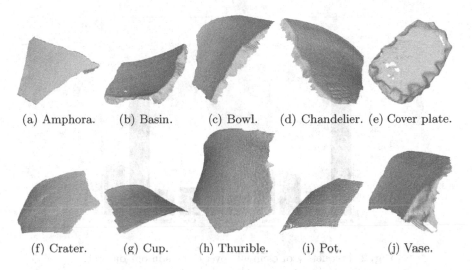

(a) Amphora.   (b) Basin.   (c) Bowl.   (d) Chandelier. (e) Cover plate.

(f) Crater.   (g) Cup.   (h) Thurible.   (i) Pot.   (j) Vase.

Fig. 1. Examples of single view 3D surfaces scanned from the collection of potsherds. Each example corresponds to one of the classes in our dataset.

Dataset Statistics. The dataset used in this work consists of 282 3D surfaces scanned from potsherds of 10 different ceramic objects, i.e., classes. Figure 2 shows the distribution of 3D surfaces over the 10 ceramic classes. Note that this dataset is not well balanced, i.e., there are much more instances in classes amphora and pot (over 60 in each), while there are only a few instances in classes basin, cover plate, and thurible (only above 10 in each). This behavior results as a consequence of the frequency at which the potsherds are found in the field. Nevertheless, in Sect. 5.2, we conducted an analysis using a balanced subset of this data.

3 Model Description

Given the promising results reported in previous works [6,12], we use 3D SIFT descriptors and the bag-of-words model (BOW) [13] to represent the 3D surfaces.

3D SIFT. The 3D Scale Invariant Feature Transform (SIFT) [6] is the 3D extension of the well known SIFT descriptor [7].

Given that points on a 3D surface (i.e., mesh vertices) might be non-uniformly spaced, the repeatability of the traditional difference of Gaussian (DoG) [7] detector of points of interest might be negatively affected [6]. Nevertheless, the Gaussian scale space can be constructed (approximated) by using an alternative method which is invariant to distance between vertices but not to their location. Namely, the smoothing of a given vertex v_i^{s+1} at scale $(s+1)$ is given by,

$$v_i^{s+1} = \frac{1}{|V_i^s|} \sum_{v_j^s \in V_i^s} v_j^s, \tag{1}$$

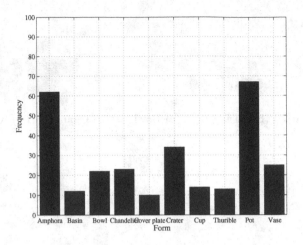

Fig. 2. Frequency of elements over classes in our dataset.

where, V_i^s is the set of first order neighbors of $\boldsymbol{v}_i^s$, i.e., the set of vertices sharing an edge with $\boldsymbol{v}_i^s$, $|\cdot|$ denotes the cardinality operator, and the summation of the vertices is performed component-wise on the x, y, and z components of $\boldsymbol{v}_i$.

After computing the Gaussian smoothing of the 3D surface using Eq. (1), the difference of Gaussians is approximated by,

$$d_i^s = \frac{1}{(s \cdot \sigma_{i,0}^2)} \left(\boldsymbol{v}_i^s - \boldsymbol{v}_i^{s+1} \right), \tag{2}$$

where, σ_0 denotes the initial variance of the integration parameter, which is independently estimated for each vertex as,

$$\sigma_{i,0} = \frac{1}{|V_i^s|} \sum_{\boldsymbol{v}_j^s \in V_i^s} abs \left(\boldsymbol{v}_i^s - \boldsymbol{v}_j^s \right), \tag{3}$$

where, $abs\,(\cdot)$ indicates absolute.

Implementing this methodology, allows to select those 3D vertices maximizing Eq. (2) as points of interest for which a 3D SIFT descriptor will be computed.

In turn, the computation of the 3D SIFT descriptor requires the computation of a depth map [6], which is estimated by projecting the 3D vertices onto a 2D surface, thus generating a kind of image on which the traditional SIFT descriptor [7] is computed. More precisely, a dominant plane P_i is estimated for each vertex $\boldsymbol{v}_i$. Namely, P_i corresponds to the tangent plane of $\boldsymbol{v}_i$, which can be found by its normal $\boldsymbol{n}_i$ and the point $\boldsymbol{v}_i$ itself.

Once the dominant plane P_i is computed, all the neighboring points to $\boldsymbol{v}_i$ are mapped onto a 2D array, for which the SIFT descriptor is computed. This is done by filling in the 2D array the distance that exists from the 3D surface to the dominant plane P_i for each neighboring point. In turn, the neighboring points of v_i correspond to those vertices within a distance D_i from v_i, where D_i is defined as,

$$D_i = C\sqrt{s_i}\sigma_{i,0}, \tag{4}$$

where, $\sigma_{i,0}$ is computed as defined in Eq. (3), s_i is the scale at which the point of interest attains its maximum value according to Eq. (2), and C is a parameter to control the level of locality [6].

After computing the sets of 3D SIFT descriptors for each 3D surface, we quantized them to construct bag-of-words representation (BOW), using dictionaries of different sizes. Later, we used the bag representations to compute distances between 3D surfaces and to conduct our similarity analysis of potsherds.

4   Analysis of Potsherds

We performed two types of analysis of the Aztec potsheds using their 3D surfaces. Namely, experiments of automatic classification, and the analysis of similarity.

Classification Performance. By relying on the k-NN classification approach [14], with $k = 1$, we evaluated the classification performance achieved by different BOW representations [13] of 3D-SIFT descriptors [6].

To this end, we estimated vocabularies of different sizes using the k-means clustering algorithm [15]. More precisely, vocabularies of 100, 250, 500, 1000, 2500, and 5000 words, and then compared the resulting bag representations using the Euclidean distance. In Sect. 5.1, we present the results of this evaluation along with the confusion matrix generated by the dictionary that achieved the best classification rate.

Similarity Analysis. To better understand the numeric results of the previously described classification experiments, we estimated the intra-class and inter-class distance that can be expected when using our approach, i.e., BOW [13] of 3D-SIFT descriptors [6] computed over surfaces. We computed these values as:

- Intra-class distance: The average of the pairwise distance between all instances within the class of interest.
- Inter-class distance: The average of the pairwise distance from each instance of the class of interest to all instances on the remaining classes.

Section 5.2 presents the intra-class and inter-class average distance of our dataset computed using the dictionary that achieved the best classification performance. Also in that section, we present a graph-based relation of the, on average, most similar classes given a reference class.

5   Results

This section presents the results of both of our analysis: classification performance and similarity analysis.

5.1 Classification Performance

The first result from the classification performance test indicates that small vocabularies are better to achieve good classification rates. Table 1 shows that 58.45 % of the potsherds are correctly classified by using only 100 and 250 words, and that this rate drops quickly with vocabularies of 1000 words or more. Although these results confirm previous observations regarding the impact of the vocabulary size [12], they also contradict some results of recent works [16].

Table 1. Classification performance achieved with vocabularies of different sizes.

vocabulary size	100	250	500	1000	2500	5000
classification rate (%)	58.45	58.45	54.26	47.52	39.36	36.17

By visual inspection, we realized the reason for which small vocabularies work well for our particular 3D structures, i.e., surfaces. Namely, besides of having instances with large amount of variations, these variations can be captured by very local descriptors, i.e., descriptors that capture information only within the closest neighborhood of the point of interest. This particularity of our data allows to represent all important local variations with only a small set of prototype descriptors, i.e., words. See Fig. 1 for visual examples of the dataset.

Figure 3 shows examples of the two most different words in the 100 words vocabulary. Namely, words number 67 and 95, where the distance between words is estimated as the distance between centroids of their corresponding clusters. More precisely, Fig. 3 shows 3D surfaces, each with highlighted sections that correspond to the 3D SIFT descriptors assigned as instances of those particular words, i.e., closest to the centroid computed by k-means.

More precisely, instances of word 95 (orange circles in Fig. 3) describe a slightly curved section, while instances of word 67 (blue circles) describe roughly flat sections near an abrupt change on the surface, like a hole (Fig. 3a and b) or the starting point of the base of the ceramic (Fig. 3c). However, note that given the nature of the 3D data, a small set of local descriptors suffices for accurate description, as well as for competitive classification performance.

Figure 4 shows the confusion matrix obtained with the vocabulary of 100 words. As shown by its main diagonal, our approach is able to achieve competitive classification performance for most of the classes, and specially for classes Amphora, Chandelier, and Pot, i.e., 81 %, 74 %, and 64 % of the times, a potsherd from those classes is properly assigned to the correct class label. On the other hand, potsherd of classes Basin, Thurible, and Vase seem to be more challenging.

5.2 Similarity Analysis

As explained in Sect. 4, we evaluated the intra-class and inter-class distance of the 3D surfaces, to acquire a better understanding of the classification potential that

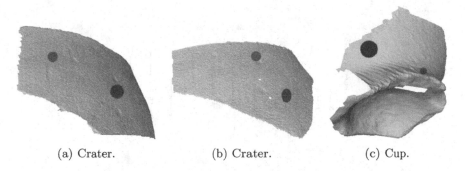

(a) Crater.   (b) Crater.   (c) Cup.

Fig. 3. 3D surfaces of potsherds from crater and cup classes. Highlighted sections correspond to the spatial locality of 3D SIFT descriptors. Blue sections indicate instances of word 67, and orange sections indicate instances of word 95, which are the two most different words in the 100 words vocabulary (Color figure online).

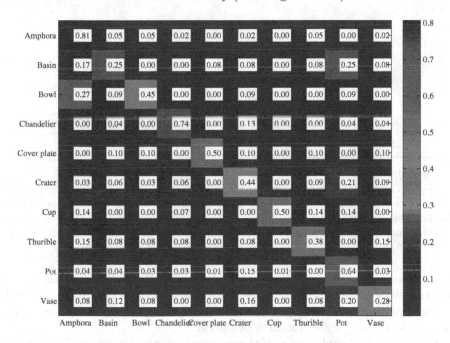

	Amphora	Basin	Bowl	Chandelier	Cover plate	Crater	Cup	Thurible	Pot	Vase
Amphora	0.81	0.05	0.05	0.02	0.00	0.02	0.00	0.05	0.00	0.02
Basin	0.17	0.25	0.00	0.00	0.08	0.08	0.00	0.08	0.25	0.08
Bowl	0.27	0.09	0.45	0.00	0.00	0.09	0.00	0.00	0.09	0.00
Chandelier	0.00	0.04	0.00	0.74	0.00	0.13	0.00	0.00	0.04	0.04
Cover plate	0.00	0.10	0.10	0.00	0.50	0.10	0.00	0.10	0.00	0.10
Crater	0.03	0.06	0.03	0.06	0.00	0.44	0.00	0.09	0.21	0.09
Cup	0.14	0.00	0.00	0.07	0.00	0.00	0.50	0.14	0.14	0.00
Thurible	0.15	0.08	0.08	0.08	0.00	0.08	0.00	0.38	0.00	0.15
Pot	0.04	0.04	0.03	0.03	0.01	0.15	0.01	0.00	0.64	0.03
Vase	0.08	0.12	0.08	0.00	0.00	0.16	0.00	0.08	0.20	0.28

Fig. 4. Confusion matrix obtained from the automatic classification of 3D surfaces scanned from Potshards of the ancient Aztec culture.

the 3D SIFT descriptor has to deal with potsherds. For this analysis we used the vocabulary of 100 words, which achieved the highest classification performance, as shown in Sect. 5.1.

For each class in our dataset, Fig. 5 shows the intra-class distance alongside its inter-class distance counterpart. As one could expect for a well behaved scenario, the distance between elements of the same class is shorter than the distance

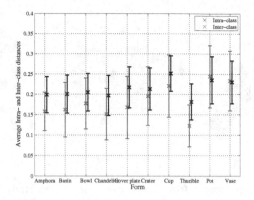

Fig. 5. Intra-class and Inter-class similarity of Potsherds.

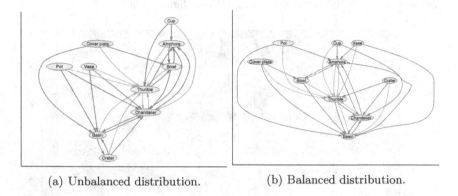

(a) Unbalanced distribution.   (b) Balanced distribution.

Fig. 6. Inter-class similarity in our dataset. Green arrows indicate the most similar reference class to each other class (class of interest). Gray arrows point to the second and thirds most similar reference classes to each other class (Color figure online).

between elements of different classes, i.e., the average intra-class distance is lower than the average inter-class distance. This observation is true for 8 of the 10 classes of potsherd, with classes Pot and Vase having slightly larger intra-class distance, as well as larger standard deviation.

A more detailed explanation of the average distance is shown in Fig. 6, where more precisely, the average similarity between pairs of classes is presented. For this purpose, a class of interest c_i is considered similar to a reference class c_r if the average inter-class distance, as explained in Sect. 4, between c_i and c_r is smaller than the average inter-class distance from c_i to any other class.

For each class in our dataset, Fig. 6a shows its three most similar classes. Note that the class Thurible is indicated as the most similar for 8 out of the 9 remaining classes, which might explain the low classification rate shown in the confusion matrix of Fig. 4. As previously shown in Fig. 2, the distribution of instances is not well balanced across classes of potsherds. To verify whether or not this unbalanced characteristic of the dataset has an impact on our analysis, we randomly selected a subset of 10 instances per class, thus generating a balanced dataset. Figure 6b shows the three most similar classes for each reference class using this balanced datset. Although a few arrows changed their end node, not important differences were observed when using a balanced dataset. This suggests that this is the natural behavior of the potsherds, and it does not depend on the amount of instances available for each class.

6   Conclusions

We presented a novel approach for automatic analysis of archaeological potsherds, based on the use of 3D surfaces scanned from a single viewpoint. More precisely, potsherds of ceramic artifacts from the ancient Aztec culture. The proposed scanning approach is much faster than the traditional method that takes several shots from different viewpoints and then registers them to construct a complete 3D model. Yet, our methodology produces data with enough information that can be described using 3D SIFT descriptors and bag representations.

Using the proposed methodology, we evaluated the impact that the size of different visual dictionaries has on classification tasks. Our results show that small vocabularies suffice to obtain competitive classification performance, which becomes harmed as the size of the dictionary increases. In particular, we obtained the best classification performance by using dictionaries of 100 and 250 words. Namely, our methodology is able to attain competitive classification performance in most cases, with the few exceptions of certain classes of potsherd that are more difficult to categorize, i.e., basin, thurible, and vase. More over, thurible and basin are the two classes that were ranked the most common classes, i.e., the classes whose instances are most similar to instances of other classes, as shown by the quantitative similarity analysis that we conducted.

Overall, this type of analysis is of interest for archaeologists, as it shows the potential that patter recognition techniques have to deal with some of the mos challenging problems in Archaeology.

Acknowledgments. This work was founded by the Swiss-NSF through the Tepalcatl project (P2ELP2-152166).

References

1. Cervantes Rosado, J., Fournier, P., Carballal, M.: La cerámica del Posclásico en la Cuenca de México. In: La producción Alfarera en el México Antiguo, vol. V, Colección Científica. Insituto Nacional de Antropología e Historia (2007)
2. Karasik, A., Smilanski, U.: 3D scanning technology as a standard archaeological tool for pottery analysis: practice and theory. J. Archaeol. Sci. **35**, 1148–1168 (2008)
3. Kampel, M., Sablatnig, R.: Virtual reconstruction of broken and unbroken pottery. In: Proceedings of the International Conference on 3-D Digital Imaging and Modeling (2003)

4. Razdan, A., Liu, D., Bae, M., Zhu, M., Farin, G.: Using geometric modeling for archiving and searching 3D archaeological vessels. In: Proceedings of the International Conference on Imaging Science, Systems, and Technology, CISST (2001)
5. Larue, F., Di-Benedetto, M., Dellepiane, M., Scopigno, R.: From the digitization of cultural artifacts to the web publishing of digital 3D collections: an automatic pipeline for knowledge sharing. J. Multimedia **7**, 132–144 (2012)
6. Darom, T., Keller, Y.: Scale-invariant features for 3-D mesh models. IEEE Trans. Image Process. **21**, 2758–2769 (2012)
7. Lowe, D.G.: Distinctive image features from scale-invariant keypoints. Int. J. Comput. Vis. **60**, 91–110 (2004)
8. Sanders, W.T., Parsons, J., Santley, R.S.: The Basin of Mexico: Ecological Processes in the Evolution of a Civilization. Academic Press, New York (1979)
9. Parsons, J.R., Brumfiel, E.S., Parsons, M.H., Wilson, D.J.: Prehistoric settlement patterns in the southern valley of Mexico: the Chalco-Xochimilco regions. In: Memoirs of the Museum of Anthropology, vol. 14. University of Michigan (1982)
10. Hodge, M.G., Minc, L.D.: The spatial patterning of Aztec ceramics: implications for prehispanic exchange systems in the valley of Mexico. J. Field Archaeol. **17**, 415–437 (1990)
11. Hodge, M.G., Neff, H., Blackman, M.J., Minc, L.D.: Black-on-orange ceramic production in the Aztec empire's heartland. Lat. Am. Antiq. **4**, 130–157 (1993)
12. Roman-Rangel, E., Jimenez-Badillo, D., Aguayo-Ortiz, E.: Categorization of Aztec Potsherds using 3D local descriptors. In: Jawahar, C.V., Shan, S. (eds.) ACCV 2014 Workshops. LNCS, vol. 9009, pp. 567–582. Springer, Heidelberg (2015)
13. Sivic, J., Zisserman, A.: Video Google: a text retrieval approach to object matching in videos. In: Proceedings of the IEEE International Conference on Computer Vision (2003)
14. Cover, T., Hart, P.: Nearest neighbor pattern classification. IEEE Trans. Inf. Theor. **13**(1), 21–27 (1967)
15. Lloyd, S.: Least squares quantization in PCM. IEEE Trans. Inf. Theor. **28**, 129–137 (1982)
16. Boyer, E., Bronstein, A.M., Bronstein, M.M., Bustos, B., Darom, T., Horaud, R., Hotz, I., Keller, Y., Keustermans, J., Kovnatsky, A., Litman, R., Reininghaus, J., Sipiran, I., Smeets, D., Suetens, P., Vandermeulen, D., Zaharescu, A., Zobel, V.: SHREC 2011: robust feature detection and description benchmark. In: Proceedings of the 4th Eurographics Conference on 3D Object Retrieval (2011)

Automatic Segmentation of Regions of Interest in Breast Thermographic Images

Adrian J. Villalobos-Montiel[1], Mario I. Chacon-Murguia[1(✉)],
Jorge D. Calderon-Contreras[1], and Leticia Ortega-Maynez[2]

[1] Visual Perception Applications on Robotics Lab,
Chihuahua Institute of Technology, Chihuahua, Mexico
{avillalobos,mchacon,jdcalderon}@itchihuahua.edu.mx
[2] Universidad Autonoma de Ciudad Juarez, Ciudad Juarez, Mexico
lortega@uaoj.mx

Abstract. Breast thermography is a promising technique allowing breast cancer detection with the aid of infrared technology. However, the automatic segmentation of the regions of interest (ROI) to be analyzed is a difficult task and, thus, it is not commonly performed. In this paper we propose an automated technique for ROI extraction. The algorithm uses Canny edge detection with automatic threshold and symmetry inspection in order to correct tilt errors and to define a symmetry axis in the image. Furthermore, the Hough transform is used to provide circles giving a favorable approximation to the ROI. A region-based active contours technique and an edge smoothing technique are also used to improve the previous segmentation, taking in consideration the energies of the images. Experimental results indicate that the ROI were extracted more accurately and with higher precision than a state-of-the-art method.

Keywords: Breast cancer · Thermography · Breast segmentation · Active contours

1 Introduction

Breast cancer is a disease in which a group of malignant cells located in the breast grow uncontrollably and invade surrounding tissues or spread (metastasize) to distant areas of the body. It figures among the leading causes of death worldwide. The World Health Organization reported that in 2012, there were 521,000 deaths due to this disease [1]. Early cancer detection methods not only increase the chances of survival, but also allow treatments that will avoid disfiguring surgery.

Breast thermography is a promising noninvasive technique specializing in the early detection of breast cancer through thermal images. Due to the chemical and physiological principles, in precancerous, cancerous and surrounding tissues, there is a greater blood supply and an increased cellular activity which causes an increase in temperature equal to or greater than 0.5 °C compared to other tissues. Thermographic images are captured with infrared (IR) cameras, and can be digitally analyzed in order to detect any abnormalities within the breast tissue [2]. Breast thermography is a safe technique because it emits no radiation, and it is more comfortable and less expensive compared to other screening methods. With this technique, it is possible to detect small (<0.8 cm)

© Springer International Publishing Switzerland 2015
J.A. Carrasco-Ochoa et al. (Eds.): MCPR 2015, LNCS 9116, pp. 135–144, 2015.
DOI: 10.1007/978-3-319-19264-2_14

non-palpable lesions, that cannot be detected by physical examination or with a mammography [3]. Adding a thermography study to mammography provides a 95 % of accuracy in the diagnosis, whereas mammography by itself has only an 80 % of accuracy in the early stages of cancer [4]. Additionally, breast thermography allows tumors to be detected even 10 years earlier than with mammography [5].

Segmentation of the ROI allows delimitation of the data to be analyzed by the computer aided diagnosis system (CAD). The ROI must include all the breast tissue and the near ganglion groups since cancerous cells usually appear in the glands that produce milk and the ducts that carry it to the nipples [6]. Most authors perform semi-automatic or manual ROI extraction because it is a hard task due to the inherent characteristics of each breast that make them amorphous, and to the lack of clear limits in this kind of images [6].

In reference [6], a study of several algorithms found in the literature that perform ROI extraction is presented. These algorithms are usually based on a combination of some of the following methods: borders detection, parabolic Hough transform, active contours, anisotropic filters and interpolation methods [6]. However, most algorithms fail to perform a correct segmentation in images where the borders are not clear due to camouflage. So the main change on segmenting the ROI, is usually detecting the lower bounds of the breasts.

In this study we present an algorithm that performs the ROI segmentation from breast thermographic images. The proposed technique can perform this task correctly even when there is a camouflage problem. The algorithm also performs a tilt correction through symmetry analysis. Figure 1 summarizes the entire process in a block diagram.

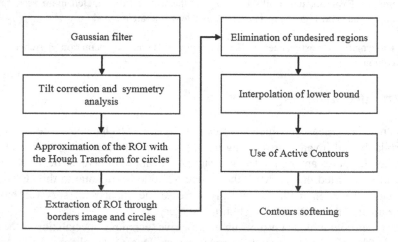

Fig. 1. Block diagram of the proposed algorithm.

Forty thermographic images from the database [7] (only available database found) were used in the development of the ROI extraction algorithm, and other different 100 images from the same database were used to evaluate the performance of the proposed algorithm. The proposed algorithm is then compared to the algorithm of Marques [8] which, through thresholding, obtains some of the lower edges of the breasts, and then with different interpolation methods completes the lower bounds of the ROI. However, the thresholding method of Marques fails when processing breast thermographies with unclear bounds.

2 Methods

Forty thermographic images from the database from [7] were used in the development of the ROI extraction algorithm. Figure 1 summarizes the entire process in a block diagram.

In order to avoid detection of false borders, the first step of the method is noise reduction through a Gaussian filter ($\sigma = 0.5$) within the breast thermography to generate the image $I_G(x,y)$.

2.1 Tilt Error Correction

Due to difficulty to keep a patient still during the image capture, patients tend to incline toward a side. Tilt error correction is necessary due that it can affect the diagnosis [9]. Symmetry analysis enables to correct tilt errors, as well to obtain important information that can be used to identify the ROI. To perform this correction, the algorithm identifies the symmetry axis in the proper position through an iterative process. A border image, $I_B(x,y)$, is obtained using the Canny operator to the filtered image $I_G(x,y)$. The Canny threshold, C_{TH}, is fixed to a value that allows detecting the lower borders of the breasts on at least 90 % of the input images. Then, a dilation operation (performed 4 times) is used over each edge pixel $p_e(x,y)$ to obtain thicker edges

$$for\ each\ p_e(x, y)$$
$$N_8(p_e(x, y)) = 1 \tag{1}$$

This image is rotated initially $-10°$, but in future iterations it is rotated with $1°$ interval increments until $10°$ is reached (values can be modified to desired range). The rotation operation is defined by

$$x_2 = \cos(\theta)(x_1 - x_0) - \sin(\theta)(y_1 - y_0) + x_0 \tag{2}$$

$$y_2 = \sin(\theta)(x_1 - x_0) + \cos(\theta)(y_1 - y_0) + y_0 \tag{3}$$

where θ is the angle of rotation in a clockwise direction and $(x0,y0)$ are the coordinates of the center of rotation [10]. The obtained image is copied and pasted in the center of a blank image with greater size with same proportions (recommended ratio of 1.5) denoted as $I_R(x,y)$. This step allows increasing the working space for the future steps. Afterwards, the size of a window W is defined based on the biggest size of breasts that can be found in any breast thermography. This window will continuously be placed in different positions over $I_R(x,y)$ in order to find the symmetry axis. Initially, it is centered in the y axis at the left border of $I_R(x,y)$. But in future iterations it will be horizontally shifted to the right in intervals of 1 pixel until reaching the right border of $I_R(x,y)$. Then two window images of the same size, W_R and W_L, are obtained by cutting W vertically.

The obtained image on the right side, W_R, is then reflected horizontally through a morphological operation, and denoted as W_{RR}. In order to find the similarities between the two images obtained from W, an image W_{AND} is generated by

$$W_{AND} = W_L(x, y)\, W_{RR}(x, y) \tag{4}$$

W_{AND} is a binary image indicating the pixels that W_{RR} and W_L have in common at same positions. The pixels with a value of 1 in W_{AND} are counted and denoted as the symmetry level. The algorithm uses 3 conditions of higher symmetry given by the variables γ_S, θ_A and α_P, which correspond respectively to the symmetry level, the angle of rotation and the position of W in the x axis of $I_R(x,y)$. Iteratively, the algorithm tests the different angles of rotation, and the different horizontal shift positions of W, calculating on every combination the symmetry level. If the symmetry level of the current iteration is higher than γ_S, the variables are overwritten with the current level of symmetry, angle of rotation, and shift position. Once the algorithm has tested all the possible combinations, a W_{AND} image is generated with the values stored in the higher symmetry variables. Ideally, the resulting W_{AND} image will show the borders of a breast and the body. Furthermore, the symmetry axis can be easily and correctly placed as a vertical line over the rotated thermography by an angle of θ_A, and at position λ in the x axis based on α_P and the width of W_{AND}, ω_{AND}:

$$\lambda = \alpha_P + \omega_{AND} \tag{5}$$

The rotated thermography will have no tilt error. The symmetry analysis steps are illustrated in Fig. 2. A more extended description of this method can be found in [11].

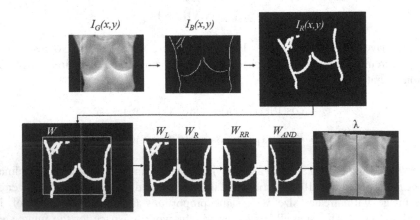

Fig. 2. Symmetry analysis and tilt error correction.

2.2 Breast Location Approximation

In order to approximate the location of the breasts on the tilt corrected thermographic image, the Hough transform is used. Because breasts tend to be circular/oval shaped, one circle C is located over W_{AND} with the Hough transform. In order to detect the desired circles, it is important to constrain the identification of circles to possible expected breast sizes. Also, the position of circles must be limited by providing a minimum distance of the circle's origin to the upper image limit to avoid detection of false circles on neck/armpit. Due that the desired threshold value for the Canny edge detector, C_{TH}, may vary for different thermographies, the algorithm has an automatic threshold adjustment to ensure a desired amount of detected edges. In order to perform this adjustment, a threshold value for minimum symmetry level is defined experimentally. C_{TH} is initialized at a value permitting detection of few edges, and it is continuously reduced until reaching the desired symmetry level threshold in W_{AND}. This process is exemplified in Fig. 3. In (a) the threshold C_{TH} is adjusted until the y component of the circle C, y_C, is greater or equal to 37 (defined threshold for this size of thermography) making sure that the circle is not in the armpit. In (b) the value of C_{TH} is also continuously decremented until the symmetry level, L_S, is greater than 600 (defined threshold for this size of thermography) allowing a correct C circle detection over the breast on W_{AND}.

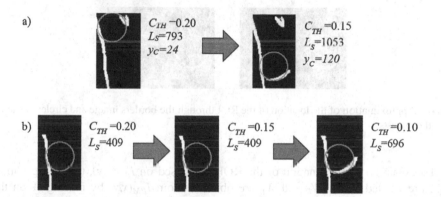

Fig. 3. Automatic threshold adjustment to make sure that (a) the circle C is not in the armpit, and that (b) the symmetry level L_S is above the desired threshold.

With the origin coordinates (x_C, y_C) of the circle C, and its radius r_C, it is easy to generate an image with circles located over the breasts positions by drawing 2 circles of radius r_c at the coordinates (x_{C1}, y_C) and (x_{C2}, y_C) where

$$x_{C1}, x_{C2} = \lambda \pm (\omega_{AND} - x_C) \qquad (6)$$

The following condition is used to draw the circles in an image denominated $I_P(x,y)$

$$\text{if } \sqrt{(x_{C1} - x_p)^2 + (y_C - y_p)^2} \approx r_C \vee \sqrt{(x_{C2} - x_p)^2 + (y_C - y_p)^2} \approx r_C \qquad (7)$$

$$I_P(x_p, y_p) = 1$$

evaluating every pixel $I_P(x_p, y_p) \in I_P(x,y)$.

The image $I_P(x,y)$ is then added to the border image of the rotated thermography obtained with the automatically adjusted threshold C_{TH}. The resulting image $I_{BM}(x,y)$ shows the edges of the body and two circles over the breasts. Since false edges can appear below the breasts, all the pixels below the circles (plus a small tolerance) are cleared. If there is a camouflage problem that does not allow the correct detection of the lower bounds of the breasts, the found circles provide a closed area and a good approximation of the breasts, considering that the circle C was placed correctly on the W_{AND} image. Figure 4 illustrates these steps.

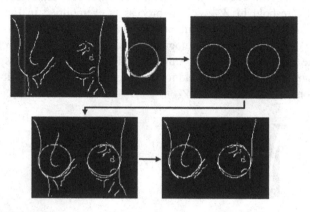

Fig. 4. Approximation of the location of the ROI through the borders image and circles' location over the breasts.

To obtain an approximation of the ROI area based on $I_{BM}(x,y)$, the next group of steps are needed. First, W_R and W_L are obtained from $I_{BM}(x,y)$ by placing W on the conditions of higher symmetry, and dilating the borders as previously explained. The dilation allows connection of separated borders. Then, a process of search and elimination of small areas takes place, removing small borders caused by noise that do not belong to the ROI. In order to provide closed ROI, through vertical and horizontal scans, the two closing pixels (endings) on the edges of every breast are detected. Having the coordinates of these points, it is possible to close the breast borders by drawing straight lines. Afterwards, the closed regions in W_R and W_L with greater areas are filled. If it is desired to obtain two separate ROI for each breast (recommended), the pixels in the last column of W_L are set to zero in order to create a separation line between the breasts. Since the generated areas are thickened, an erosion procedure is performed, allowing having borders closer to where initially the borders were detected. Additionally, with this process, edges that are not part of the ROI are deleted. Next, these images are placed on a blank image of the same size as $I_R(x,y)$, in the same positions where W_R and W_L were initially extracted. The resulting image $I_E(x,y)$ is a mask containing the ROI. This group of steps is illustrated on Fig. 5.

It is very possible that the ROI generated in $I_E(x,y)$ have an error of shape where the outer body edges may be connected to the lower edges of the breasts due to the strength

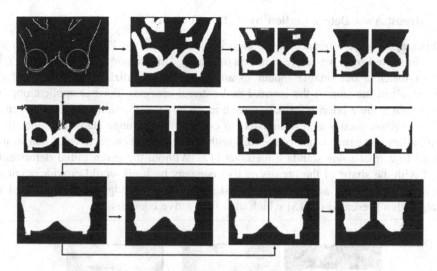

Fig. 5. Extraction of ROI through approximation of ROI.

of the outer body edges. In order to reduce this error, as shown in Fig. 6, a function to eliminate the undesired regions given by the following code:

$$\textit{for all white pixels } m(i,j) \textit{ in } I_E(x,y)$$

$$\textit{if } m(i,j-5) = 0 \textbf{ AND } m(i,j+5) = 0 \textbf{ THEN } m(i,j-4) = 0; \ m(i,j-3) = 0; \ \cdots$$
$$m(i,j+4) = 0;$$

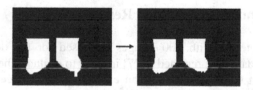

Fig. 6. Elimination of undesired regions in $I_E(x,y)$.

In order to generate a single ROI, as the ground truths provided in [7], an interpolation is needed. For this task, the coordinates of the lower bounds of the breasts in the ROI are obtained through a vertical scan. Additionally, the edge pixels in $I_{BM}(x,y)$ between the two extracted regions are considered and filtered, given that the interpolation will consider only coordinates from $I_{BM}(x,y)$ forming a monotonically incremental function before the symmetry axis, and a monotonically decremental function after the symmetry axis. The interpolation algorithm is cubic spline ($k = 3$), as it provides an excellent fit to the tabulated points and is not unduly complex. Subsequently, $I_E(x,y)$ is filled with pixels with a value of 1 in the middle region from the upper bound to the interpolated function generating a single ROI.

2.3   Breast Area Determination by Active Contours

At this stage, the performance of the algorithm is satisfactory. However, the accuracy can be improved by using active contours or snakes. This technique, shown in Fig. 7, is used to match a deformable model to an image by minimizing the energy which corresponds to the sum of the internal and external energies. For our application, the current ROI is the a priori information used as the initial deformable model. The active contour method used was reformulated to consider local image statistics rather than global, allowing segmentation of the ROI with heterogeneous characteristics that would be hard to extract using standard methods [12]. Without the given initial deformable model with the shape of the breasts, active contours by itself would provide accurate ROI. Rather it would amorphously adjust to the detected edges as it happened in the algorithm presented in [13] which also uses active contours.

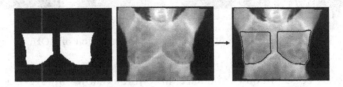

Fig. 7. Active contours adjust the borders of the ROI.

Since the edges of the ROI obtained from the active contour technique can be somewhat irregular, a contour softening is performed as the final process of the algorithm.

3   Discussion and Experimental Results

The algorithm was tested with 100 images (not used for the development of the algorithm) from the data base reported in [7] in order to evaluate the proposed method. It was observed that the algorithm had an excellent performance for the symmetry analysis, correcting tilt errors and providing accurate symmetry axes.

The approximation of the breasts was also very precise since in 100 % of the processed images, the circles were placed correctly over the breasts.

Given the ground truths from [7], the proposed method was compared to the algorithm of Marques [8] since he uses the same database and was found as the algorithm with best performance from all the methods found in the literature. The evaluation showed that the proposed algorithm had an accuracy and sensibility of 0.987 and 0.984, respectively, while Marques' algorithm showed an accuracy of 0.982 and sensibility of 0.974. Additionally, the proposed method had lower standard deviations in accuracy (0.007) and sensibility (0.017) compared to Marques (0.027 for accuracy and 0.066 for sensibility), showing that the suggested method is more reliable. Results also indicated that the proposed algorithm had specificity, positive and negative predictivity values of 0.988, 0.979 and 0.991, respectively.

Due to the characteristics of the proposed method, the algorithm detects breasts of all sizes and of all asymmetry conditions, even when there is a camouflage problem. Figure 8 shows the final extracted ROI for several thermographies. In its first image (upper left), the edges of the patient's left breast are not very visible (as in many other thermographies), and the algorithm presented in [8] failed to incorporate this breast into the ROI. However, the proposed method was able to estimate where this breast was more likely to be located, and included it into the ROI.

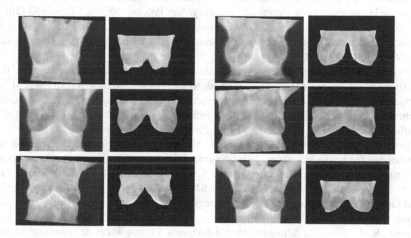

Fig. 8. Tilt corrections and extracted ROI for different breasts thermographies.

4 Conclusions

A novel algorithm for breast segmentation in digital thermographic images, with potential to be used in a CAD system, was presented.

The developed technique is able to correct the tilt error, providing a symmetry axis and detecting the breasts' positions even when their borders are not very visible. Results showed that the algorithm had an excellent performance, providing ROI with higher precision, accuracy, specificity and predictivity metrics compared to another algorithm found in the literature performing the same task.

Acknowledgements. The authors thanks to TNM, for the support of this research. Dr. Aura Conci and the personal of the Visual Lab at the Federal Fluminense University are also acknowledged for providing the used data base. Without their support, this research work would not have been possible.

References

1. World Health Organization: Cancer: fact sheet N°297 (2014). http://www.who.int/mediacentre/factsheets/fs297/en/

2. Drosu, O., Hantila, F., Maricaru, M.: Non-invasive method for screening and early detection of breast tumors using thermal field analysis. J. Electr. Electron. **2**(2), 21–25 (2009)
3. Yokoe, T., Yuichi, I., Yoshiki, T., Hidetada, A., Noritaka, S., Tohru, K., Susumu, O., Yasuo, M.: Breast cancer only detected by thermography: a case report. Ann. Cancer Res. Ther. Jpn. Soc. Strat. Cancer Res. Ther. **4**(2), 89–90 (1995)
4. Hobbins, W., Amalu, W.: Beating breast cancer. Original Internist **16**(1), 46 (2009)
5. Ng, E.Y.K., Kee, E.C.: Advanced integrated technique in breast cancer thermography. J. Med. Eng. Technol. Singap. **32**(2), 103–114 (2008)
6. Borchartt, T.B., Concia, A., Limab, R.C.F., Resminia, R., Sanchez, A.: Breast thermography from an image processing viewpoint: a survey. Sig. Process. **93**(10), 2785–2803 (2013)
7. Silva, L.F., Saade, D.C.M., Sequeiros-Olivera, G.O., Silva, A.C., Paiva, A.C., Bravo, R.S., Conci, A.: A new database for breast research with infrared image. J. Med. Imaging Health Inf. **4**(1), 92–100 (2014)
8. Marques, R.S.: Segmentação automática das mamas em imagens térmicas. M.S. thesis, Instituto de Computação, Universidade Federal Fluminense, Niterói, RJ, Brasil (2012)
9. Borchartt, T.B.: Análise de imagens termográficas para a classificação de alterações na mama. Ph.D. thesis, Universidade Federal Fluminense, Icaraí, Brasil (2013)
10. Ballard, D., Brown, C.: Computer Vision. Prentice-Hall, Englewood Cliffs (1982)
11. Chacon-Murguia, M.I., Villalobos-Montiel, A.J., Calderon-Contreras, J.D.: Thermal image processing for breast symmetry detection oriented to automatic breast cancer analysis. In: Martínez-Trinidad, J.F., Carrasco-Ochoa, J.A., Olvera-Lopez, J.A., Salas-Rodríguez, J., Suen, C.Y. (eds.) MCPR 2014. LNCS, vol. 8495, pp. 271–280. Springer, Heidelberg (2014)
12. Lankton, S., Tannenbaum, A.: Localizing region-based active contours. IEEE Trans. Image Process. **17**(11), 2029–2039 (2008)
13. Jaimes, R., Osorio, J.E.: Segmentación de Imágenes Termográficas de Glándulas Mamarias Utilizando Técnicas de Detección de Discontinuidades. Bachelor thesis. Universidad Industrial de Santander, Santander, Spain (2008)

Automatic Detection of Clouds from Aerial Photographs of Snowy Volcanoes

Carolina Chang[1,2]([⊠]) and Fernando Vaca[3]

[1] Grupo de Inteligencia Artificial, Universidad Simón Bolívar, Caracas, Venezuela
[2] PROMETEO Researcher, Instituto Geográfico Militar, Quito, Ecuador
cchang@usb.ve
[3] Instituto Geográfico Militar, Quito, Ecuador
fernando.vaca@mail.igm.gob.ec

Abstract. We propose a method for cloud detection from RGB aerial photographs of snow-capped volcanoes of Ecuador. For cartography purposes, clouds are undesired objects that occlude the terrain, while snow-covered areas are valid regions of a map. The traditional approach of image thresholding does not suffice when snowy areas cannot be dismissed from the image in advanced. We combine image thresholding with region growing and neural networks classification to detect clouds at the object level. We show that there is overlap at the pixel level of clouds and snow. At the classification task a fuzzy ARTMAP neural net achieves 91.4 % of success in fast learning mode and 95.5 % of success in slow learning mode at the same vigilance level, for 32×32 pixel images. Incremental learning is achieved at a loss of 0.4 % of the network performance.

Keywords: Cloud detection · Object-based image analysis · Fuzzy artmap neural network

1 Introduction

The Military Geographic Institute of Ecuador is in charge of generating the base cartography of the country. The land is mapped from aerial photographs taken from an airplane by means of a high resolution digital camera. Through the technique of orthophotography the images are corrected to produce a photographic presentation of the land without perspectives effects and other distortions. It is well known that clouds hinder the production of maps, as the clouds occlude the ground. The problem of automatic cloud detection has been extensively researched, but the existing solutions do not suite the Institute's task for two reasons:

a. The Institute's Maps are Produced from Aerial Digital RGB Photographs: Several of the existing techniques use multi-temporal information from satellite images of the same spot. The difference in brightness between two images of a region provides a starting point for identifying clouds. Additionally,

© Springer International Publishing Switzerland 2015
J.A. Carrasco-Ochoa et al. (Eds.): MCPR 2015, LNCS 9116, pp. 145–155, 2015.
DOI: 10.1007/978-3-319-19264-2_15

the movement of the clouds in the image sequence helps identify them from other areas of high brightness. In the case of aerial photography, the flight plan covers some overlap between pictures, but not high enough or with sufficient time span to apply the techniques of image comparison.

b. The Ecuadorian Andes have Perpetual Snow: The Continental Ecuador has a mountain range known as *"Volcanoes Avenue"*, about 300 Km long and 50 Km wide. In this *"Avenue"* there are many volcanoes and most of the highest mountains of the country. Due to the height, there are many snowy-capped peaks. Among them there is the Chimborazo (6,268 m Earth's closest point to the sun) and the Cotopaxi (5,897 m, one of the highest active volcanoes in the world). High brightness of perpetual snow mountains in the photos difficult to distinguish clouds through the technique of brightness of clouds. Unlikely clouds, snow-covered areas are valid areas of land to be retained in cartography.

It is necessary to differentiate the clouds from the snow-capped peaks in the Institute's photos. Currently, the process of cloud detection is performed entirely manually, which means a large investment of time.

The goal of this research is to detect the clouds by using only the RGB bands of a given photo and taking into account the possibility of the existence of snow-capped peaks in the photography. In order to guarantee this, linear transformations techniques, region growing and pattern classification through neural networks will be combined.

2   Related Work

Cloud detection has been an important issue since the beginning of remote sensing because half of the planet is covered with clouds at any given instant of time [12]. Several methods are based on the assumptions that clouds are bright and cold regions, therefore the RGB bands and infrared bands such as near-infrared (NIR), short wave infrared (SWIR) or mid-infrared (MIR) from satellites are widely used [4,6,9,13,14]. Since many images of a scene can be captured by satellite, cloud detection approaches often compare several images of the scene, at different times of the day, or over several days. For example, Champions [3] uses a pile of 6 overlapping images. Pixel-to-pixel comparison is done to detect clouds, based on the assumptions that clouds produce a significant increase of reflectance, and that the clouds should move over time (have different locations across the image time series).

Other methods use information such as the solar zenith angle or the solar irradiance. The D transform uses the the normalized difference vegetation index (NDVI) [5]. Pixels from vegetation have a positive NDVI, from water negative, and from clouds approximately zero. However, soil and snow also have NDVI close to zero as well.

Tasseled-Cap linear transformation (TC) is effective for detection of thin clouds and haze, and uses only information from the red (R) and blue (B) channels of the Landsat satellite:

$$TC = 0.846B - 0.464R \tag{1}$$

Variations of the tasseled-cap transformation such as the Haze Optimised Transform (HOT) use the image data to calculate the weights of the blue and red channels [15]. HOT calculates a linear regression between B (independent variable) and R (dependent variable). ϕ is the angle of the adjusted line:

$$HOT = sin(\phi)B - cos(\phi)R \qquad (2)$$

A similar transformation is proposed by Le Hégarat-Mascle and André, [9], but using the green and the MIR channels.

These conversions from multi-channel images to a single channel image (greyscale) that can be thresholded for cloud detection is a commonly used method [11] because it is simple, fast, and cheap. Marais et al. [10] studied how to transform a four-dimensional image into a greyscale image. They computed optimal thresholds for the D transform, the HOT transform and the Heteroscedastic Discriminant Analysis transfomation (HDA). They showed that HDA discriminates the clouds better, yet they are aware that HDA is slower that HOT, and that it will fail in the detection of clouds over snow or ice. It is quite clear to us that any method based on a single threshold will fail to discriminate clouds from snowy mountain peaks. Instead of using thresholds Jang et al. trained multi-layer perceptrons to classify pixels of SPOT Vegetation images [8].

Based on the work of Le Hégarat-Mascle and André, [9], Champion [3] states that clouds are connected objects and are brighter than the underlying landscape. This is a simple way of understanding the underlying process of approaches based on region growing algorithms. The brightest pixels are chosen as seeds for the algorithm. Neighboring pixels that satisfy certain criteria are aggregated with the region. All the neighbors of a pixel that belong to the region are evaluated. While Champion grows regions based on the brightness of pixels of panchromatic images, Sedano et al. implemented the procedure on the SWIR band after cloud patches were segmented by comparing cloud-free and cloudy images [14]. The cloud free images were obtained within a 17-day window from the date the cloudy image was obtained. Notice that in the multi-temporal approach most of the bright pixels that belong to the terrain will not be chosen since their brightness will not vary significantly across the multi-temporal images, and if so, they will not move across images as clouds do.

3   Input Data

The data set consists of 105 photos from 6 different regions of Ecuador: Antisana, Balzar, Chaguarpamba, Chillanes, Cotopaxi and Ilinizas. The photos are high-resolution RGB images in TIFF format. The photos were shoot from a *Cessna Citation II IGM-62* airplane equipped with an *UltraCam XP* aerial digital camera installed on a gyrostabilized mount. The mount compensates for abrupt movements of the airplane, and combined with a GPS/IMU system provides information for image correction through the process of orthophotography.

Antisana, Cotopaxi and Ilinizas are volcanoes of Ecuador. 44 of the photos of the data set include parts of the snowy peaks. As compared to related works, this is a large dataset of high resolution images. For example, Sedano and colleagues worked on 7 satellite images [14], while Marais et al. worked on 13 images from which 32 sub-scenes were extracted [10]. Each sub-scene measured 1000×1000 pixels. Le Héhart & Mascle worked with 39 images of the African Monsoon Multidisciplinary Analysis (AMMA) dataset [9].

17 additional photos of the Guayas region were obtained at the end of this study, so they were not used in experiments I and II (Sect. 5). This is an urban region that is very different from the other six. Therefore, these photos were ideal for testing the incremental learning capability of the system.

Table 1 shows a description of the input images. The ground sample distance (GSD) is the distance between pixel centers measured on the ground. For example adjacent pixels in the images of the Antisana region are 30 cm apart on the ground. We take into account that images have different GSD, as explained in Sect. 4.

Table 1. Description of the input data

Region	Number of images	Size (pixel x pixel)	GSD (cm)
Antisana	42	11310 × 17310	34
Balzar	5	11310 × 17310	14
Chaguarpamba	14	11310 × 17310	30
Chillanes	8	11310 × 17310	34
Cotopaxi	17	7680 × 13824	40
Ilinizas	19	11310 × 17310	30
Guayas	17	11310 × 17310	14

4 Methods

Our method for cloud detection consists of three steps: Preprocessing, Classification and Postprocessing. In this section we describe each of these steps.

4.1 Preprocessing

The high-resolution RGB images are preprocessed to segment bright objects. Preprocessing consists of Cloud Masking, Region Growing, and GSD image normalization.

Cloud Masking: In Sect. 2 we described several filters for cloud detection. From the RGB images we only have access to the color channels, so our best options for cloud masking are the Tasseled-Cap transformation and the HOT transformation. Our goal is neither to evaluate which transform is better nor to

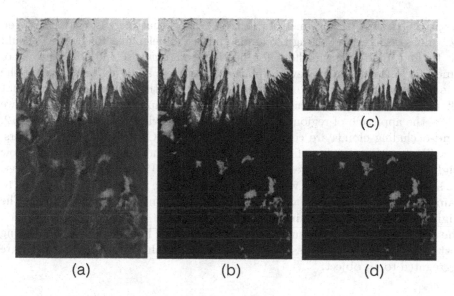

Fig. 1. Tasseled-Cap transformation of an photograph of the Cotopaxi region. (a) Original hophoto. (b) Pixels masked with the Tasseled-Cap transformation. (c) Region containing only pixels from snow. (d) Region containing only pixels of the Cloud class.

find the optimal thresholds, but to test whether a neural network can generalize what clouds are from object features.

The simplest and lowest-cost transformation is the Tasseled-Cap (TC) transformation (Eq. 1). This transformation was proposed for agricultural applications. Some vegetation can pass the filter as the green channel is not weighted in the equation. This transformation does not take into account the overall brightness of the image as the HOT transformation does, and this is the reason why we chose the TC transform. We decided to challenge the neural network approach with objects segmented by sub-optimal cloud masks. If the network can learn to recognize clouds from very variable input examples, then a better image preprocessing stage should improve the overall performance of the system. We used the following cloud mask for all images:

$$(TC > 35) \text{ AND } (G > 200) \tag{3}$$

Setting a threshold for the green channel (G) allows us to discard dark green pixels that have high TC values. The threshold of the Tasseled-Cap transform (TC) and the green channel (G) are smart guesses that we must improve in the future. Approaches for choosing these thresholds involve a process of manually labelling each pixel of the image in order to evaluate the filter's threshold accuracy. This is a time-consuming process that we cannot afford for the time being. The size of our images and data set are too large for this endeavour. For example Marais and colleagues invested two weeks labelling 32 1000 × 1000 pixel images [10]. This whole set is about 1/5 of the size of a single high-resolution image from our data set.

Region Growing: Experiment I (Sect. 5.1) shows that there is overlap between the values of pixels belonging to clouds and pixels belonging to snow. We could try to find an upper threshold for pixels belonging to clouds at the cost of some misclassification. However, we are interested in studying the clouds at the object level instead of at the pixel level. We believe that there are features that differentiate clouds from other objects of a photograph. To segment objects we follow the approach of region growing ([3, 9, 14]) previously discussed in Sect. 2. Land-occluding clouds are connected bright pixels. Contrary to those authors, we will encounter regions of snow in the images. Snow-covered land will grow bright regions as well. Images were compressed to 10 % of their size to increase the speed of the process. We will need to compress the regions anyway later on to train the neural network, so we would rather add some speed at this point. The brightest pixel of a masked image is the seed for the region-growing algorithm. The bounding-box of the resulting region is saved for labelling the clouds during post processing. The process is repeated until all the pixels of the image are aggregated to an object.

Ground Sampling Distance (GSD) Normalization: Images in our data set have different sizes and GSD (see Sect. 3). For this reason, the sizes of the segmented objects are not normalized. We would like to feed the neural network with normalized objects, even though we are aware that there are clouds of many sizes (Table 2).

The least common multiple (lcm) of all the GSDs is 14280 cm. Since images are compressed to 10 % of their size, the lcm of the compressed GSDs is 142800 cm. Hence, we created squared windows from the regions of 1428×1428 m.

Large regions are divided into several squared images of the size of the window. Tiny objects are discarded. They may belong to scattered snow, or other shiny objects. If they were indeed clouds, they will not invalidate the photo as there is some tolerance on the amount of occlusion allowed in the maps. The minimum object size was set to 2 % of the window. The other regions are completed to a square image.

4.2   Classification

There are many options of neural networks that can be used in this classification task. However, we chose to use fuzzy ARTMAP neural networks because

Table 2. Size of the regions window according to the image GSD

Compressed image GSD (cm)	Window size (pixel x pixel)
140	1040 × 1040
300	476 × 476
340	420 × 420
400	357 × 357

they are capable of incremental learning and fast learning ([1,2]). Incremental learning allows the network to learn new inputs after an initial training has been completed. Many other architectures would require a complete retraining of the network in order to learn the new inputs (*i.e.* catastrophic learning). With fast learning, inputs can be encoded in a single presentation while traditional networks may require thousands of training epochs. Usually fast learning is not as accurate as slow learning, but it has a greater capability for real-time applications and exploratory experimentation.

Fuzzy ARTMAP is a biologically inspired neural network architecture that combines fuzzy logic and adaptive resonance theory (ART). The parameter ρ (*vigilance*) determines how much generalization is permitted ($0 \leq \rho \leq 1$). Small vigilance values lead to greater generalization (*i.e.* fewer recognition categories or nodes are generated) while larger values lead to more differentiation among inputs. The parameter β controls the speed of learning ($0 \leq \beta \leq 1$). At $\beta = 1$ the network is set to fast learning.

4.3   Postprocessing

Recognized clouds are marked back in the original image using the coordinates of the regions, which were saved during preprocessing. The percentage of pixels of the image that belong to clouds is computed and reported.

5   Experiments and Results

We show the results of 3 experiments conducted on the image data set. In the first experiment we compare the pixels of regions that belong to clouds with those that belong to snow. In the second experiment we train ARTMAP neural networks with the square images of the regions. In the third experiment we test the incremental learning capability of the network.

5.1   Experiment I: Image Thresholding Counterexample

Some related works focus on finding an optimal threshold for the image transformations applied to detect clouds. These works do not deal with the problem of having brighter-than-cloud pixels. We hypothesize that there is no error-free way to classify clouds at the pixel level because some snow pixels and cloud pixels may have the same brightness value.

To confirm this hypothesis we conducted the following experiment. We applied the Tasseled-Cap transformation to a photo that has well defined snow regions (Fig. 1c) and cloud regions (Fig. 1d). The threshold was set to $\tau > 35$ manually. We compared the histograms of both regions. Results show that for every single brightness level of cloud pixels there were also snow pixels with the same value. In this single image, 7,846,843 pixels of the snow class have the same brightness as clouds. Therefore, it is not possible to classify pixels by setting a brightness threshold without producing classification errors.

5.2   Experiment II: ARTMAP Neural Network
for Cloud Recognition

We obtained 1302 images after preprocessing the 105 initial photos (Sect. 4.1).
The images were classified manually as belonging to the *Cloud Class* or to the
Not Cloud Class. Examples of the images are shown in Fig. 2 (upper and middle
rows). There are 794 images in the *Cloud Class* and 508 images in the *Not Cloud
Class*. We used a 7-fold cross-validation approach to train and test the neural
networks. Thus, the data set is partitioned into 7 disjoint subsets. Each subset
contains 186 images. Training and validation is performed 7 times, each time
using a different subset as the validation set, and using the remaining 6 subsets
as the training set. We report on the average performance of the networks over
the 7 runs of the cross-validation procedure.

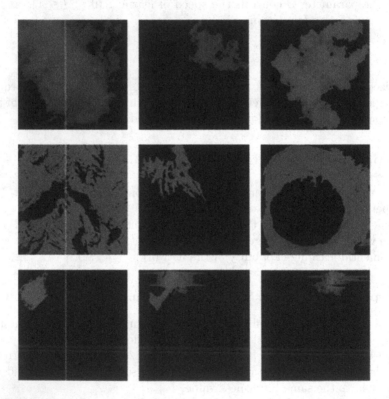

Fig. 2. Examples of high-resolution images obtained. **Upper Row:** Clouds. **Middle
Row:** Snow-covered volcanoes. **Bottom Row:** Objects from the Guayas region.

We run the cross-validation procedure for 3 different input image sizes trained
with fast learning a at different base vigilance values. ρ was varied form 0 to 1
at increments of 0.01. Table 3 shows the minimum vigilance value and number
of category nodes required to achieve a target performance.

The best results are obtained for all image sizes at success rates between 80 % and 90 %. Better performances are achieved at the cost of little network generalization, *i.e.* too many category nodes are created.

For the three image sizes we compare the performance of the network during fast learning and slow learning. We choose the vigilance value obtained at the 90 % success value. We vary the learning rate β starting at 0.5 at increments of 0.05 and let the neural network learn during 100 epochs. Results of the best performances are presented in Table 4. In all cases we obtained a better performance at slow learning, keeping the vigilance value obtained during fast learning. The number of category nodes increased in all cases, but the number of nodes are about half of the nodes obtained during fast learning at the 95 % of performance (see Table 3). For example for 32×32 images, we obtained 95.9 % of success with 429 category nodes, while 992 category nodes were created during fast learning. Thus, slow learning lets the network achieve a better generalization.

Table 3. ARTMAP neural network performance in fast learning for 3 different image sizes

Target %	16 × 16				24 × 24				32×32			
	Test	Train	ρ	Nodes	Test	Train	ρ	Nodes	Test	Train	ρ	Nodes
min	75.8	73.9	0.00	20	**78.0**	75.8	0.00	**20**	67.7	75.1	0.00	18
80	80.1	78.0	0.32	83	82.2	78.4	0.20	73	83.9	77.6	0.24	66
85	85.5	85.2	0.59	188	85.5	82.9	0.52	163	85.5	84.8	0.42	124
90	91.9	90.1	0.82	418	90.1	90.1	0.79	362	91.4	89.9	0.74	317
95	-	-	-	-	95.2	96.2	0.99	**980**	95.6	96.7	0.99	992
Max	94.6	95.8	0.99	975	**96.2**	96.9	1.00	**1114**	96.2	97.1	1.00	1114

Table 4. ARTMAP neural network performance after slow learning

		Fast Learning		Slow Learning		
Image size	ρ	Test %	Nodes	β	Test %	Nodes
16×16	0.82	91.9	418	0.70	95.7	488
24×24	0.79	90.1	362	0.65	95.7	454
32×32	0.74	91.4	317	0.65	**95.9**	**429**

Table 5 shows the confusion matrix of the network for 32×32 images after slow learning. We note a slightly better prediction of the Cloud Class. Recall however that there are less examples in the Not Cloud Class, and that this class include a variety of objects.

Table 5. Average confusion matrix of the ARTMAP network for 32×32 input images and slow learning. Recall that there are 794 images in the Cloud Class and 508 images in the Not Cloud Class

Slow Learning 32×32		Predicted	
		Cloud	Not Cloud
Actual	Cloud	770 (97.0 %)	24 (3.0 %)
	Not Cloud	22 (4.3 %)	486 (95.7 %)

5.3   Experiment III: Incremental Learning from Urban Scenes

We used an ARTMAP neural network trained previously at slow learning ($\beta=0.65$) with 32×32 image inputs. At a vigilance level of $\rho=0.74$ the network has a success rate of 95.5 % on all the 1302 images. Then, we obtained 41 objects (20 Clouds, and 21 Not Clouds) from the Guayas region photos. The not cloud objects are building roofs and image distortions not seen by the network before (see the bottom row of Fig. 2). We tested the 41 images. 80 % of the new misclassified images were presented at the network for fast learning.

Table 6. Incremental Learning. The network is tested on a total of 1343 images

Guayas objects	20 Clouds	21 Not Clouds	Network Total Performance
Before fast learning	17 (85 %)	2 (1.5 %)	95.2 %
After fast learning	19 (95 %)	17 (80 %)	94.8 %

Table 6 shows the network performance for incremental learning. As expected, the network performed poorly on the not cloud objects. After 17 randomly chosen images are presented to the network for fast learning, classification of the new images improved significantly. However, we observe that new learning occurs at the cost of a small loss of the previously acquired knowledge.

6   Discussion

Results show that an object-based image analysis approach is suitable for a cloud recognition task. We have focused on exploring whether bright regions of the images encode enough information for a neural network to learn the differences between clouds and other objects of the scene. We achieved over 95 % of accuracy in the classification task, which is a promising result. We must address now the issue of image preprocessing optimization and multi-resolution learning. If we can depure the inputs, learning should improve. We could evaluate other neural networks architectures, voting mechanisms, and a variety of classification techniques as well. However we would like to preserve the incremental learning capability that the ARTMAP neural network has to some extent.

Acknowledgement. This work has been financed by the PROMETEO Project from Secretaría de Educación Superior, Ciencia, Tecnología e Innovación (SENESCYT) of the Republic of Ecuador.

We would like to thank Mireya Chuga and Lenin Jaramillo for their help collecting the input data set.

References

1. Carpenter, G.A., Grossberg, S., Markuzon, N., Reynolds, J.H., Rosen, D.B.: Fuzzy ARTMAP: a neural network architecture for incremental supervised learning of analog multidimensional maps. IEEE Trans. Neural Netw. **3**, 698–713 (1992)
2. Carpenter, G.A.: Default ARTMAP. In: Proceedings of the International Joint Conference on Neural Networks (IJCNN 2003), Portland, Oregon (2003)
3. Champion, N.: Automatic cloud detection from multi-temporal satellite images: towards the use of pléaides time series. International Archives of the Photogrammetry, Remote Sensing and Spatial Information Sciences. vol. XXXIX-B3, pp. 559–564 (2012)
4. Chen, P.Y., Srinivasan, R., Fedosejevs, G., Narasimhan, B.: An automated cloud detection method for daily NOAA-14 AVHRR data for Texas, USA. Int. J. Remote Sens. **23**(15), 2939–2950 (2002)
5. Di Girolamo, L., Davies, R.: The image navigation cloud mask for the multi-angle imaging spectroradiometer (MISR). J. Atmos. Ocean. Technol. **12**, 1215–1228 (1995)
6. Gutman, G.G.: Satellite daytime image classification for global studies of earth's surface parameters from polar orbiters. Int. J. Remote Sens. **13**(2), 209–234 (1992)
7. Hagolle, O., Huc, M., Pascual, D.V., Dedieu, G.: A multi-temporal method for cloud detection, applied to FORMOSAT-2, VENS, LANDSAT and SENTINEL-2 images. Remote Sens. Environ. **114**(8), 1747–1755 (2010)
8. Jang, J.D., Viau, A., Anctil, F., Bartholomé, E.: Neural network application for cloud detection in SPOT VEGETATION images. Int. J. Remote Sens. **27**, 719–736 (2006)
9. Le Hégarat-Mascle, S., André, C.: Use of markov random fields for automatic cloud/shadow detection on high resolution optical images. ISPRS J. Photogramm. Remote Sens. **64**(4), 351–366 (2009)
10. Marais, I.V.Z., Du Preez, J.A., Steyn, W.H.: An optimal image transform for threshold-based cloud detection using heteroscedastic discriminant analysis. Int. J. Remote Sens. **32**(6), 1713–1729 (2011)
11. Rossow, W.B.: Measuring cloud properties from space: a review. J. Clim. **2**, 201–213 (1989)
12. Saunders, R.W.: An automated scheme for the removal of cloud contamination from AVHRR radiances over western europe. Int. J. Remote Sens. **7**(7), 867–886 (1986)
13. Saunders, R.W., Kriebel, K.T.: Improved method for detecting clear sky and cloudy radiances from AVHRR data. Int. J. Remote Sens. **9**, 123–150 (1988)
14. Sedano, F., Kempeneers, P., Strobl, P., Kucera, J., Vogt, P., Seebach, L., San-Miguel-Ayanz, J.: A cloud mask methodology for high resolution remote sensing data combining information from high and medium resolution optical sensors. ISPRS J. Photogramm. Remote Sens. **66**(5), 588–596 (2011)
15. Zhang, Y., Guindon, B., Cihlar, J.: An image transform to characterize and compensate for spatial variations in thin cloud contamination of landsat images. Remote Sens. Environ. **82**(2), 173–187 (2002)

A Comparative Study of Robust Segmentation Algorithms for Iris Verification System of High Reliability

Mireya S. García-Vázquez[1(✉)], Eduardo Garea-Llano[2],
Juan M. Colores-Vargas[3], Luis M. Zamudio-Fuentes[1],
and Alejandro A. Ramírez-Acosta[4]

[1] Instituto Politécnico Nacional - CITEDI, Tijuana, Mexico
{msarai,lzamudiof}@ipn.mx
[2] Advanced Technologies Application Center - CENATAV, Habana, Cuba
egarea@cenatav.co.cu
[3] Universidad Autónoma de Baja California - CITEC, Tijuana, Mexico
dr.jcolores@gmail.com
[4] MIRAL R&D&I, Houston, CA, USA
ramacosl0@hotmail.com

Abstract. Iris recognition is being widely used in different environments where the identity of a person is necessary. Therefore, it is a challenging problem to maintain high reliability and stability of this kind of systems in harsh environments. Iris segmentation is one of the most important process in iris recognition to preserve the above-mentioned characteristics. Indeed, iris segmentation may compromise the performance of the entire system. This paper presents a comparative study of four segmentation algorithms in the frame of the high reliability iris verification system. These segmentation approaches are implemented, evaluated and compared based on their accuracy using three unconstraint databases, one of them is a video iris database. The result shows that, for an ultra-high security system on verification at FAR = 0.01 %, segmentation 3 (Viterbi) presents the best results.

Keywords: Iris recognition · Segmentation · Uncontrolled environments

1 Introduction

Nowadays, iris recognition is being widely used in different environments where the identity of a person is necessary. So, it is a challenging problem to maintain a stable and reliable iris recognition system which is effective in unconstrained environments. Indeed, in difficult conditions, the person to recognize usually moves his head in different ways giving rise to non-ideal images (with occlusion, off-angle, motion-blur and defocus) for recognition [1, 2]. Most iris recognition systems achieve recognition rates higher than 99 % under controlled conditions. These cases are well documented in [3–6]. However, the iris recognition rates may significantly decrease in unconstrained environments if the capabilities of the key processing stages are not developed accordingly. For this reason in [7–9], some improvements have been incorporated on

© Springer International Publishing Switzerland 2015
J.A. Carrasco-Ochoa et al. (Eds.): MCPR 2015, LNCS 9116, pp. 156–165, 2015.
DOI: 10.1007/978-3-319-19264-2_16

the acquisition module. As it is mentioned, iris segmentation is one of the most important process in iris recognition to preserve a high reliable and stable iris verification systems in harsh environments. It may compromise the performance of the entire system. Thus, this paper presents a comparative study of four segmentation algorithms in the frame of the high reliability iris verification system. These segmentation approaches are implemented, evaluated and compared based on their accuracy using three unconstraint databases, one of them is a video iris database. The remainder of this paper is organized as follows. Section 2 presents the iris verification scheme and segmentation approaches in the frame of the high reliability. Section 3 presents the experimental results, and Sect. 4 gives the conclusion of this work.

2 Iris Verification Scheme and Segmentation Approaches

The iris verification scheme (Fig. 1) comprises the following steps. From capturing one or more images of the iris of a person or persons either for the same type of sensor or multiple sensors, preprocessing of the input image is performed. In this step the inner and outer boundaries of the iris is extracted using at least one of the four previously selected segmentation algorithms implemented [10, 17]. Once the iris segmentation was obtained, the transformation of coordinates is performed to obtain the normalized iris image.

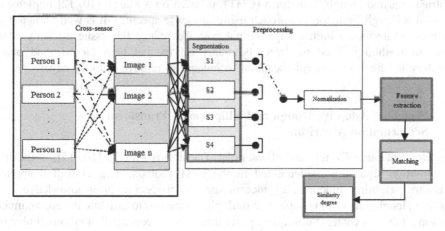

Fig. 1. Iris verification scheme.

Finally, the characteristic extraction, matching and similarity degree steps [17] are performed as it is shown in Fig. 1.

2.1 Viterbi-Based Segmentation Algorithm

The first step of this segmentation approach [10, 11] consists in a rough localization of the pupil area. First, filling the white holes removes specular reflections due to illuminators. Then, a morphological opening removes dark areas smaller than the disk-shaped structuring element. Then, the pupil area is almost the biggest dark area, and is surrounded by the iris, which is darker than the sclera and the skin. Consequently the sum of intensity values in large windows in the image is computed, and the minimum corresponds to the pupil area. The pupil being roughly located, a morphological reconstruction allows estimating a first center, which is required for exploiting the Viterbi algorithm. The second step consists in accurately extracting the pupil contour and a well estimated pupil circle for normalization. Relying on the pupil center, the Viterbi algorithm is used to extract the accurate pupil contour. This accurate contour will be used to build the iris mask for recognition purposes. Viterbi algorithm is exploited at two resolutions, corresponding to the number of points considered to estimate the contour. At a high resolution (all pixels are considered), it allows finding precise contours, while at a low resolution; it retrieves coarse contours that will be used to improve the accuracy of normalization circles. One advantage of this approach is that it can be easily generalized to elliptic normalization curves as well as to other parametric normalization curves in polar coordinates. Also, a clear interest of this approach is that it does not require any threshold on the gradient map. Moreover, the Viterbi algorithm implementation is generic (the system can be used on different databases presenting various degradations, without any adaptation).

2.2 Contrast-Adjusted Hough Transform Segmentation Algorithm

Contrast-adjusted Hough Transform (CHT), is based on a Masek [10, 12] implementation of a Hough Transform approach using (database-specific). It is well known that this method implies a high computational cost. To reduce this issue a Canny edge detection method is used to detect boundary curves that help the iris and pupil boundary localization, and enhancement techniques to remove unlikely edges.

2.3 Weighted Adaptive Hough and Ellipsopolar Transform Segmentation Algorithm

Weighted Adaptive Hough and ellipsopolar Transforms (WHT) [10, 13], is the iris segmentation algorithm implemented in the USIT toolbox. This algorithm applies Gaussian weighting functions to incorporate model-specific prior knowledge. An adaptive Hough transform is applied at multiple resolutions to estimate the approximate position of the iris center. Subsequent polar transform detects the first elliptic limbic or pupillary boundary, and an ellipsopolar transform finds the second boundary based on the outcome of the first. This way, both iris images with clear limbic (typical for visible wavelength) and with clear pupillary boundaries (typical for near infrared) can be processed in a uniform manner.

2.4 Modified Hough Transform Segmentation Algorithm

Modified Hough Transform (MHT), uses the circular Hough transform initially employed by Wildes et al. [14] combined with a Canny edge detector [15–17]. From the edge map, votes are cast in Hough space for the parameters of circles passing through each edge point. These parameters are the centre coordinates and the radius, for the iris and pupil outer boundaries. These parameters are the centre's coordinates $\left[(x_p, y_p), (x_i, y_i)\right]$ and radius $\left[r_p, r_i\right]$, for the iris and pupil outer boundaries respectively.

3 Experimental Results

The aim of this research was oriented to explore the capacity of the robust methods at level of segmentation stage for unconstrained environments to increase the recognition rates, in the frame of the high reliability iris verification system.

3.1 Databases

To develop a robust iris image preprocessing, feature extraction and matching methods in unconstrained environments, it is necessary to use a database collected with different iris cameras and different capture conditions. We highlight the fact for this research is been used two image still databases and one video database. CASIA-V3-INTERVAL (images) [18] all iris images are 8 bit gray-level JPEG files, collected under near infrared illumination, with 320 × 280 pixel resolution (2639 images, 395 classes). Almost all subjects are Chinese. CASIA-V4-THOUSAND (images) [19], which were contains 20,000 iris images from 1,000 subjects. The main sources of intra-class variations in CASIA-Iris-Thousand are eyeglasses and specular reflections. The MBGC-V2 (video) [20] database provided 986 near infrared eye videos. All videos were acquired using an LG2200 EOU iris capture system [21]. This database presents noise factors, especially those relative to reflections, contrast, luminosity, eyelid and eyelash iris obstruction and focus characteristics. These facts make it the most appropriate for the objectives of real iris systems for uncontrolled. A sample set of three database images are shown in Fig. 2.

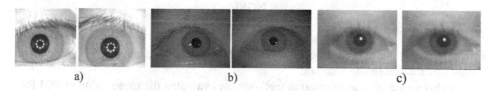

a)   b)   c)

Fig. 2. Examples of databases images (a) CASIA-V3-Interval, (b) CASIA-V4-Thousand, (c) MBGC-V2.

The Table 1 shows the used iris database with detailed specifications.

Table 1. Iris database.

Database[a]	Type	Size	Format image	Format video	Iris sensor
Db1	NIR	320 × 280	JPG	–	Casia-CAM
Db2	NIR	640 × 480	JPG	–	IrisKing IKEMB-100
Db3	NIR	640 × 480	–	MP4	LG2200 EOU

[a]Db1 - Casia-V3-Interval, Db2 - Casia-V4-Thousand, Db3 - MBGC-V2.

3.2 Quality for Segmented Images

In this part we considered two categories of quality for segmented images: good segmented and bad segmented images. Good segmented images contain more than 60 % of the iris texture and less than 40 % of eyelids or eyelashes or elements that do not belong to the eye (noise elements). Bad segmented images contain more than 40 % of noise elements (see Fig. 3).

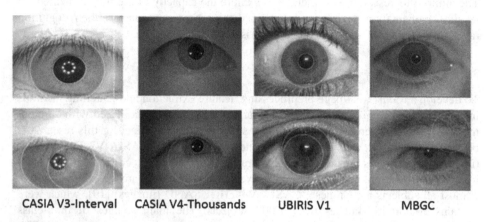

CASIA V3-Interval   CASIA V4-Thousands   UBIRIS V1   MBGC

Fig. 3. Some examples of segmented images. Above: good segmented, below: bad segmented.

Table 2 shows the obtained segmentation results on the analyzed databases. The process of evaluation was manually assessed by comparing the segmented iris images. As measure of segmentation performance we computed the percentage of good segmented images for each evaluated database by the expression 1:

$$PGI = \frac{NGSI}{NTI} \times 100 \qquad (1)$$

where: NGSI, is the number of good segmented images in the database; NTI is the total number of images in the database.

To choose the best segmentation methods we evaluated the mean value of PGI for each segmentation method in all databases by expression 2:

$$MS = \frac{\sum_{i=1}^{3} PGIi}{3} \qquad (2)$$

From Table 2 it is possible to see that taking into account the MS values obtained for each segmentation method the first two best performances were obtained by Viterbi and Weighted Adaptive Hough transform. These methods obtained stable results on the three evaluated databases.

Table 2. Quality for segmentation images, WHT: Weighted adaptive Hough Transform; CHT: Contrast adjusted Hough Transform; MHT: Modified Hough Transform.

	Viterbi	WHT	CHT	MHT
CASIA V3- Interval				
NGSI	2639	2639	2639	2600
PGI %	100	100	100	98.5
CASIA V4-Thousands				
NGSI	3196	3704	2639	2365
PGI %	80.7	93.5	66.6	59.7
MBGC				
NGSI	1736	1663	1747	1764
PGI %	86.8	83.1	87.3	88.2
MS %	**89.2**	**92.2**	**84.6**	**82.1**

3.3 Experimental Results in Verification Task

The recognition tests were conducted using the experimental design presented in Fig. 1. All these processes were implemented in C language.

The matching probes generate two distributions: Inter-Class and Intra-Class (Hamming distances for Clients and Impostors) useful to compare the performance of segmentation algorithms. To evaluate any identity verification system, it is necessary to determine the point in which the FAR (false accept rate) and FRR (false reject rate) have the same value, which is called EER (equal error rate), because it allows the user to determine the appropriate threshold Th, for a given application. The Table 3 contains the above mentioned values for the verification scheme (Fig. 1).

Table 3. Results in verification task at EER (FAR = FRR).

	Viterbi	WHT	CHT	MHT
CASIA V3- Interval				
EER	7.4	6.45	7.59	8.48
CASIA V4-Thousands				
EER	10.73	7.34	10.81	8.44
MBGC				
EER	3.73	3.74	3.56	13.47

3.3.1 ROC Curve Analysis

The receiver operating characteristics (ROC) curve was used to obtain the optimal threshold decision. In a ROC curve the false accept rate is plotted in function of the false reject rate for different threshold points. With this type of curve, we assure that the high security systems are interested in not allow unauthorized users to access restricted places. Otherwise, the system would allow a false accepted user to access. Therefore, these biometric systems work at low FAR/high FRR values. In this scene, the system have better to reject a valid user than allow an unauthorized user to access. The FAR and FRR evaluation measurements are the most common indicators of recognition accuracy when the biometric system is meant for verification mode, see Fig. 4.

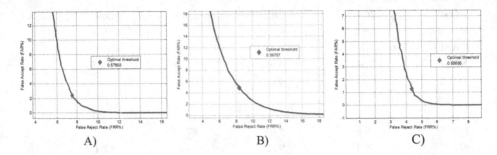

Fig. 4. ROC curves together with EER threshold value. (A) CASIA-V3- Interval. WHT segmentatidson, (B) CASIA-V4-Thousands. WHT segmentation, (C) MBGC-V2. CHT segmentation.

The Table 4 contains the obtained results that choose the optimal decision threshold for discrimination between classes (Intra-Class and Inter-Class). This improvement was described using ROC curves; FAR and GAR (GAR = 1-FRR) [22].

Table 4. Results in verification task with optimal threshold for security system applications.

Algorithm	Masek								
	GAR	FAR	FRR	GAR	FAR	FRR	GAR	FAR	FRR
	CASIA V3- Interval			CASIA V4-Thousands			MBGCv2		
CHT	91.36	3.21	8.63	86.82	5.03	13.17	95.83	1.21	4.16
WHT	92.47	2.39	7.52	91.6	4.85	8.39	95.63	1.27	4.36
MHT	90.03	4.29	9.96	90.29	5.04	9.70	84.57	3.67	15.42
Viterbi	91.63	3.96	8.36	87.42	6.48	12.57	95.71	1.83	4.28

Under conditions of CASIA-V3-Interval which is the database that was captured in more controlled conditions, the best performance is obtained for WHT segmentation method with GAR = 92.47 % at FAR = 2.39 % (see Fig. 3A).

Under conditions of CASIA4-V4-Thousands database, also the highest rating was obtained by WHT segmentation method with GAR = 91.6 % at FAR = 4.85 % (see Fig. 3B). For MBGC database the CHT segmentation method obtained the best results with GAR = 95.83 % at FAR = 1.21 % (see Fig. 3C). Overall the results demonstrate that WHT method behaves stable for the three databases.

The evaluation of accuracy for an ultra-high security system on verification at FAR = 0.01 % was estimated by ROC curves; False Reject Rate versus False Acceptance Rate. Table 5 reports the results of the GAR for each of automatic segmentation results.

Table 5. Results in verification task at FAR ≤ 0.01 %.

Algorithm	Masek		
	GAR		
	CASIA V3- Interval	CASIA V4-Thousands	MBGC
CHT	92.5	86.2	95.6
WHT	89.7	87.5	95.5
MHT	87.9	85.7	82.9
Viterbi	93.2	90.2	95.1

Under conditions of CASIA-V3-Interval database, the four experimented segmentation algorithms have very similar performance with GAR = 87.9–93.2 %, showing the best performance for Viterbi method at FAR ≤ 0.01 %. This shows that when image capture conditions are controlled, segmentation methods generally perform well.

Under conditions of CASIA4-V4-Thousands database, also the highest rating was obtained by Viterbi segmentation method with GAR = 90.2 % at FAR ≤ 0.01 %, but in general the three algorithms based on Hough Transform obtained stable results.

For MBGCv2 database the CHT and WHT segmentation method obtained the best results with GAR = 95.6 and 95.5 % respectively at FAR ≤ 0.01 %, the Viterbi Algorithm also obtained wood results with GAR = 95,1 % at FAR ≤ 0.01 %.

Overall the results demonstrate that Viterbi method behaves stable for the three databases. The other three methods do not achieved the best results. This fact also corresponds with the results of the evaluation of segmentation, although the Viterbi method misses the best results in terms of MS, it maintains a constant stability for the three databases evaluated with a PGI of 100, 80 and 86 %.

4   Conclusions

In this paper, we have presented a comparative study of four segmentation algorithms in the frame of the high reliability iris verification system. These segmentation approaches were implemented, evaluated and compared based on their accuracy using three unconstraint databases, one of them is a video iris database. The ability of the system to work with non-ideal iris images has a significant importance because is a common realistic scenario.

The first results test show that based on Table 2, the best segmentation method was WHT. On the other hand, based on Table 4, for optimal threshold the GAR average is for WHT 93.23 %, CHT 91.34 %, Viterbi 91.59 % and MHT 88.30 %. This shows that WHT method presents the best performance under a normal behavior of the system. However, if we raise the robustness of the system working at low FAR/ high FRR values (FAR ≤ 0.01 %), based on Table 5 the GAR average is for Viterbi 92.83 %, CHT 91.43 %, WHT 90.90 % and MHT 85.50 %. These differences could be possible due the segmentation method accuracy.

It was demonstrated that the Viterbi segmentation method has the most stable performance segmenting images taken under different conditions since the features extracted from images segmented by it contain more information than the images segmented by the others methods. It can be used in a real iris recognition system for ultra-high security. Combining this method with a set preprocessing technics to improve the image quality can produce a significant increase in the effectiveness of the recognition rates.

We believe that this problem can be also faced using combination of clustering algorithms. In particular, an algorithm of this kind may be able to face the possible high dimensionality of the image and use the existing geospatial relationship between the pixels to obtain better results.

The problem of computational cost is another line that we will continue investigating since it has to perform at least two segmentations methods, in consequence they will increase the computational time depending on the nature of the simultaneously used methods.

Aknowledgment. This research was supported by SIP2015 project grant from Instituto Politécnico Nacional from México and Iris Project grant from Advanced Technologies Application Center from Cuba.

References

1. Cao, Y., Wang, Z., Lv, Y.: Genetic algorithm based parameter identification of defocused image. In: ICCCSIT 2008, International Conference on Computer Science and Information Technology, pp. 439–442, September 2008
2. Colores, J.M., García-Vázquez, M., Ramírez-Acosta, A., Pérez-Meana, H.: Iris image evaluation for non-cooperative biometric iris recognition system. In: Batyrshin, I., Sidorov, G. (eds.) MICAI 2011, Part II. LNCS, vol. 7095, pp. 499–509. Springer, Heidelberg (2011)
3. Daugman, J.: The importance of being random: statistical principles of iris recognition. Pattern Recogn. **36**, 279–291 (2003)
4. Phillips, P., Scruggs, W., Toole, A.: FRVT 2006 and ICE 2006 large-scale results, Technical report, National Institute of Standards and Technology, NISTIR 7408 (2007)
5. Proenca, H., Alexandre, L.: The NICE.I: noisy iris challenge evaluation. In: Proceedings of the IEEE First International Conference on Biometrics: Theory, Applications and Systems, vol. 1, pp. 1–4 (2007)
6. Newton, E.M., Phillips, P.J.: Meta-analysis of third-party evaluations of iris recognition. IEEE Trans. Syst. Man Cybern. **39**(1), 4–11 (2009)
7. Kalka, N.D., Zuo, J., Schmid, N.A., Cukic, B.: Image quality assessment for iris biometric. In: SPIE 6202: Biometric Technology for Human Identification III, vol. 6202, pp. D1–D11 (2006)
8. Chen, Y., Dass, S.C., Jain, A.K.: Localized iris image quality using 2-d wavelets. In: Zhang, D., Jain, A.K. (eds.) ICB 2005. LNCS, vol. 3832, pp. 373–381. Springer, Heidelberg (2005)
9. Belcher, C., Du, Y.: A selective feature information approach for iris image quality measure. IEEE Trans. Inf. Forensics Secur. **3**(3), 572–577 (2008)

10. Sanchez-Gonzalez, Y., Chacon-Cabrera, Y., Garea-Llano, E.: A comparison of fused segmentation algorithms for iris verification. In: Bayro-Corrochano, E., Hancock, E. (eds.) CIARP 2014. LNCS, vol. 8827, pp. 112–119. Springer, Heidelberg (2014)
11. Sutra, G., Garcia-Salicetti, S., Dorizzi, B.: The viterbi algorithm at different resolutions for enhanced iris segmentation. In: 2012 5th IAPR International Conference on Biometrics (ICB), pp. 310–316. IEEE (2012)
12. Masek, L.: Recognition of human iris patterns for biometric identification. Technical report (2003)
13. Uhl, A., Wild, P.: Weighted adaptive hough and ellipsopolar transforms for realtime iris segmentation. In: 2012 5th IAPR International Conference on Biometrics (ICB), pp. 283–290. IEEE (2012)
14. Wildes, R.P., Asmuth, J.C., Green, G.L.: A system for automated recognition 0-8186-6410-X/94, IEEE (1994)
15. Canny, J.F.: Finding edges and lines in images. M.S. thesis, Massachusetts Institute of Technology (1983)
16. Hough, P.V.C.: Method and means for recognizing complex patterns. U.S. Patent 3 069 654 (1962)
17. Colores-Vargas, J.M., García-Vázquez, M., Ramírez-Acosta, A., Pérez-Meana, H., Nakano-Miyatake, M.: Video images fusion to improve iris recognition accuracy in unconstrained environments. In: Carrasco-Ochoa, J.A., Martínez-Trinidad, J.F., Rodríguez, J.S., di Baja, G. S. (eds.) MCPR 2012. LNCS, vol. 7914, pp. 114–125. Springer, Heidelberg (2013)
18. Daugman, J.G.: High confidence visual recognition of persons by a test of statistical independence. IEEE Trans. Pattern Anal. Mach. Intell. **15**(11), 1148–1161 (1993)
19. CASIA-V3-Interval. The Center of Biometrics and Security Research, CASIA Iris Image Database. http://biometrics.idealtest.org/
20. CASIA-V4-Thousands. The Center of Biometrics and Security Research, CASIA Iris Image Database. http://biometrics.idealtest.org/
21. Multiple Biometric Grand Challenge. http://face.nist.gov/mbgc/
22. Bowyer, K.W., Hollingsworth, K., Flynn, P.J.: Image understanding for iris biometrics: a survey. Comput. Vis. Image Underst. **110**(2), 281–307 (2008)
23. Zweig, M., Campbell, G.: Receiver-operating characteristic ROC plots: a fundamental evaluation tool in clinical medicine. Clin. Chem. **39**, 561–577 (1993)

Robotics and Computer Vision

Vision-Based Humanoid Robot Navigation in a Featureless Environment

Julio Delgado-Galvan, Alberto Navarro-Ramirez, Jose Nunez-Varela[✉],
Cesar Puente-Montejano, and Francisco Martinez-Perez

College of Engineering, Universidad Autonoma de San Luis Potosi,
78290 San Luis Potosi, Mexico
{cesardelgado,albertonavarro}@alumnos.uaslp.edu.mx,
{jose.nunez,cesar.puente,eduardo.perez}@uaslp.mx

Abstract. One of the most basic tasks for any autonomous mobile robot is that of safely navigating from one point to another (e.g. service robots should be able to find their way in different kinds of environments). Typically, vision is used to find landmarks in that environment to help the robot localise itself reliably. However, some environments may lack of these landmarks and the robot would need to be able to find its way in a featureless environment. This paper presents a topological vision-based approach for navigating through a featureless maze-like environment using a NAO humanoid robot, where all processing is performed by the robot's embedded computer. We show how our approach allows the robot to reliably navigate in this kind of environment in real-time.

Keywords: Navigation · Humanoid robot · Vision

1 Introduction

One of the most basic tasks that any autonomous mobile robot is expected to perform is that of safely navigating from one point to another. For instance, service robots should be capable of bringing and taking objects from one place to another in different kinds of environments, such as homes, offices, factories, etc. To solve the navigation problem two other problems need to be taken into account, namely, localization and mapping. *Localization* is the problem of knowing where the robot is in the environment, whilst *mapping* is the process of creating a map of such environment in order to know where to go [1,2]. Sensors such as cameras, sonars and laser range finders are used, individually or in combination, to tackle these problems. Visual information is commonly used for robot localization, where the autonomous detection of visual landmarks (e.g. posters on walls, signs, etc.) in the environment help the robot to discriminate between different places, and so the robot is able to know approximately where

J. Delgado-Galvan—We gratefully acknowledge the undergraduate scholarships given by CONACYT-Mexico (No. 177041) and the support given by FAI-UASLP (SA06-0031).

© Springer International Publishing Switzerland 2015
J.A. Carrasco-Ochoa et al. (Eds.): MCPR 2015, LNCS 9116, pp. 169–178, 2015.
DOI: 10.1007/978-3-319-19264-2_17

it is [3,4]. However, these landmarks might not be static and could disappear or move at some point in time. Furthermore, for some environments, such as office buildings, is difficult to find reliable visual landmarks. The contributions of this paper are twofold: (i) it presents a topological vision-based approach for navigating through a featureless environment in form of a maze, and (ii) it shows how a humanoid robot is able to navigate through this environment in real-time.

Our work is implemented using the NAO humanoid robot, a standard robotic platform developed by Aldebaran Robotics. NAO has 25 degrees of freedom, two video cameras, two sonars, two bumpers on its feet, tactile sensors on its head and arms, force sensors and an inertial unit (Fig. 1A). It is important to note that the video cameras are not located in the "eyes" of the robot, it has a top camera on its forehead and a bottom camera on its "mouth". Furthermore, the cameras' field of view do not intersect, i.e. it is not possible to use stereo vision. For processing, the robot has an embedded computer with an Atom microprocessor and a Linux-based operating system. The only sensor used in this work is the bottom camera, as it is able to provide the robot with visual information about both, the walls and the floor of the environment. The top camera's field of view barely obtains information about the floor, so it is not used at the moment. All processing is performed by the robot's embedded computer in real-time.

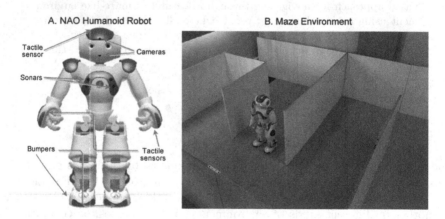

Fig. 1. A. The NAO humanoid robot and the location of some of its sensors. B. Photo of the NAO robot solving the maze.

The robot has the task to navigate and successfully exit a maze with featureless walls and floor (Fig. 1B). The robot knows it has reached the end of the maze by detecting a black strip on the floor located at the exit. The colour of the walls is white and the floor is made of wood. The robot is placed at the entrance of the maze and the goal is to finish the maze in the shortest amount of time. To perform the task our approach builds a topological map (i.e. builds a model of the environment as a graph) using vision, then the robot uses this graph in order to know where it is and to decide which path to follow.

The paper is structured as follows. Section 2 analyses related work, Sect. 3 describes our vision-based navigation approach, Sect. 4 discusses some experiments performed to test our approach, and Sect. 5 presents conclusions and future work.

2 Related Work

Navigating in an unknown environment is a hard problem. The first thing to take into account is the kind of environment. Well-structured environments, such as assembly lines, are relatively easy to navigate since things are not expected to move. On the other hand, unstructured enviroments, such as highways or homes, are more difficult because they are dynamic and non-deterministic [1]. Moreover, we can refer to indoor or outdoor environments. Another thing to take into account is the type of map that the robot could use for navigation[1]. Metric maps represent the environment using a set of geometrical features and raw data obtained from the robot's sensors, this allows the localization and mapping to be more precise, although it requires more memory and difficult calculations. In contrast, topological maps represent the environment by means of a graph structure, which is not as precise as metric maps but it requires less memory and demand less calculations. Finally, hybrid maps combine the previous two types of maps [4]. It is important to note that the map could be created on-line (i.e. while the robot is navigating) or off-line (i.e. the map is given to the robot a priori). In terms of the environment, our work deals with an indoor semi-structured environment. The maze does not have static or dynamic obstacles, but it is non-deterministic. Our work makes use of topological maps created on-line.

An example of navigation in a well-structured environment is presented in [6]. Here, a Lego robot navigates inside a maze in order to reach designated locations. Because the robot uses photoelectric sensors for navigation, black strips are placed in the maze so that the robot can follow them. The robot knows it has encountered a junction because a white square is placed in the path. Although we also use a maze, we do not place reliable marks in the environment so that the robot stays in the middle of the path or to know when it has reached a junction. More complex approaches allow mobile robots to localise themselves and build maps simultaneously by detecting salient visual landmarks in the environment (e.g. [3,5]). These approaches have proved to be succesful in indoor and outdoor environments. Our problem at the moment is to deal with a featureless environment, thus no landmarks are available to us.

To successfully navigate in a featureless environment the robot should be able to detect walls, floors, doors and corridor lines. In [7], the authors determine the corridor lines location and the vanishing point by using three stages of image processing: linear feature extraction, line generation by subtractive clustering, and confirmation of corridor lines by counting the number of vertical

---

[1] The robot does not need to follow a map, some works make use of reactive behaviors in order to navigate through an environment [13].

lines that fall onto the proposed corridor lines. In [8] a technique based on stereo homography is proposed for robust obstacle/floor detection. A fast stereo matching technique is used to find corresponding points in the right and left images, and then for each pair of matched points, its symmetric transfer distance from the ground homography is computed. Regions with enough points close to the ground plane homography are classified as floor, and the remaining regions are labeled as obstacles. [9] explores the usefulness of a omni-directional camera for identifying corridor status when a mobile robot has to map an unknown maze. Their work is divided in two parts: (1) algorithm for processing image data and (2) algorithm for identifying corridor condition. The image processing algorithm delivers two digital pictures of the front of the robot, labeled as left side and right side. Those pictures are processed separately by the second algorithm which reports any possible condition of the corridor: (a) straight-way (b) end-corner (c) end-turning (d) crossing-point. Our work follow a similar approach to these three works in the sense that we also detect walls, floor and types of junctions. Nevertheless, we restrict ourselves to what the NAO robot offers, so stereo vision or omni-directional cameras cannot be considered.

In [10], it is described an approach to generate walking commands that enable a NAO robot to walk at the center of the corridor using only visual information with no a priori knowledge of the environment. To do so, they extend a technique used in unicycle robots that can cope with turns and junctions as well. The experiments are performed on just one corridor, not a complete maze. An approach for navigation in a real office environment using the NAO robot is described in [11]. However, they assume that the metric map is given to the robot and landmarks are place around the environment for localization.

Finally, [2] presents a survey for visual navigation mobile robots, where they explore approaches using metric and topological maps. On the other hand, [4] present a survey where only topological approaches to simultaneous localization and mapping are described.

In this paper, we propose a technique based on the watershed segmentation algorithm and the k-means clustering for the wall/floor classification of the scene in order to exit a maze using the NAO humanoid robot. Technical details of our proposal are shown in the next section.

3   Visual-Based Navigation Approach

The goal of our approach is to allow a humanoid robot to reliably navigate through a maze-like featureless environment in the shortest amount of time. To achieve the task the robot follows a visual-based approach which could be divided into the following tasks:

- **Visual processing:** The robot captures an image using its bottom camera every time it is not moving (no visual information is gathered whilst walking). This image is processed in order to determine which regions are walls and floor. Once this is done, those detected regions are classified into different classes of intersections or junctions.

- **Mapping:** According to the class of junction currently detected, the robot builds a map in real-time by adding nodes into a graph structure which represents the paths where the robot could go.
- **Navigation:** Based on the topological map being created and the identified junctions, the robot decides which node to visit next based on its current pose (position and orientation) and a wall-following algorithm.

3.1 Visual Processing

This task has two main goals: (i) detect and identify which regions of the image are walls and floor, and (ii) determine what kind of intersection or junction the robot is facing. This task starts by acquiring the image using the NAO's bottom camera and applying a smoothing filter in order to reduce part of the noise present in the image. Visual processing was implemented using OpenCV.

Wall and Floor Detection. The first step is to segment the image using the marker-controlled *watershed* algorithm [12]. Basically, this algorithm tries to find edges based on the image's gradient and by determining different markers within the image. The algorithm then simulates filling with water regions of the image starting from each of the previously defined markers. This filling ends when a border is found between regions. Because each region is filled according to its gradient, the borders or edges found by the algorithm commonly represent the contours of objects. Figure 2B shows the result of applying the watershed algorithm to the original image. The problem with this technique is to determine how many and where to place the markers within the image. The number of markers will determine the number of regions in which the image will be segmented. For the maze problem, around 4 regions could represent walls and floor, however if they are not placed in the correct part of the image the algorithm could not return the correct regions. We divided the original image into a 6×6 grid, which results in 36 cells. A marker was randomly placed inside each of these cells, thus 36 regions resulted after segmentation. In order to guarantee a correct segmentation, the watershed algorithm is repeated 2 more times.

The following step is to merge all 10 segmented images in order to find the edges that separate walls from the floor. To achieve this a voting mechanism is followed, where those edges found in at least five of the segmented images are taken as valid edges. At the end of this step we end up with an image showing the borders and regions that form walls and the floor (Fig. 2C).

Next, we need to label each of the found regions as "floor" or "wall". To do this we make use of the k-means algorithm for clustering data. K-means is fed with four values that come from each of the found regions, i.e. for each region found after segmentation four values are calculated that represent such region. Three of these values are the average of the region's pixels in each HSV channel, and the fourth value is the average of the blue RGB channel. In order to find the pixels for each region and calculate the averages, masks are defined for each region using the contours generated from the segmentation (Fig. 2D).

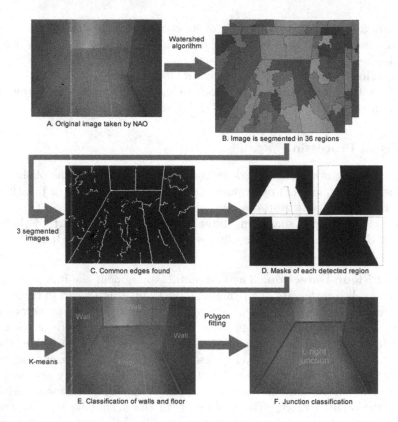

Fig. 2. The goal of the visual processing task is to determine what type of junction the robot is currently facing. A. The robot acquires an image. B. The original image is segmented and this process is repeated 3 times. C. The 3 segmented images are used to find the wall and floor edges. D. Masks are obtained for each region delimited by the found edges. E. K-means classifier is used to determine which regions are walls and floor. F. The angles formed by the edges are calculated which allow the robot to know the type of junction.

After clustering these values we end up with two groups of regions. These two groups still need to be labeled. Because the floor should have a higher level in the red RGB channel (because is a woodfloor), the group with the higher level of red is labeled as "floor", whilst the other is labeled as "wall". If it is the case, all regions belonging to the "floor" group are merged together (Fig. 2E).

Junction Classification. Once visual processing has detected which parts of the image are walls and the floor, the robot needs to determine which type of junction appears in the image in order to make a decision about where to go. We have defined four types of junctions: *L-right*, *L-left*, *T-junction* and *dead-end* (Fig. 3). To find the type of junction the angles between each edge forming the floor are calculated (from the segmented image as explained above). The vertices

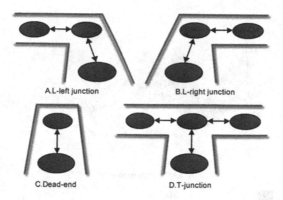

A.L-left junction   B.L-right junction

C.Dead-end   D.T-junction

Fig. 3. The four types of junctions defined in this paper, and the nodes used to represent each junction. These nodes are added to the graph (map) as the robot faces each type of junction.

forming an edge are calculated by approximating a polygon with the shape of the floor. With the obtained angles the robot is able to classify the left, right and front walls. So, for instance, if the left wall intersects the front wall and the right wall does not, then the robot has encountered an L-right junction (Fig. 2F). If the right nor the left wall intersect the front wall, then the robot is facing a T-junction.

3.2 Mapping

Having identified the type of junction that the robot is facing then it is time to build the map of the maze. The physical maze is a rectangular area of 3.5 m in length, 2.1 m in width and 0.6 m in height. The width of all corridors is 0.7 m. To simplify the mapping process the maze is discretised into 3 × 5 cells of approximantely 0.7 × 0.7 m. The map is represented by a graph, where each node represents a cell of the maze, and each edge represents a safe path where the robot can navigate. A node in the graph could have up to four neighboring nodes, one for each direction the robot could go. Initially, the map only contains the node where the robot starts. Then the robot builds the map as it moves through the maze. Every time the robot makes a turn or walks forward around 0.5 m, an image is acquired, processed and the map is updated. Each update adds a number of nodes according to the type of junction detected. Figure 3 shows different types of junctions and the nodes and edges related to each type.

3.3 Navigation

Finally, the robot turns or walks according to the map being built. The nodes in the graph show the robot those parts where it is safe to move. At this point, the robot follows a simple maze-solving strategy called "wall follower" algorithm [13].

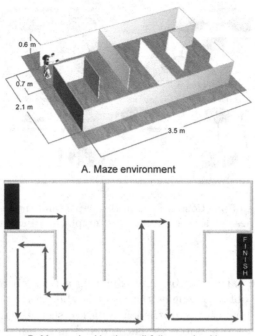

A. Maze environment

B. Maze solved by the wall-follower algorithm

Fig. 4. A. Example of a maze solved by the wall-follower algorithm using the right turn. B. Virtual view of the maze environment.

The idea is to always follow the left or right wall in order to find the exit. It is shown that the algorithm guarantees to solve the maze if and only if the maze has all its walls connected. For this paper, the robot always tries to go right, if there is no path available it chooses to go forward, otherwise it goes left. The last option is to go backwards, which sometimes is needed if there is a dead-end. At the end of the maze a black strip was placed on the floor so that the robot could know it has finished the maze (Fig. 4). It is important to note that once the map is built, the wall-following strategy is no longer required and a more intelligent path could be found. The robot uses both, the wall-following strategy to take quick decisions about where to go next, and the built map to consider other paths not previously available. Furthermore, once the robot has built the map of a particular maze, it can just use the map to successfully navigate the maze again.

Another problem during navigation is that the robot must center itself within the corridors. This problem was solved using the calculated angles found after a junction was classified. The angles determine if the robot has drifted away from the center, if so it moves back so that the horizon remains approximately horizontal.

4 Experiments

Our vision-based navigation approach was tested in two ways, in simulation and using the real robot. In both cases the goal of the robot is to find the exit in the shortest amount of time. The robot knows it has reached the end of the maze by detecting a black strip on the floor located at the exit. The colour of the walls is white (the physical walls are made of melamine), whilst the floor is made of wood. The robot is placed at the entrance of the maze and the goal is to finish the maze in the shortest amount of time. We recorded the number of times the robot successfully finished the maze and the times that it could not finish.

For the simulation we used Webots for NAO, which allows the creation of virtual environments (Fig. 4B shows a screenshot of the simulator). Here, the robot was tested in 3 maze designs (Fig. 4B shows one of these designs), where there were 18 attempts for each design. Taking into account all 54 attempts, the robot successfully finished the maze in 82 % of the cases taking 9 min on average on each attempt. The 18 % of the attempts the robot did not find the exit due to falls or failures in the algorithm. Even in simulation the robot drifts away from the center and could hit the walls. In other cases the visual processing could fail in detecting the correct junction.

Experiments with the real robot were performed in just one maze design, as shown in Fig. 1B. The robot made 35 attempts to solve the maze, where 88 % of the cases the robot successfully finished the maze, taking an average time of 7.30 min. In 5 % of the cases the robot would go back to the entrance, and in 7 % did not find the exit. This is mainly due to the noise in the images (e.g. illumination, shadows, depth of the walls etc.) and noise in the actuators (e.g. the robot would drift whilst walking).

5 Conclusions and Future Work

This paper presented a vision-based approach for robot navigation through a maze-like featureless environment. This approach was implemented in a NAO humanoid robot performing the task in real-time, where all processing was done in the robot's embedded computer. Navigating in a featureless environment is important since some real-world places such as office buildings do not have landmarks which the robot could use to localise itself. Our experiments demostrated how our vison-based approach is, in general, reliable, although improvements and more experiments need to be done. The main problem is that there is uncertainty at different levels in the system. In particular, one of the main problems is the noise due to illumination changes. For future work, wall and floor detection could be improved by defining a more efficient segmentation algorithm, since the number of images used to determine the regions was obtained empirically. Also, junction classification is a critical sub-task which needs further testing. We are planning to combine sonar and visual information in order to better determine the depth of where some walls really are. Finally, our goal is to have the NAO navigating in a real office environment where landmarks will also be considered in order to increase the reliability of the system.

References

1. Thrun, S.: Robotic mapping: a survey. In: Lakemeyer, G., Nebel, B. (eds.) Exploring Artificial Intelligence in the New Millennium, pp. 1–35. Morgan Kaufmann, San Francisco (2002)
2. Bonin-Font, F., Alberto, O., Gabriel, O.: Visual navigation for mobile robots: a survey. J. Intell. Robot. Syst. **53**(3), 263–296 (2008)
3. Frintrop, S.: VOCUS: A Visual Attention System for Object Detection and Goal-Directed Search. Springer-Verlag, New York (2006)
4. Boal, J., Sanchez-Miralles, A., Arranz, A.: Topological simultaneous localization and mapping: a survey. Robotica **32**(5), 803–821 (2014)
5. Sim, R., Little, J.J.: Autonomous vision-based exploration and mapping using hybrid maps and Rao-Blackwellised particle filters. In: IEEE International Conference on Intelligent Robots and Systems, pp. 2082–2089 (2006)
6. Shuying, Z., Wenjun, T., Shiguang, W., Chongshuang, G.: Research on robotic education based on LEGO bricks. In: International Conference on Computer Science and Software Engineering. vol. 5, pp. 733–736 (2008)
7. Shi, W., Samarabandu, J.: Investigating the performance of corridor and door detection algorithms in different environments. In: International Conference on Information and Automation, pp. 206–211 (2006)
8. Fazl-Ersi, E., Tsotsos, J.K.: Region classification for robust floor detection in indoor environments. In: Kamel, M., Campilho, A. (eds.) ICIAR 2009. LNCS, vol. 5627, pp. 717–726. Springer, Heidelberg (2009)
9. Bagus, A., Mellisa, W.: The development of corridor identification algorithm using omni-directional vision sensor. In: International Conference on Affective Computation and Intelligence Interaction, pp. 412–417 (2012)
10. Faragasso, A., Oriolo, G., Paolillo, A., Vendittelli, M.: Vision-based corridor navigation for humanoid robots. In: IEEE International Conference on Robotics and Automation, pp. 3190–3195 (2013)
11. Wei, C., Xu, J., Wang, C., Wiggers, P., Hindriks, K.: An approach to navigation for the humanoid robot nao in domestic environments. In: Natraj, A., Cameron, S., Melhuish, C., Witkowski, M. (eds.) TAROS 2013. LNCS, vol. 8069, pp. 298–310. Springer, Heidelberg (2014)
12. Beucher, S.: The watershed transformation applied to image segmentation. Scanning Microsc. Suppl. **6**, 299–314 (1992)
13. Sharma, M., Robeonics, K.: Algorithms for micro-mouse. In: IEEE International Conference on Future Computer and Communication (2009)

Evaluation of Local Descriptors for Vision-Based Localization of Humanoid Robots

Noé G. Aldana-Murillo$^{(\boxtimes)}$, Jean-Bernard Hayet, and Héctor M. Becerra

Centro de Investigación en Matemáticas (CIMAT),
C.P. 36240 Guanajuato, GTO, Mexico
{noe.aldana,jbhayet,hector.becerra}@cimat.mx

Abstract. In this paper, we address the problem of appearance-based localization of humanoid robots in the context of robot navigation using a visual memory. This problem consists in determining the most similar image belonging to a previously acquired set of key images (visual memory) to the current view of the monocular camera carried by the robot. The robot is initially kidnapped and the current image has to be compared with the visual memory. We tackle the problem by using a hierarchical visual bag of words approach. The main contribution of the paper is a comparative evaluation of local descriptors to represent the images. Real-valued, binary and color descriptors are compared using real datasets captured by a small-size humanoid robot. A specific visual vocabulary is proposed to deal with issues generated by the humanoid locomotion: blurring and rotation around the optical axis.

Keywords: Vision-based localization · Humanoid robots · Local descriptors comparison · Visual bag of words

1 Introduction

Recently, the problem of navigation of humanoid robots based only on monocular vision has raised much interest. Many research has been reported for this problem in the context of wheeled mobile robots. In particular, the visual memory approach [1] has been largely studied. It mimics the human behavior of remembering key visual information when moving in unknown environments, to make the future navigation easier.

Robot navigation based on a visual memory consists of two stages [1]. First, in a learning stage, the robot creates a representation of an unknown environment by means of a set of key images that forms the so-called visual memory. Then, in an autonomous navigation stage, the robot has to reach a location associated to a desired key image by following a visual path. That path is defined by a subset of images of the visual memory that topologically connects the most similar key image compared with the current view of the robot with the target image.

Few work has been done for humanoids navigation based on a visual memory [2,3]. In both works, the robot is not initially kidnapped but it starts the navigation from a known position. In this context, the main interest in this paper

© Springer International Publishing Switzerland 2015
J.A. Carrasco-Ochoa et al. (Eds.): MCPR 2015, LNCS 9116, pp. 179–189, 2015.
DOI: 10.1007/978-3-319-19264-2_18

is to solve an appearance-based localization problem, where the current image is matched to a known location only by comparing images [4]. In particular, we address the localization of humanoid robots using only monocular vision.

This paper addresses the problem of determining the key image in a visual memory that is the most similar in appearance to the current view of the robot (input image). Figure 1 presents a general diagram of the problem. Consider that the visual memory consists of n ordered key images $(\mathcal{I}_1^*, \mathcal{I}_2^*, ..., \mathcal{I}_n^*)$. The robot is initially kidnapped and the current view $\mathcal{I}$ has to be compared with the n key images and the method should give as an output the most similar key image $\mathcal{I}_o^*$ within the visual memory.

Since a naive comparison might take too much time depending on the size of the visual memory, we propose to take advantage of a method that compresses the visual memory into a compact, efficient to access representation: the visual bag of words (VBoW) [5]. A bag of words is a structure that represents an image as a numerical vector, allowing fast images comparisons. In robotics, the VBoW approach has been used in particular for loop-closure in simultaneous localization and mapping (SLAM) [6,7], where re-visited places have to be recognized.

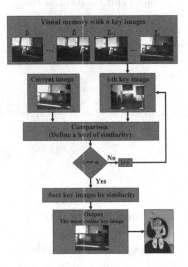

Fig. 1. General diagram of the appearance-based localization from a visual memory.

Fig. 2. Example of images from an onboard camera of a humanoid NAO robot.

In this paper, a quantitative evaluation of the VBoW approach using different local descriptors is carried out. In particular, we evaluate the approach on real datasets captured by a camera mounted in the head of a small-size humanoid robot. The images are affected by issues related to the sway motion introduced by the humanoid locomotion: blurring and rotation around the optical axis. A specific visual vocabulary is proposed to tackle those issues. Figure 2 shows two examples of images captured from our experimental platform: a NAO humanoid robot. These images are of 640 × 480 pixels.

The paper is organized as follows. Section 2 introduces the local descriptors included in our evaluation. Section 3 details the VBoW approach as implemented. Section 4 presents the results of the experimental evaluation and Sect. 5 gives some conclusions.

2 Local Descriptors

Local features describe regions of interest of an image through descriptor vectors. In the context of image comparison, groups of local features should be robust against occlusions and changes in view point, in contrast to global methods. From the existing local detectors/descriptors, we wish to select the best option for the specific task of appearance-based humanoids localization. Hereafter, we introduce the local descriptors selected for a comparative evaluation.

2.1 Real-Valued Descriptors

A popular keypoint detector/descriptor is SURF (Speeded Up Robust Features) [8]. It has good properties of invariance to scale and rotation. SURF keypoints can be computed and compared much faster than their previous competitors. Thus, we selected SURF as a real-valued descriptor to be compared in our localization framework. The detection is based on the Hessian matrix and uses integral images to reduce the computation time. The descriptor combines Haar-wavelet responses within the interest point neighborhood and exploits integral images to increase speed. In our evaluation, the standard implementation of SURF (vector of dimension 64) included in the OpenCV library is used.

2.2 Binary Descriptors

Binary descriptors represent image features by binary strings instead of floating-point vectors. Thus, the extracted information is very compact, occupies less memory and can be compared faster. Two popular binary descriptors have been selected for our evaluation: Binary Robust Independent Elementary Features (BRIEF [9]) and Oriented FAST and Rotated BRIEF (ORB [10]). Both use variants of FAST (Features from Accelerated Segment Tests) [11], i.e. they detect keypoints by comparing the gray levels along a circle of radius 3 to the gray level of the circle center. In average, most pixels can be discarded soon, hence the detection is fast. BRIEF uses the standard FAST keypoints while ORB uses oFAST keypoints, an improved version of FAST including an adequate orientation component. The BRIEF descriptor is a binary vector of user-choice length where each bit results from an intensity comparison between some pairs of pixels within a patch around keypoints. The patches are previously smoothed with a Gaussian kernel to reduce noise. They do not include information of rotation or scale, so they are hardly invariant to them. This issue can be overcome by using the rotation-aware BRIEF descriptor (ORB), which computes a dominant orientation between the center of the keypoint and the intensity centroid of its patch. The BRIEF comparison pattern is rotated to obtain a descriptor that should not vary when the image is rotated in the plane. In our evaluation, we use oFAST keypoints given by the ORB detection method as implemented in OpenCV along with BRIEF with size of patches 48 and descriptor length 256. The ORB implementation is the one of OpenCV with descriptors of 256 bits.

2.3   Color Descriptors

We also evaluate the image comparison approach by using only color information. To do so, we use rectangular patches and a color histogram is associated to each patch as a descriptor. We select the color space HSL (Hue-Saturation-Lightness) because its three components are more natural to interpret and less correlated than in other color spaces. Also, only the H and S channels are used, in order to achieve robustness against illumination changes. The color descriptor of each rectangular patch is formed by a two-dimensional histogram of hue and saturation and the length of the descriptor was set experimentally to 64 bits. Three different alternatives are evaluated using color descriptors:

- Random patches: 500 patches of size 48×64, randomly selected. This option is referred to as Color-Random.
- Uniform grid: A uniform grid of 19×19 patches covering the image, with patches overlapped a half of their size. This option is referred to as Color-Whole.
- Uniform grid on half of the image: Instead of using the whole image, only the upper half is used. This is because the inferior parts, when taken by the humanoid robot, are mainly projections of the floor and do not discrimante well for localization purposes. This option is referred to as Color-Half.

3   Visual Bag of Words for Humanoid Localization

As mentioned above, this work relies on the hierarchical visual bag of words approach [12] to combine the high descriptive power of local descriptors with the versatility and robustness of histograms. In Sect. 3.1, we recall the main characteristics of [12], and then in Sect. 3.2, we introduce a novel use of the BRIEF descriptor suited within a VBoW approach in the context of humanoid robots.

3.1   Hierarchical Visual Bag of Words Approach

The visual bag of words approach first discretizes the local descriptors space in a series of words, i.e., clusters in the local descriptors space. Here, we followed the strategy of Nister et al. [12], who perform this step in a hierarchical way: in the set of n key images $\mathcal{I}_1^*, \mathcal{I}_2^*, ..., \mathcal{I}_n^*$ forming the visual memory, a pool of D

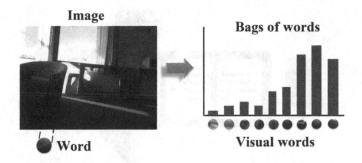

Fig. 3. Representation of an image in visual words.

local descriptors is detected, as illustrated in Fig. 3, left. The local descriptors can be extracted by any of the methods mentioned before. Given a branch factor k (a small integer, typically 3 or 4), the idea is to form k clusters among the D descriptors by using the kmeans++ algorithm. Then, the sets of descriptors associated to these k clusters are recursively clustered into other k clusters, and so on, up to a maximum depth of L levels. At each level, the formed clusters are associated to a representative descriptor chosen randomly (by the kmeans++ algorithm) that will be compared with new descriptors. The leaves of this tree of recursively refined clusters correspond to the visual words, i.e., the clusters in the local descriptors space. The advantage is that, when faced with descriptors found in new images, it is computationally efficient to associate them to a visual word, namely with kL distance computations, i.e., k at each level. Since we obtain $W = k^L$ leaves (i.e., words), characterizing a descriptor as a word is done in $O(k \log_k W)$ operations, where W is the number of words, instead of the W computations with a naive approach. This principle is illustrated in Fig. 4.

When handling a new image $\mathcal{I}$, d descriptors are extracted, and each of these is associated to a visual word as explained. This way, we obtain an empirical distribution of the visual words in $\mathcal{I}$, in the form of a histogram of visual words $v(\mathcal{I})$ (see Fig. 3, right). Now the content of $\mathcal{I}$ can be compared with the one of any of the

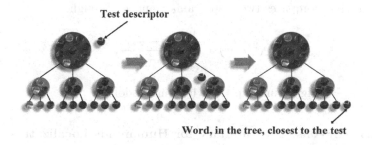

Fig. 4. Hierarchical approach of visual words search: When a new descriptor is found in some image $\mathcal{I}$, it is recursively compared to representatives of each cluster.

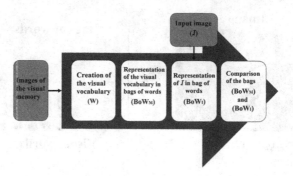

Fig. 5. Complete method of image comparison based on visual bag of words.

key images $\mathcal{I}_i^*$ by comparing their histograms. Of course, because n may be very high, it is out of question to compare the histogram of $\mathcal{I}$ with the n histograms of the key images. That is why an important element in this representation is the notion of inverse dictionary: for each visual word, one stores the list of images containing this word. Then, on a new image, we can easily determine, for each visual word it contains, the list of key images also containing this word. To limit the comparisons, we restrain the search for the most similar images to the subset of key images having at least 5 visual words in common. For an image $\mathcal{I}$, each histogram entry v_i (where i refers to the visual word) is defined as:

$$v_i(\mathcal{I}) = \frac{c_i(\mathcal{I})}{c(\mathcal{I})} \log(\frac{n}{n_i})$$

where $c(\mathcal{I})$ is the total number of descriptors present in $\mathcal{I}$, $c_i(\mathcal{I})$ the numbers of descriptors in $\mathcal{I}$ classified as word i, and n_i the number of key images where the word i has been found. The log term allows to weight the frequency of word i in $\mathcal{I}$ in function of its overall presence: When a word is present everywhere in the database, then the information of its presence is not that pertinent.

Last, we should choose how to compare histograms. After intensive comparisons made among the most popular metrics for histograms, we have chosen the χ^2 distance, that compares two histograms v and w through:

$$\chi^2(v, w) = \sum_{i=0}^{W} \frac{(\hat{v}_i - \hat{w}_i)^2}{\hat{v}_i + \hat{w}_i},$$

where $\hat{v} = \frac{1}{\|v\|_1} v$. Fig. 5 sums up the whole methodology.

3.2   A BRIEF-based Vocabulary for Humanoids Localization

We introduce a novel use of the BRIEF descriptor suited within a VBoW approach in the context of humanoid robots. This is a specific vocabulary that we called BRIEFROT, which deals with the issues generated by the humanoid locomotion. BRIEFROT possesses three independent internal vocabularies, two of which are rotated a fixed angle: one anti-clockwise, the other clockwise. Through experimentation, we found that for the NAO humanoid platform a suitable value for the rotation of the vocabularies is 10 degrees. These rotated vocabularies were implemented with the idea of settling the slight variations in the rotation caused by the locomotion of these robotic systems. The rotated vocabularies represent to the images of the visual memory as rotated images. The third vocabulary is identical to the normal BRIEF. The idea of using three vocabularies is that if the input image is rotated with respect to any image of the visual memory, then the image is detected by any of the rotated vocabularies; if the input image is not rotated with respect to any image of the visual memory, then it is detected with the vocabulary without rotation. Additionally, the detected local patches are smoothed with a Gaussian kernel to reduce the blur effect.

4   Experimental Evaluation

We evaluated the local descriptors mentioned in Sect. 2 on 4 datasets. We used three datasets in indoor environments (CIMAT-NAO-A, CIMAT-NAO-B and Bicocca) and one outdoors (New College). The tests were done in a laptop using Ubuntu 12.04 with 4 Gb of RAM and 1.30 GHz processor.

4.1   Description of the Evaluation Datasets

The *CIMAT-NAO-A* dataset was acquired with a NAO humanoid robot inside CIMAT. This dataset contains 640 × 480 images of good quality but also blurry ones. Some images are affected by rotations introduced by the humanoid loco-motion or by changes of lighting. We used 187 images, hand-selected, as a visual memory and 258 images for testing. The *CIMAT-NAO-B* dataset was also cap-tured indoors at CIMAT with the humanoid robot. It also contains good quality and blurry 640 × 480 images, but it does not have images with drastic light changes, as in the previous dataset. We used 94 images as a visual memory and 94 images for testing. Both datasets *CIMAT-NAO-A* and *CIMAT-NAO-B* are available in http://personal.cimat.mx:8181/~hmbecerra/CimatDatasets.zip.

The *Bicocca 2009-02-25b* dataset is available online [13] and was acquired by a wheeled robot inside a university. The 320 × 240 images have no rotation around the optical axis nor blur. We used 120 images as a visual memory and 120 images for testing. Unlike the three previous datasets that were obtained indoors, the *New College* dataset was acquired outside the Oxford University by a wheeled robot [14], with important light changes. The 384 × 512 images are of good quality with no rotation nor blur. For this dataset, 122 images were chosen as a visual memory and 117 images for testing.

4.2   Evaluation Metrics

Since the goal of this work is to evaluate different descriptors in VBoW approaches, it is critical to define corresponding metrics to assess the quality of the result from our application. We propose two metrics; the first one is:

$$\mu_1(\mathcal{I}) = \text{rank}(\bar{k}(\mathcal{I}))$$

where $\bar{k}(\mathcal{I})$ is defined as the ground truth index of the key image associated to $\mathcal{I}$. In the best case, the rank of the closest image to ours should be one, so $\mu_1(\mathcal{I}) = 1$ means that the retrieval is perfect, whereas higher values correspond to worse evaluations. The second metric is:

$$\mu_2(\mathcal{I}) = \sum_l \frac{z_l(\bar{k}(\mathcal{I}))}{\sum_{l'} z_{l'}(\bar{k}(\mathcal{I}))} \text{rank}(l)$$

where the $z_l(k)$ is the similarity score between the key images k and l inside the visual memory. This metric is proposed to handle similar key images within the dataset; hence, with this metric, the final score integrates weights (normalized by $\sum_{l'} z_{l'}(\bar{k}(\mathcal{I}))$ to sum to one) from the key images l similar to the closest ground truth image $\bar{k}(\mathcal{I})$; this ensures that all the closest images are well ranked.

4.3 Parameters Selection

There are three free parameters: the number of clusters k, the tree depth L and the measure of similarity. Tests were performed by varying k from $k = 8$ to $k = 10$ and varying L from $L = 4$ to $L = 8$. Also, different similarity measures between histograms were tested: L1-Norm, L2-Norm, χ^2, Bhattacharyya and dot product. To do the tests, we generated ground truth data, by defining manually the most similar key image to the input image. The parameters were selected so that the confidence levels μ_2 were close to 1. We obtained the best results with $k = 8$, $L = 8$ and the metric χ^2. The dataset used for the parameters selection was the CIMAT-NAO-A, since it is the most challenging dataset for the type of images it contains.

4.4 Analysis of the Results Obtained on the Evaluation Datasets

We present the results in the following tables for the seven vocabularies created. On the one hand, the efficiency of the vocabularies is observed using the confidence μ_1. In this case, the threshold chosen for the test to be classified as correct was 1. On the other hand, for the level of confidence μ_2 we choose a threshold of 2.5, all tests below 2.5 were considered correct. This level of confidence takes into account the possible similarity between the images on the visual memory.

In Table 2, we present the results obtained for the *CIMAT-NAO-A* dataset. In this case, the BRIEFROT vocabulary obtained the best behavior for both levels of confidence. For the case of μ_1, it has an efficiency of 60.85 % and for μ_2 of 75.19 %. Also, the ORB vocabulary offered good performance for μ_2 and was the second best for μ_1. The Color-Half vocabulary obtained the worst results. On the other hand, for the *CIMAT-NAO-B* dataset, the SURF vocabulary behaved better than BRIEFROT, but with higher computation times. The times reported were measured from the stage of features extraction to the stage of comparison. ORB, again, behaved well and was the second best vocabulary. The Color-Random vocabulary obtained the worst performance for μ_1, but for μ_2 it was

Table 1. Percentages of correct results for the dataset CIMAT-NAO-A.

Descriptor	Number of tests	Correct tests μ_1	Effectiveness μ_1 (%)	Correct tests μ_2	Effectiveness μ_2 (%)	Average time comparison (ms)
BRIEF	258	132	51.16	185	71.71	122.6
BRIEFROT	258	157	60.85	194	75.19	132.4
Color-Random	258	110	42.64	117	45.35	129.4
Color-Half	258	104	40.31	160	62.01	93.1
Color-Whole	258	110	42.64	162	62.79	101.8
ORB	258	144	55.81	194	75.19	107.5
SURF	258	135	52.32	187	72.48	296.5

Table 2. Percentages of correct results for the dataset CIMAT-NAO-B.

Descriptor	Number of tests	Correct tests μ_1	Effectiveness μ_1 (%)	Correct tests μ_2	Effectiveness μ_2 (%)	Average time comparison (ms)
BRIEF	94	63	67.02	82	87.23	87.23
BRIEFROT	94	65	69.14	81	86.17	112.0
Color-Random	94	62	65.96	86	91.49	109.9
Color-Half	94	64	68.09	78	82.98	63.2
Color-Whole	94	68	72.34	83	88.3	73.1
ORB	94	69	73.40	83	88.3	77.7
SURF	94	70	74.46	86	91.49	267.9

Table 3. Percentages of correct results for the dataset Bicocca25b.

Descriptor	Number of tests	Correct tests μ_1	Effectiveness μ_1 (%)	Correct tests μ_2	Effectiveness μ_2 (%)	Average time comparison (ms)
BRIEF	120	111	92.5	116	96.67	73.4
BRIEFROT	120	111	92.5	116	96.67	98.5
Color-Random	120	60	50	69	57.50	72.7
Color-Half	120	57	47.5	67	55.83	36.0
Color-Whole	120	59	49.17	65	54.17	79.4
ORB	120	110	91.67	114	95.00	60.2
SURF	120	111	92.5	114	95.00	120.0

Table 4. Percentages of correct results for the dataset New College.

Descriptor	Number of tests	Correct tests μ_1	Effectiveness μ_1 (%)	Correct tests μ_2	Effectiveness μ_2 (%)	Average time comparison (ms)
BRIEF	117	70	59.83	85	72.65	105.9
BRIEFROT	117	70	59.83	87	74.36	134.6
Color-Random	117	48	41.02	69	58.97	110.9
Color-Half	117	34	29.06	64	54.70	60.4
Color-Whole	117	44	37.61	74	63.25	110.5
ORB	117	69	58.97	85	72.65	107.0
SURF	117	74	63.25	89	76.07	302.3

one of the best vocabulary; this means that it tends to put the correct key image in the second rank. Color-Half had the worst results for μ_2.

In the *Bicocca 2009-02-25b* dataset, three vocabularies obtained the best results for μ_1: BRIEFROT, Color-Random and SURF. The difference between these three vocabularies is in the computation time: SURF consumes much more time. For μ_2, BRIEFROT was the best. In the *New College* dataset, the SURF vocabulary obtained the best behavior for both levels of confidence. In both cases the BRIEFROT vocabulary obtained a good behavior, close to SURF, but BRIEFROT consumes less than half the time required by SURF (Tables 1, 3 and 4).

5 Conclusions

This paper addresses the problem of vision-based localization of humanoid robots, i.e., determining the most similar image among a set of previously acquired images (visual memory) to the current robot view. To this end, we use a hierarchical visual bag of words (VBoW) approach. A comparative evaluation of local descriptors to use to feed the VBoW is reported: Real-valued, binary and color descriptors were compared on real datasets captured by a small-size humanoid robot. We presented a novel use of the BRIEF descriptor suited to the VBoW approach for humanoid robots: BRIEFROT. According to our evaluation, the BRIEFROT vocabulary is very effective in this context, as reliable as SURF to solve the localization problem, but in much less time. We also show that keypoints-based vocabularies performed better than color-based vocabularies.

As future work, we will explore the combination of visual vocabularies to robustify the localization results. We will implement the method onboard the NAO robot using a larger visual memory. We also wish to use the localization algorithm in the construction of the visual memory to identify revisited places.

References

1. Courbon, J., Mezouar, Y., Martinet, P.: Autonomous navigation of vehicles from a visual memory using a generic camera model. IEEE Trans. Intell. Transp. Syst. **10**(3), 392–402 (2009)
2. Ido, J., Shimizu, Y., Matsumoto, Y., Ogasawara, T.: Indoor navigation for a humanoid robot using a view sequence. Int. J. Robot. Res. **28**(2), 315–325 (2009)
3. Delfin, J., Becerra, H.M., Arechavaleta, G.: Visual path following using a sequence of target images and smooth robot velocities for humanoid navigation. In: IEEE International Conference on Humanoid Robots, pp. 354–359 (2014)
4. Ulrich, I., Nourbakhsh, I.: Appearance-based place recognition for topological localization. In: IEEE International Conference on Robotics and Automation, pp. 1023–1029 (2000)
5. Sivic, J., Zisserman, A.: Video google: a text retrieval approach to object matching in videos. In: IEEE International Conference on Computer Vision, pp. 1–8 (2003)
6. Botterill, T., Mills, S., Green, R.: Bag-of-words-driven, single-camera simultaneous localization and mapping. J. Field Robot. **28**(2), 204–226 (2011)
7. Galvez-Lopez, D., Tardos, J.D.: Bags of binary words for fast place recognition in image sequences. IEEE Trans. Robot. **28**(5), 1188–1197 (2012)
8. Bay, H., Ess, A., Tuytelaars, T., Van Gool, L.: Speeded-up robust features (SURF). Comput. Vis. Image Underst. **110**(3), 346–359 (2008)
9. Calonder, M., Lepetit, V., Strecha, C., Fua, P.: BRIEF: binary robust independent elementary features. In: Daniilidis, K., Maragos, P., Paragios, N. (eds.) ECCV 2010, Part IV. LNCS, vol. 6314, pp. 778–792. Springer, Heidelberg (2010)
10. Rublee, E., Rabaud, V., Konolige, K., Bradski, G.: ORB: an efficient alternative to SIFT or SURF. In: IEEE International Conference on Computer Vision, pp. 2564–2571 (2011)
11. Rosten, E., Drummond, T.W.: Machine learning for high-speed corner detection. In: Leonardis, A., Bischof, H., Pinz, A. (eds.) ECCV 2006, Part I. LNCS, vol. 3951, pp. 430–443. Springer, Heidelberg (2006)
12. Nister, D., Stewenius, H.: Scalable recognition with a vocabulary tree. In: IEEE International Conference on Computer Vision and Pattern Recognition, pp. 2161–2168 (2006)
13. Bonarini, A., Burgard, W., Fontana, G., Matteucci, M., Sorrenti, D.G., Tardos, J.D.: Rawseeds: robotics advancement through web-publishing of sensorial and elaborated extensive data sets. In: International Conference on Intel, Robots and Systems (2006)
14. Smith, M., Baldwin, I., Churchill, W., Paul, R., Newman, P.: The new college vision and laser data set. Int. J. Robot. Res. **28**(5), 595–599 (2009)

NURBS Based Multi-objective Path Planning

Sawssen Jalel[1,2]([✉]), Philippe Marthon[2], and Atef Hamouda[1]

[1] LIPAH Research Laboratory, Faculty of Sciences of Tunis,
Tunis El Manar University, 2092 Tunis, Tunisia
atef_hammouda@yahoo.fr

[2] Site ENSEEIHT de l'Institut de Recherche en Informatique de Toulouse (IRIT),
University of Toulouse, 2 rue Charles Camichel, BP 7122, Toulouse, France
{sawssen.jalel,philippe.marthon}@enseeiht.fr

Abstract. Path planning presents a key question for an autonomous robot to evolve in its environment. Hence, it has been largely dealt in recent years. Actually, finding feasible paths and optimizing them for different objectives is computationally difficult. In this context, this paper introduces a new mobile robot path planning algorithm by introducing an optimized NURBS (Non Uniform Rational B-Spline) curve modelling using Genetic Algorithm to represent the generated path from the specified start location to the desired goal. Thus, given an a priori knowledge of the environment, an accurate fitness function is used to compute a curvature-constrained and obstacles-avoiding smooth path, with minimum length and low variations of curvature. The performance of the proposed algorithm is demonstrated through extensive MATLAB simulation studies.

Keywords: Robot path planning optimization · Genetic algorithm · NURBS curves parameterization · Weight parameter

1 Introduction

The autonomous mobile robotics aims, more specifically, to develop systems able to move independently. Direct applications are particularly in the fields of automotive, planetary exploration and service robotics. Which is why the problem of motion planning with obstacle avoidance has been extensively studied over the last decade since one of the crucial tasks for an autonomous robot is to navigate intelligently from a starting node to a target node.

The path planning environment can be categorized into two major classes which are static and dynamic. In the static environment, the whole solution must be found before starting execution. However, for dynamic or partially observable environments, replannings are required frequently and more update time is needed. Depending on environment type, path planning algorithms are divided into two categories, which are local and global methods. They might also be divided into traditional and intelligent methods. Eminent traditional path planning methods include potential field [1], visibility graph [2] and cell decomposition approaches [3]. As to intelligent methods, they include particle swarm

© Springer International Publishing Switzerland 2015
J.A. Carrasco-Ochoa et al. (Eds.): MCPR 2015, LNCS 9116, pp. 190–199, 2015.
DOI: 10.1007/978-3-319-19264-2_19

optimization [4], neural networks [5], ant clony algorithms [6], fuzzy logic [7] and genetic algorithms. Obviously, each one of them has its own strengths and weaknesses which encourage researchers to search alternative and more efficient methods.

This paper deals with a novel path planning algorithm that consists of two steps. The first agrees to calculate the shortest, obstacles-avoiding polyline path from a start to a goal position. While the second undertakes to use both the wealth of genetic algorithms and the flexibility of NURBS curves to meet the path planner constraints.

The remainder of this paper is organized as follows: In Sect. 2, we present the theoretical background describing the theory of NURBS curves. The approach proposed is outlined in Sect. 3. Section 4 shows the simulation studies made to verify and validate the effectiveness of the presented method. Conclusions are finally presented in Sect. 5.

2   Theoretical Background

A pth degree NURBS curve is a vector valued piecewise rational polynomial function of the form:

$$C(u) = \frac{\sum_{i=0}^{n} N_{i,p}(u)w_i P_i}{\sum_{i=0}^{n} N_{i,p}(u)w_i} \tag{1}$$

where $\{P_i\}$ are the n control points, $\{w_i\}$ are the corresponding weights and the $\{N_{i,p}(u)\}$ are the pth B-spline basis functions defined on the non-uniform knot vector U, by DeBoor-Cox Calculation as follows :

$$N_{i,p}(u) = \frac{u - u_i}{u_{i+p} - u_i} N_{i,p-1}(u) + \frac{u_{i+p+1} - u}{u_{i+p+1} - u_{i+1}} N_{i+1,p-1}(u) \tag{2}$$

$$N_{i,0}(u) = \begin{cases} 1 & \text{if } u_i \leq u < u_{i+1} \\ 0 & \text{else} \end{cases}$$

The degree, number of knots, and number of control points are related by $m = n + p + 1$. The knot vector is defined by $U = \{a, ..., a, u_{p+1}, ..., u_{mp-1}, b, ..., b\}$ with a multiplicity of a and b which is equal to the order of the curve. This constraint ensures the interpolation of the two endpoints. Throughout this paper, we assume that the parameter lies in the range $u \in [0, 1]$. For more details we refer the reader to [9].

NURBS have various useful properties which are ideal for the design of complicated geometry, making them a standard tool in the CAD/CAM and graphics community. In fact, they offer a common mathematical form for representing and designing both standard analytic shapes and free-form curves and surfaces. Furthermore, by manipulating the control points as well as the weights or the knot vector, NURBS provide the flexibility to design a large variety of shapes. Another useful property is the local modification, for example, if a control point P_i is moved or a weight w_i is changed, it will affect the curve only in $[u_i, u_{i+p+1})$.

Also not to overlook the fact that evaluation is reasonably and computationally stable. As well they have clear geometric interpretations and are invariant under scaling, rotation, translation and shear as well as parallel and perspective projection.

A considerable amount of research has been carried out in the parametrization domain. Indeed, finding a correct parameterization and the weight of the control points when calculating the curve have been the main issues in curve fitting techniques. Many evolutionary optimization techniques have been successfully applied [10,11]. Through this paper, we aim to exploit the effect of the weight parameter. For this, we recall that the weight of a point P_k determines its influence on the associated curve. Therefore, increasing (decreasing) w_k pulls (pushes) the curve toward (away from) this point (Fig. 1). It should also be noted that a control point with zero weight removes the contribution of this point on the generated curve.

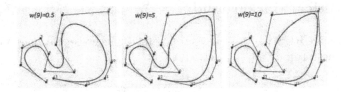

Fig. 1. NURBS Curve with w(9) Varying

3   Proposed Path Planning Optimization Algorithm

In this section, an optimized NURBS curve modelling using Genetic Algorithm for mobile robot navigation is presented. Thus, given a map with a priori knowledge of the locations of obstacles and positions of departure and arrival, the objective is to compute an optimal trajectory. In other words, a curvature-constrained smooth path avoiding collisions, with minimum length, and having low variations of curvature. For this, we assume that the robot is represented by a circle of diameter d (taking into account the robot's dimensions) and minimum radius of curvature ρ_{min}. As for the workspace Ω, it is considered as the union of the subspace of abstacles-free configuration Ω_{free} and the subspace of abstacles configuration Ω_{obst}.

Algorithm 1 describes the procedure of shortest polyline path computation. This algorithm starts by extracting the skeleton of the obstacle-free space, to which, it connects the start and target point giving, accordingly, the extended skeleton X upon which the robot may move (Fig. 2(a)). Then, a morphological process is applied to classify the pixels of X into End Points (EP), Junction Points (JP), Critical Curve Points (CCP) and Curve Points (CP). Afterward, a weighted graph is constructed by associating $EP(X)$, $JP(X)$ and $CCP(X)$ with the vertices, and the set of $CP(X)$ with the edges. The weights of edges are generated by considering the Euclidean distances between vertices. Before applying Bellman Ford Algorithm to calculate the shortest path between S and T, the algorithm applies a graph reduction step to remove the edges that represent inaccessible corridors by the robot [8].

Algorithm 1. Shortest polyline path computation

Input: Ω, S, T, d
Output: Shortest polyline path S_{opt}

1 **begin**
2     $X = SkeletonExtraction(\Omega_{free}, S, T)$;
3     $EP(X) = [\bigcup_i \epsilon_{\theta_i(\overline{A})}(\overline{X})] \cap X$;
4     $JP(X) = [\bigcup_i \epsilon_{\theta_i(B)}(X)] \cup [\bigcup_i \epsilon_{\theta_i(C)}(X)]$;
5     $CCP(X) = CurveCriticalPoint(X)$;
6     $CP(X) = X\backslash(EP(X) \cup JP(X) \cup CCP(X))$;
7     $G = GraphConstruction(EP(X), JP(X), CCP(X), CP(X))$;
8     $G' = GraphReduction(G, d, \Omega)$;
9     $S_{opt} = BellmanFordAlgorithm(G', S, T)$;
10   **return** (S_{opt});

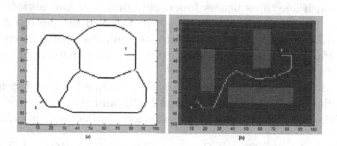

Fig. 2. Shortest path finding process. (a) Extended skeleton. (b) Shortest path between S and T.

As shown Fig. 2(b), the shortest path S_{opt} is modeled by a set of connected edges. This trajectory is not smooth and don't respect the curvature constraint. Then, it needs a mathematical model to refine it. Hence, the importance of using NURBS curves given their properties previously cited. Thus, a good parameterisation is needed to obtain the desired result. In fact, generating a NURBS curve requires the specification of the set of control points as well as their associated weights. In our algorithm, the control points are extracted from the selected polyline path S_{opt}:

$$P = \{P_i\}_{i=1..n} = JP(S_{opt}) \cup EP(S_{opt}) \cup CCP(S_{opt}) \tag{3}$$

where $JP(S_{opt})$ are the junction points of S_{opt}, $EP(S_{opt})$ are the end points of S_{opt} and $CCP(S_{opt})$ are the critical curve points corresponding to variations of directions of S_{opt}. Regarding the weight factor, we propose a novel parameterization method based on genetic algorithm.

3.1 Genetic Representation

The chromosome structure must have sufficient information about the entire path from the start to the end point in order to be able to represent it. In our context, an individual is a NURBS curve representing a path candidate. Such a curve is defined by a set of n weighted points for which P_1 and P_n are always the starting and destination configurations. Each gene represents the location of a control point (x and y coordinate) and its associated weight. It is noteworthy that location of this point is fixed while its associated weight is varying.

3.2 Fitness Function

During each generation, the population of paths is evaluated to quantify their degree of elitism which depends on how suitable the solution (path) is according to the problem. Consequently, an individual's fitness value should be proportional to its survival ability. Most GA-based path planning existing methods consider the path length as fitness function to minimize and neglect feasibility. Consequently, the resulted path may not be followed by the robot if it does not respect the curvature limit related to its minimum radius of curvature. In addition, having low variation of curvatures along the path is of considerable importance.

In this study, we propose an accurate evaluation function that depends on four parameters: the safety, the feasibility, the length and the curvatures's standard deviation of the path. The objective function for a path C is then defined as:

$$f(C) = \alpha.P_{safety}(C) + \beta.P_{feasibility}(C) + \gamma.P_{length}(C) + \delta.P_\sigma(C) \qquad (4)$$

where $P_{safety}(C)$ is the path safety constraint and which denotes the number of collisions:

$$P_{safety}(C) = |P_{col}| \qquad (5)$$

where:

$$P_{col} = \{p_i \in C, i = 1..k1\} = C \cap \Omega_{obst} \qquad (6)$$

$P_{feasibility}(C)$ is the path feasibility related to the robot's minimum radius of curvature and determined as the number of curve points that exceeds the absolute value of the curvature limit:

$$P_{feasibility}(C) = |P_{cur}| \qquad (7)$$

where:

$$P_{cur} = \{p_j \in C, j = 1..k2 | |curvature(p_j)| > \frac{1}{\rho_{min}}\} \qquad (8)$$

$P_{length}(C)$ is the path length, determined as:

$$P_{length}(C) = \int_0^1 \|C'(u)\| \, du \qquad (9)$$

And finally $P_\sigma(C)$ denotes the standard deviation of curvatures along the path which measures the dispersion of the of the curvatures's values around their arithmetical mean c_{mean}:

$$P_\sigma(C) = \sqrt{\frac{1}{m} \sum_{i=1}^m (c_i - c_{mean})^2} \qquad (10)$$

α, β, γ and δ are weighting factors. Their values will be fixed through experimentation. It is abvious, in our method, that the best individual will have the minimum fitness value.

3.3   Genetic Operators

Classical GA algorithms primarily use three genetic operators: selection, crossover and mutation that simulate the process of natural selection and gives them their powerful search ability.

Selection Method. There are many selection strategies such as rank-based selection, roulette wheel selection, elitist selection and tournament selection. In this study, the rank-based fitness assignment is employed.

Crossover Operator. The fundamental role of crossover is to allow the recombination of information in the genetic heritage of the population by combining the features of two parents to form two offsprings. Conventional crossover methods include single-point crossover and two-point crossover which is proven more effective. To enhance the population diversity, an n-point crossover is used where n is the number of crossing points. The number and sites of the crossover are randomly generated in each iteration. Therefore, the two offsprings are obtained by copying in the first offspring the genes of $Parent1$ up to the first crosspoint then supplementing with the genes of $Parent2$ up to the second crosspoint and so on until the nth crosspoint as shown in Fig. 3, where n equals 3 and the crossover sites are 2, 5 and 7.

Mutation Operator. For the purpose of increasing the population's diversity, mutaion operation is required in genetic algorithms despite its low probability. In this paper, the classical mutation operator is used which takes as input a selected individual for mutation and returns a mutant individual obtained by local transformation of one of its genes. In fact, a gene is replaced with a new gene not included in the individual. Thus, the mutation operator modifies randomly the characteristics of a solution, which allows to introduce and maintain the diversity within our population.

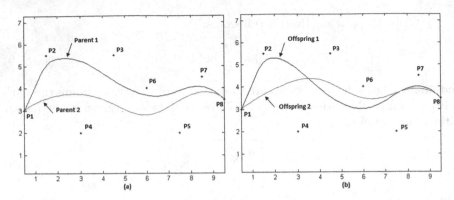

Fig. 3. (a) Two chromosomes (NURBS curves approximation of 7 control points) before the crossover: weight vector of Parent 1(0.1, 3.4, 4.5, 1.2, 2.3, 3, 4, 2), weight vector of Parent 2(5.2, 0.4, 3.2, 3.6, 5.1, 2, 2, 2) (b) Generated offsprings after the crossover: weight vector of offspring 1(0.1, 3.4, 3.2, 3.6, 5.1, 3, 4, 2), weight vector of offspring 2(5.2, 0.4, 4.5, 1.2, 2.3, 2, 2, 2).

4   Experimental Results

To investigate the feasibility and effectiveness of the proposed algorithm, three simulation experiments were conducted. Various input parameters related to the environment and the robot are summarized in Table 1. Likewise, different values of the GA parameters are considered as listed in Table 2. We notice that Blue denotes an accessible area while Red denotes an impenetrable area. S and T represent the starting point and the goal point. All generated NURBS curves are of degree 4.

Table 1. Input parameters of environment and robot

Parameters	Map I	Map II	Map III
Space size	50*60	100*100	80*80
Start position	(28,46)	(5,80)	(33,48)
Target position	(21,7)	(67,12)	(47,8)
Robot diameter	2	1	2
Minimum radius of curvature	0.4	0.33	2

For Map I, after calculating the shortest polyline path from S to T, a set of 12 control points (red points in Fig. 4(a)) is determined. Initial population is created by generating randomly values of weights between 0.1 and 5. This population is evolved generation by generation to calculate the optimum path with respect to the system requirements. We note that with a crossover probability equal to 0.6 and a mutation probability of 0.2, the algorithm succeeded to provide, within 5

Table 2. Input parameters of GA

Parameters	Map I	Map II	Map III
Chromosome length	12	16	23
Population size	20	100	100
Crossover probability	0.6	0.07	0.7
Mutation probability	0.2	0.19	0.4
Max generations	50	100	40

iterations (Fig. 4(b)), the best solution. In fact, the best individual of the first generation has a fittness value of 22.25, a length equal to 73.75 and a standard deviation of curvatures equal to 0.19. As seen in Table 3, these values have been optimized to provide an optimal path of length equal to 69.44, with 20.95 fittness value and 0.18 as final measure of curvatures dispersion. The weight vector of the computed path is $W = (0.92, 2.56, 2.77, 1.09, 3.21, 0.23, 2.09, 4.53, 0.94, 0.99, 1.31, 3.85)$.

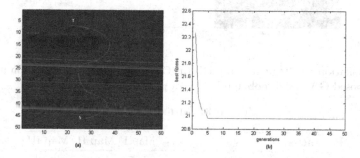

Fig. 4. Simulation results under Map I. (a) Generated path. (b) Evolution procedure of the proposed GA based mobile robot path planning algorithm (Color figure online).

As the robot's minimum radius of curvature of the first experiment was set to 0.4, the generated path must have 2.5 as the maximum allowable value of curvature. This constraint was ensured and the maximum value of curvature is 0.54.

Figure 5 shows the simulation results of the proposed algorithm under Map II. The obtained path is of length 105.95 computed from the approximation of 16 control points and it has the value 0.28 as standard deviation of curvatures. The corresponding weight vector is $W = (3.03, 2.48, 3.18, 0.50, 3.03, 4.92, 0.30, 1.24, 3.45, 1.10, 0.17, 3.91, 4.45, 1.13, 0.57, 1.03)$. For Map III, the planned path is modeled by a NURBS curve which approximates a set of 23 control points (Fig. 6). The length of this curve is 143.39 and the standard deviation of curvatures equals 0.24. The corresponding weight vector is $W = (1.75, 2.20, 2.50, 4.64, 2.96, 2.22, 0.27, 4.85, 3.95, 1.80, 4.55, 0.20, 0.15, 3.35, 4.18, 0.51, 0.15, 1.92, 2.16, 2.18, 2.41, 2.43, 2.84)$. Experimental results are summarized in Table 3.

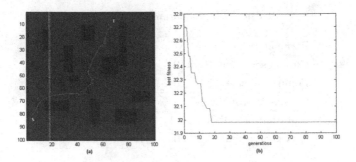

Fig. 5. Simulation results under Map II. (a) Generated path. (b) Evolution procedure of the proposed GA based mobile robot path planning algorithm.

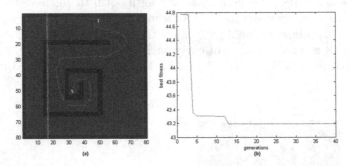

Fig. 6. Simulation results under Map III. (a) Generated path. (b) Evolution procedure of the proposed GA based mobile robot path planning algorithm.

Table 3. Output results

Parameters	Map I	Map II	Map III
Best fitness	20.95	31.98	43.19
Standard deviation of curvatures	0.18	0.28	0.24
Path length	69.44	105.95	143.39
|Max curvature|	0.54	1.49	0.48

As shown, the presented algorithm can be applied to mobile robot global path planning under different environments and is able to guide a mobile robot while achieving an optimal or near-optimal collision-free path thanks to the proposed multi-objective fitness function used in the parameterization step of the NURBS curve modelling the generated solution.

5   Conclusion

This paper has presented a novel approach to the path planning problem based on an optimized NURBS curve using genetic algorithm. It has introduced an accurate fitness function showing that the proposed algorithm is capable of efficiently generating a smooth and obstacle avoiding path with minimal length and minimal variations of curvature. Future studies include investigating the path planning task with different optimization methods. Furthermore, it will be interesting to adopt our algorithms into real world applications. Also, it involves to research the problem of path planning in an unknown dynamic environments.

References

1. Khatib, O.: Real time obstacle avoidance for manipulators and mobile robots. Int. J. Robot. Res. **5**, 90–98 (1986)
2. Nilsson, N.J.: A Mobile automaton: an application of artificial intelligence techniques. In: Proceedings of the 1st International Joint Conference on Artificial Intelligence, Washington, pp. 509–520 (1969)
3. Lingelbach, F.: Path planning using probabilistic cell decomposition. In: Proceedings of the IEEE International Conference on Robotics and Automation (ICRA), pp. 467–472 (2004)
4. Zhang, Y., Gong, D.W., Zhang, J.: Robot path planning in uncertain environment using multi-objective particle swarm optimization. Neurocomputing **103**, 172–185 (2013)
5. Bin, N., Xiong, C., Liming, Z., Wendong, X.: Recurrent neural network for robot path planning. In: Liew, K.-M., Shen, H., See, S., Cai, W. (eds.) PDCAT 2004. LNCS, vol. 3320, pp. 188–191. Springer, Heidelberg (2004)
6. Cong, Y.Z., Ponnambalam, S.G.: Mobile robot path planning using ant colony optimization. In: IEEE/ASME International Conference on Advanced Intelligent Mechatronics (AIM), pp. 851–856 (2009)
7. Hassanzadeh, I., Sadigh, S.M.: Path planning for a mobile robot using fuzzy logic controller tuned by GA. In: 6th International Symposium on Mechatronics and its Applications (ISMA), pp. 1–5 (2009)
8. Jalel, S., Marthon, P., Hamouda, A.: Optimum path planning for mobile robots in static environments using graph modelling and nurbs curves. In: 12th WSEAS International Conference on Signal Processing, Robotics and Automation(ISPRA), pp. 216–221 (2013)
9. Piegl, L., Tiller, W.: The NURBS book, 2nd edn. Springer, Heidelberg (1997)
10. Adi, D.I.S., Shamsuddin, S.M., Ali, A.: Particle swarm optimization for NURBS curve fitting. In: Proceedings of the Sixth International Conference on Computer Graphics, Imaging and Visualization: New Advances and Trends(CGIV), pp. 259–263 (2009)
11. Jing, Z., Shaowei, F., Hanguo, C.: Optimized NURBS curve and surface modelling using simulated evolution algorithm. In: Second International Workshop on Computer Science and Engineering (WCSE), pp. 435–439 (2009)

Natural Language Processing
and Recognition

Sampled Weighted Min-Hashing
for Large-Scale Topic Mining

Gibran Fuentes-Pineda[(⊠)] and Ivan Vladimir Meza-Ruíz

Instituto de Investigaciones en Matemáticas Aplicadas y en Sistemas,
Universidad Nacional Autónoma de México, Mexico city, Mexico
gibranfp@turing.iimas.unam.mx

Abstract. We present Sampled Weighted Min-Hashing (SWMH), a randomized approach to automatically mine topics from large-scale corpora. SWMH generates multiple random partitions of the corpus vocabulary based on term co-occurrence and agglomerates highly overlapping interpartition cells to produce the mined topics. While other approaches define a topic as a probabilistic distribution over a vocabulary, SWMH topics are ordered subsets of such vocabulary. Interestingly, the topics mined by SWMH underlie themes from the corpus at different levels of granularity. We extensively evaluate the meaningfulness of the mined topics both qualitatively and quantitatively on the NIPS (1.7 K documents), 20 Newsgroups (20 K), Reuters (800 K) and Wikipedia (4 M) corpora. Additionally, we compare the quality of SWMH with Online LDA topics for document representation in classification.

Keywords: Large-scale topic mining · Min-Hashing · Co-occurring terms

1 Introduction

The automatic extraction of topics has become very important in recent years since they provide a meaningful way to organize, browse and represent large-scale collections of documents. Among the most successful approaches to topic discovery are directed topic models such as Latent Dirichlet Allocation (LDA) [1] and Hierarchical Dirichlet Processes (HDP) [15] which are Directed Graphical Models with latent topic variables. More recently, undirected graphical models have been also applied to topic modeling, (e.g., Boltzmann Machines [12, 13] and Neural Autoregressive Distribution Estimators [9]). The topics generated by both directed and undirected models have been shown to underlie the thematic structure of a text corpus. These topics are defined as distributions over terms of a vocabulary and documents in turn as distributions over topics. Traditionally, inference in topic models has not scale well to large corpora, however, more efficient strategies have been proposed to overcome this problem (e.g., Online LDA [8] and stochastic variational inference [10]). Undirected Topic Models can be also trained efficient using approximate strategies such as Contrastive Divergence [7].

© Springer International Publishing Switzerland 2015
J.A. Carrasco-Ochoa et al. (Eds.): MCPR 2015, LNCS 9116, pp. 203–213, 2015.
DOI: 10.1007/978-3-319-19264-2_20

Table 1. SWMH topic examples.

NIPS	introduction, references, shown, figure, abstract, shows, back, left, process, ... (51)
	chip, fabricated, cmos, vlsi, chips, voltage, capacitor, digital, inherent, ... (42)
	spiking, spikes, spike, firing, cell, neuron, reproduces, episodes, cellular, ... (17)
20 Newsgroups	algorithm communications clipper encryption chip key
	lakers, athletics, alphabetical, pdp, rams, pct, mariners, clippers, ... (37)
	embryo, embryos, infertility, ivfet, safetybelt, gonorrhea, dhhs, ... (37)
Reuters	prior, quarterly, record, pay, amount, latest, oct
	precious, platinum, ounce, silver, metals, gold
	udinese, reggiana, piacenza, verona, cagliari, atalanta, perugia, ... (64)
Wikipedia	median, householder, capita, couples, racial, makeup, residing, ... (54)
	decepticons', galvatron's, autobots', botcon, starscream's, rodimus, galvatron
	avg, strikeouts, pitchers, rbi, batters, pos, starters, pitched, hr, batting, ... (21)

In this work, we explore the mining of topics based on term co-occurrence. The underlying intuition is that terms consistently co-occurring in the same documents are likely to belong to the same topic. The resulting topics correspond to ordered subsets of the vocabulary rather than distributions over such a vocabulary. Since finding co-occurring terms is a combinatorial problem that lies in a large search space, we propose Sampled Weighted Min-Hashing (SWMH), an extended version of Sampled Min-Hashing (SMH) [6]. SMH partitions the vocabulary into sets of highly co-occurring terms by applying Min-Hashing [2] to the inverted file entries of the corpus. The basic idea of Min-Hashing is to generate random partitions of the space so that sets with high Jaccard similarity are more likely to lie in the same partition cell.

One limitation of SMH is that the generated random partitions are drawn from uniform distributions. This setting is not ideal for information retrieval applications where weighting have a positive impact on the quality of the retrieved documents [3,14]. For this reason, we extend SMH by allowing weights in the mining process which effectively extends the uniform distribution to a distribution based on weights. We demonstrate the validity and scalability of the proposed approach by mining topics in the NIPS, 20 Newsgroups, Reuters and Wikipedia corpora which range from small (a thousand of documents) to large scale (millions of documents). Table 1 presents some examples of mined topics and their sizes. Interestingly, SWMH can mine meaningful topics of different levels of granularity.

The remainder of the paper is organized as follows. Section 2 reviews the Min-Hashing scheme for pairwise set similarity search. The proposed approach for topic mining by SWMH is described in Sect. 3. Section 4 reports the experimental evaluation of SWMH as well as a comparison against Online LDA. Finally, Sect. 5 concludes the paper with some discussion and future work.

2   Min-Hashing for Pairwise Similarity Search

Min-Hashing is a randomized algorithm for efficient pairwise set similarity search (see Algorithm 1). The basic idea is to define MinHash functions h with the property that the probability of any two sets A_1, A_2 having the same MinHash value is equal to their Jaccard Similarity, i.e.,

$$P[h(A_1) = h(A_2)] = \frac{|A_1 \cap A_2|}{|A_1 \cup A_2|} \in [0, 1]. \tag{1}$$

Each MinHash function h is realized by generating a random permutation π of all the elements and assigning the first element of a set on the permutation as its MinHash value. The rationale behind Min-Hashing is that similar sets will have a high probability of taking the same MinHash value whereas dissimilar sets will have a low probability. To cope with random fluctuations, multiple MinHash values are computed for each set from independent random permutations. Remarkably, it has been shown that the portion of identical MinHash values between two sets is an unbiased estimator of their Jaccard similarity [2].

Taking into account the above properties, in Min-Hashing similar sets are retrieved by grouping l tuples $g_1, \ldots, g_l$ of r different MinHash values as follows

$$g_1(A_1) = (h_1(A_1), h_2(A_1), \ldots, h_r(A_1))$$
$$g_2(A_1) = (h_{r+1}(A_1), h_{r+2}(A_1), \ldots, h_{2 \cdot r}(A_1))$$
$$\ldots$$
$$g_l(A_1) = (h_{(l-1) \cdot r+1}(A_1), h_{(l-1) \cdot r+2}(A_1), \ldots, h_{l \cdot r}(A_1))$$

where $h_j(A_1)$ is the j-th MinHash value. Thus, l different hash tables are constructed and two sets A_1, A_2 are stored in the same hash bucket on the k-th hash table if $g_k(A_1) = g_k(A_2), k = 1, \ldots, l$. Because similar sets are expected to agree in several MinHash values, they will be stored in the same hash bucket with high probability. In contrast, dissimilar sets will seldom have the same MinHash value and therefore the probability that they have an identical tuple will be low. More precisely, the probability that two sets A_1, A_2 agree in the r MinHash values of a given tuple g_k is $P[g_k(A_1) = g_k(A_2)] = sim(A_1, A_2)^r$. Therefore, the probability that two sets A_1, A_2 have at least one identical tuple is $P_{collision}[A_1, A_2] = 1 - (1 - sim(A_1, A_2)^r)^l$.

The original Min-Hashing scheme was extended by Chum et al. [5] to weighted set similarity, defined as

$$sim_{hist}(H_1, H_2) = \frac{\sum_i w_i \min(H_1^i, H_2^i)}{\sum_i w_i \max(H_1^i, H_2^i)} \in [0, 1], \tag{2}$$

where H_1^i, H_2^i are the frecuencies of the i-th element in the histograms H_1 and H_2 respectively and w_i is the weight of the element. In this scheme, instead of generating random permutations drawn from a uniform distribution, the permutations are drawn from a distribution based on element weights. This extension allows the use of popular document representations based on weighting schemes such as *tf-idf* and has been applied to image retrieval [5] and clustering [4].

Algorithm 1. Pairwise Similarity Search by Min-Hashing

Data: Database of sets $A = A_1, \ldots, A_N$ and query set q
Result: Similar sets to q in A
Indexing
1. Compute l MinHash tuples $g_i(A_j), i = 1, \ldots, l$ for each set $A_j, j = 1, \ldots, N$ in A.
2. Construct l hash tables and store each set $A_j, j = 1, \ldots, N$ in the buckets corresponding to $g_i(A_j), i = 1, \ldots, l$.

Querying

1. Compute the l MinHash tuples $g_i(q), i = 1, \ldots, l$ for the query set q.
2. Retrieve the sets stored in the buckets corresponding to $g_i(q), i = 1, \ldots, l$.
3. Compute the similarity between each retrieved set and q and return those with similarity greater than a given threshold ϵ.

---

3   Sampled Min-Hashing for Topic Mining

Min-Hashing has been used in document and image retrieval and classification, where documents and images are represented as bags of words. Recently, it was also successfully applied to retrieving co-occurring terms by hashing the inverted file lists instead of the documents [5,6]. In particular, Fuentes-Pineda et al. [6] proposed Sampled Min-Hashing (SMH), a simple strategy based on Min-Hashing to discover objects from large-scale image collections. In the following, we briefly describe SMH using the notation of terms, topics and documents, although it can be generalized to any type of dyadic data. The underlying idea of SMH is to mine groups of terms with high *Jaccard Co-occurrence Coefficient (JCC)*, i.e.,

$$JCC(T_1, \ldots, T_k) = \frac{|T_1 \cap T_2 \cap \cdots \cap T_k|}{|T_1 \cup T_2 \cup \cdots \cup T_k|}, \tag{3}$$

where the numerator correspond to the number of documents in which terms $T_1, \ldots, T_k$ co-occur and the denominator is the number of documents with at least one of the k terms. Thus, Eq. 1 can be extended to multiple co-occurring terms as

$$P[h(T_1) = h(T_2) \ldots = h(T_k)] = JCC(T_1, \ldots, T_k). \tag{4}$$

From Eqs. 3 and 4, it is clear that the probability that all terms $T_1, \ldots, T_k$ have the same MinHash value depends on how correlated their occurrences are: the more correlated the higher is the probability of taking the same MinHash value. This implies that terms consistently co-occurring in many documents will have a high probability of taking the same MinHash value.

In the same way as pairwise Min-Hashing, l tuples of r MinHash values are computed to find groups of terms with identical tuple, which become a co-occurring term set. By choosing r and l properly, the probability that a group of k terms has an identical tuple approximates a unit step function such that

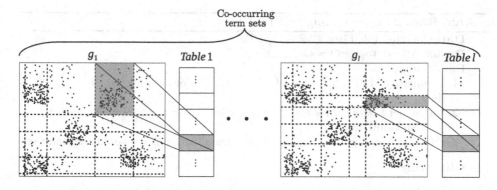

Fig. 1. Partitioning of the vocabulary by Min-Hashing.

$$P_{collision}[T_1, \ldots, T_k] \approx \begin{cases} 1 & \text{if } JCC(T_1, \ldots, T_k) \geq s* \\ 0 & \text{if } JCC(T_1, \ldots, T_k) < s* \end{cases},$$

Here, the selection of r and l is a trade-off between precision and recall. Given $s*$ and r, we can determine l by setting $P_{collision}[T_1, \ldots, T_k]$ to 0.5, which gives

$$l = \frac{\log(0.5)}{\log(1 - s*^r)}.$$

In SMH, each hash table can be seen as a random partitioning of the vocabulary into disjoint groups of highly co-occurring terms, as illustrated in Fig. 1. Different partitions are generated and groups of discriminative and stable terms belonging to the same topic are expected to lie on overlapping inter partition cells. Therefore, we cluster co-occurring term sets that share many terms in an agglomerative manner. We measure the proportion of terms shared between two co-occurring term sets C_1 and C_2 by their overlap coefficient, namely

$$ovr(C_1, C_2) = \frac{|C_1 \cap C_2|}{\min(|C_1|, |C_2|)} \in [0, 1].$$

Since a pair of co-occurring term sets with high Jaccard similarity will also have a large overlap coefficient, finding pairs of co-occurring term sets can be speeded up by using Min-Hashing, thus avoiding the overhead of computing the overlap coefficient between all the pairs of co-occurring term sets.

The clustering stage merges chains of co-occurring term sets with high overlap coefficient into the same topic. As a result, co-occurring term sets associated with the same topic can belong to the same cluster even if they do not share terms with one another, as long as they are members of the same chain. In general, the generated clusters have the property that for any co-occurring term set, there exists at least one co-occurring term set in the same cluster with which it has an overlap coefficient greater than a given threshold ϵ.

Algorithm 2. Topic mining by SWMH

Data: Inverted File Lists $T = T_1, \ldots, T_N$
Result: Mined Topics $O = O_1, \ldots, O_M$
Partitioning

1. Compute l MinHash tuples $g_i(T_j), i = 1, \ldots, l$ for each list $T_j, j = 1, \ldots, N$ in T.
2. Construct l hash tables and store each list $T_j, j = 1, \ldots, N$ in the bucket corresponding to $g_i(T_j), i = 1, \ldots, l$.
3. Mark each group of lists stored in the same bucket as a co-occurring term set.

Clustering

1. Find pairs of co-occurring term sets with overlap coefficient greater than a given threshold ϵ.
2. Form a graph G with co-occurring term sets as vertices and edges defined between pairs with overlap coefficient greater than ϵ.
3. Mark each connected component of G as a topic.

We explore the use of SMH to mine topics from documents but we judge term co-occurrence by the *Weighted Co-occurrence Coefficient (WCC)*, defined as

$$WCC(T_1, \ldots, T_k) = \frac{\sum_i w_i \min\left(T_1^i, \cdots, T_k^i\right)}{\sum_i w_i \max\left(T_1^i, \cdots, T_k^i\right)} \in [0, 1], \tag{5}$$

where $T_1^i, \cdots, T_k^i$ are the frecuencies in which terms $T_1, \ldots, T_k$ occur in the i-th document and the weight w_i is given by the inverse of the size of the i-th document. We exploit the extended Min-Hashing scheme by Chum et al.[5] to efficiently find such co-occurring terms. We call this topic mining strategy Sampled Weighted Min-Hashing (SWMH) and summarize it in Algorithm 2.

4   Experimental Results

In this section, we evaluate different aspects of the mined topics. First, we present a comparison between the topics mined by SWMH and SMH. Second, we evaluate the scalability of the proposed approach. Third, we use the mined topics to perform document classification. Finally, we compare SWMH topics with Online LDA topics.

The corpora used in our experiments were: NIPS, 20 Newsgroups, Reuters and Wikipedia[1]. NIPS is a small collection of articles ($3,649$ documents), 20 Newsgroups is a larger collection of mail newsgroups ($34,891$ documents), Reuters is a medium size collection of news ($137,589$ documents) and Wikipedia is a large-scale collection of encyclopedia articles ($1,265,756$ documents)[2].

---

[1] Wikipedia dump from 2013-09-04.
[2] All corpora were preprocessed to cut off terms that appeared less than 6 times in the whole corpus.

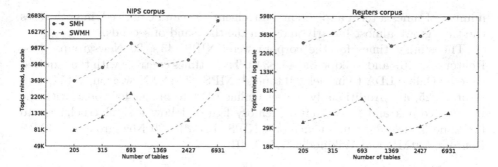

Fig. 2. Amount of mined topics for SMH and SWMH in the (a) NIPS and (b) Reuters corpora.

All the experiments presented in this work were performed on an Intel(R) Xeon(R) 2.66 GHz workstation with 8 GB of memory and with 8 processors. However, we would like to point out that the current version of the code is not parallelized, so we did not take advantage of the multiple processors.

4.1   Comparison Between SMH and SWMH

For these experiments, we used the NIPS and Reuters corpora and different values of the parameters $s*$ and r, which define the number of MinHash tables. We set the parameters of similarity ($s*$) to 0.15, 0.13 and 0.10 and the tuple size (r) to 3 and 4. These parameters rendered the following table sizes: 205, 315, 693, 1369, 2427, 6931. Figure 2 shows the effect of weighting on the amount of mined topics. First, notice the breaking point on both figures when passing from 1369 to 2427 tables. This effect corresponds to resetting the $s*$ to .10 when changing r from 3 to 4. Lower values in $s*$ are more strict and therefore less topics are mined. Figure 2 also shows that the amount of mined topics is significantly reduced by SWMH, since the colliding terms not only need to appear on similar documents but now with similar proportions. The effect of using SWMH is also noticeable in the number of terms that compose a topic. The maximum reduction reached in NIPS was 73 % while in Reuters was 45 %.

4.2   Scalability Evaluation

To test the scalability of SWMH, we measured the time and memory required to mine topics in the Reuters corpus while increasing the number of documents to be analyzed. In particular, we perform 10 experiments with SWMH, each increasing the number of documents by 10 %[3]. Figure 3 illustrates the time taken to mine topics as we increase the number of documents and as we increase an index of complexity given by a combination of the size of the vocabulary and the average

---

[3] The parameters were fixed to $s* = 0.1$, $r = 3$, and overlap threshold of 0.7.

number of times a term appears in a document. As can be noticed, in both cases the time grows almost linearly and is in the thousand of seconds.

The mining times for the corpora were: NIPS, 43 s; 20 Newsgroups, 70 s; Reuters, 4, 446 s and Wikipedia, 45, 834 s. These times contrast with the required time by Online LDA to model 100 topics[4]: NIPS, 60 s; 20 Newsgroups, 154 s and Reuters, 25, 997. Additionally, we set Online LDA to model 400 topics with the Reuters corpus and took 3 days. Memory figures follow a similar behavior to the time figures. Maximum memory: NIPS, 141 MB; 20 Newsgroups, 164 MB; Reuters, 530 MB and Wikipedia, 1, 500 MB.

4.3   Document Classification

In this evaluation we used the mined topics to create a document representation based on the similarity between topics and documents. This representation was used to train an SVM classifier with the class of the document. In particular, we focused on the 20 Newsgroups corpus for this experiment. We used the typical setting of this corpus for document classification (60 % training, 40 % testing).

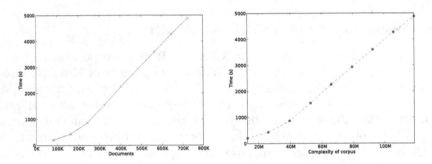

Fig. 3. Time scalability for the Reuters corpus.

Table 2. Document classification for 20 Newsgroups corpus.

Model	Topics	Accuracy	Avg. score
205	3394	59.9	60.6
319	4427	61.2	64.3
693	6090	68.9	70.7
1693	2868	53.1	55.8
2427	3687	56.2	60.0
6963	5510	64.1	66.4
Online LDA	100	59.2	60.0
Online LDA	400	65.4	65.9

---

[4] https://github.com/qpleple/online-lda-vb was adapted to use our file formats.

Table 2 shows the performance for different variants of topics mined by SWMH and Online LDA topics. The results illustrate that the number of topics is relevant for the task: Online LDA with 400 topics is better than 100 topics. A similar behavior can be noticed for SWMH, however, the parameter r has an effect on the content of the topics and therefore on the performance.

4.4 Comparison Between Mined and Modeled Topics

In this evaluation we compare the quality of the topics mined by SWMH against Online LDA topics for the 20 Newsgroups and Reuters corpora. For this we measure *topic coherence*, which is defined as

$$C(t) = \sum_{m=2}^{M} \sum_{l=1}^{m-1} \log \frac{D(v_m, v_l)}{D(v_l)},$$

where $D(v_l)$ is the document frequency of the term v_l, and $D(v_m, v_l)$ is the co-document frequency of the terms v_m and v_l [11]. This metric depends on the first M elements of the topics. For our evaluations we fixed M to 10. However, we remark that the comparison is not direct since both the SWMH and Online LDA topics are different in nature: SWMH topics are subsets of the vocabulary with uniform distributions while Online LDA topics are distributions over the complete vocabulary. In addition, Online LDA generates a fixed number of topics which is in the hundreds while SWMH produces thousands of topics. For the comparison we chose the n-best mined topics by ranking them using an ad hoc metric involving the co-occurrence of the first element of the topic. For the purpose of the evaluation we limited the SWMH to the 500 best ranked topics. Figure 4 shows the coherence for each corpus. In general, we can see a difference in the shape and quality of the coherence box plots. However, we notice that SWMH produces a considerable amount of outliers, which calls for further research in the ranking of the mined topics and their relation with the coherence.

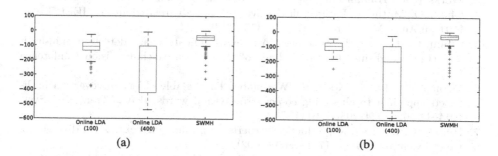

Fig. 4. Coherence of topics mined by SWMH vs Online LDA topics in the (a) 20 Newsgroups and (b) Reuters corpora.

5 Discussion and Future Work

In this work we presented a large-scale approach to automatically mine topics in a given corpus based on Sampled Weighted Min-Hashing. The mined topics consist of subsets of highly correlated terms from the vocabulary. The proposed approach is able to mine topics in corpora which go from the thousands of documents (1 min approx.) to the millions of documents (7 h approx.), including topics similar to the ones produced by Online LDA. We found that the mined topics can be used to represent a document for classification. We also showed that the complexity of the proposed approach grows linearly with the amount of documents. Interestingly, some of the topics mined by SWMH are related to the structure of the documents (e.g., in NIPS the words in the first topic correspond to parts of an article) and others to specific groups (e.g., team sports in 20 Newsgroups and Reuters, or the *Transformers* universe in Wikipedia). These examples suggest that SWMH is able to generate topics at different levels of granularity.

Further work has to be done to make sense of overly specific topics or to filter them out. In this direction, we found that weighting the terms has the effect of discarding several irrelevant topics and producing more compact ones. Another alternative, it is to restrict the vocabulary to the top most frequent terms as done by other approaches. Other interesting future work include exploring other weighting schemes, finding a better representation of documents from the mined topics and parallelizing SWMH.

References

1. Blei, D.M., Ng, A.Y., Jordan, M.I.: Latent Dirichlet allocation. J. Mach. Learn. Res. **3**, 993–1022 (2003)
2. Broder, A.Z.: On the resemblance and containment of documents. Comput. **33**(11), 46–53 (2000)
3. Buckley, C.: The importance of proper weighting methods. In: Proceedings of the Workshop on Human Language Technology, pp. 349–352 (1993)
4. Chum, O., Matas, J.: Large-scale discovery of spatially related images. IEEE Trans. Pattern Anal. Mach. Intell. **32**, 371–377 (2010)
5. Chum, O., Philbin, J., Zisserman, A.: Near duplicate image detection: min-hash and tf-idf weighting. In: Proceedings of the British Machine Vision Conference (2008)
6. Fuentes Pineda, G., Koga, H., Watanabe, T.: Scalable object discovery: a hash-based approach to clustering co-occurring visual words. IEICE Trans. Inf. Syst. **E94–D**(10), 2024–2035 (2011)
7. Hinton, G.E.: Training products of experts by minimizing contrastive divergence. Neural Comput. **14**(8), 1771–1800 (2002)
8. Hoffman, M.D., Blei, D.M., Bach, F.: Online learning for latent Dirichlet allocation. In: Advances in Neural Information Processing Systems 23 (2010)
9. Larochelle, H., Stanislas, L.: A neural autoregressive topic model. In: Advances in Neural Information Processing Systems 25, pp. 2717–2725 (2012)

10. Mimno, D., Hoffman, M.D., Blei, D.M.: Sparse stochastic inference for latent Dirichlet allocation. In: International Conference on Machine Learning (2012)
11. Mimno, D., Wallach, H.M., Talley, E., Leenders, M., McCallum, A.: Optimizing semantic coherence in topic models. In: Proceedings of the Conference on Empirical Methods in Natural Language Processing, pp. 262–272. ACL (2011)
12. Salakhutdinov, R., Srivastava, N., Hinton, G.: Modeling documents with a deep Boltzmann machine. In: Proceedings of the Conference on Uncertainty in Artificial Intelligence (2013)
13. Salakhutdinov, R., Hinton, G.E.: Replicated softmax: an undirected topic model. In: Advances in Neural Information Processing Systems 22, pp. 1607–1614 (2009)
14. Salton, G., Buckley, C.: Term-weighting approaches in automatic text retrieval. Inf. Process. Manage. **24**(5), 512–523 (1988)
15. Teh, Y.W., Jordan, M.I., Beal, M.J., Blei, D.M.: Hierarchical Dirichlet processes. J. Am. Stat. Assoc. **101**, 1566–1581 (2004)

A Graph-Based Textual Entailment Method Aware of Real-World Knowledge

Saúl León, Darnes Vilariño[✉], David Pinto, Mireya Tovar, and Beatriz Beltrán

Faculty of Computer Science, Benemérita Universidad
Autonóma de Puebla, Puebla, Mexico
{saul.leon,darnes,dpinto,mtovar,bbeltran}@cs.buap.mx
http://www.cs.buap.mx

Abstract. In this paper we propose an unsupervised methodology to solve the textual entailment task, that extracts facts associated to pair of sentences. Those extracted facts are represented as a graph. Then, two graph-based representations of two sentences may be further compared in order to determine the type of textual entailment judgment that they hold. The comparison method is based on graph-based algorithms for finding sub-graphs structures inside another graph, but generalizing the concepts by means of a real world knowledge database. The performance of the approach presented in this paper has been evaluated using the data provided in the Task 1 of the SemEval 2014 competition, obtaining 79 % accuracy.

Keywords: Textual entailment · Graph-based representation · Semantic similarity

1 Introduction

Textual entailment is defined as a process in which a directional relation between pairs of text expressions, denoted by T (the entailing "Text"), and H (the entailed "Hypothesis") is hold. We say that T entails H if the meaning of H can be inferred from the meaning of T, as would typically be interpreted by people.

Nowadays, many Natural Languaje Processing (NLP) tasks such as: Question Answering, Information Retrieval, Automatic Summary, Author Verification, Author Profiling, etc., usually require a module capable of detecting this type of semantic relation between a given pair of texts.

In this research work, we address the problem of textual entailment by proposing a methodology for extracting facts from a pair of sentences in order to discover the relevant information and use it for determining the textual entailment judgment between that pair of sentences. In this methodology, each set of facts associated to a sentence is represented as a graph (from the perspective of graph theory). Then, two graph-based representations of two sentences may be further compared in order to determine the type of textual entailment judgment that they hold. The comparison method is based on graph-based algorithms for

© Springer International Publishing Switzerland 2015
J.A. Carrasco-Ochoa et al. (Eds.): MCPR 2015, LNCS 9116, pp. 214–223, 2015.
DOI: 10.1007/978-3-319-19264-2_21

finding sub-graphs structures inside another graph, but generalizing the concepts by means of a real world knowledge database constructed on the basis of conceptnet5, wordnet and Openoffice thesaurus. The performance of the approach presented in this paper has been evaluated using the data provided in the Task 1 of the SemEval 2014 competition.

The remaining of this paper is structured as follows. In Sect. 2 we present the state of the art in textual entailment. In Sect. 3 we show our methodology based on graphs for determining the judge of textual entailment. The experimental results are presented in Sect. 4. Finally, in Sect. 5 the conclusions and further work are given.

2   Related Work

In recent years, the amount of unstructured information present in the web has overcomed our human capacity of analysis. There exist, however, a necessity of performing this process for many Natural Language Processing Tasks. With this amount of data, the analysis is a process that can only be done by using computational methods. But, the construction of automatic methods that perform this task efficiently is a big challenge that has not been solved yet.

This paper is focused in the analysis of texts for performing the particular task of textual entailment, and, therefore, we have been gathered some works reported in literature that we consider relevant for this paper.

The problem of textual entailment has been studied for more than a decade, the seminal paper know is the one published by Monz and de Rijke in 2001 [8]. This task was, thereafter, brought to a wide community in a challenge for Recognizing Textual Entailment (RTE) [2]. In general, in literature we may see that the major research works have focused their efforts in the selection of a big number of features, mainly statistical ones, that allow to build feature vectors useful in supervised learning methods for creating classification models that may further be used for determining the entailment judgment for a given new pair of sentences; some of those feature are shown in Table 1.

Presenting a comprehensive state of the art in textual entailment is not the purpose of this paper, but to present a new approach based on graph for representing information that leads to determine the entailment judgment for a pair of sentences. There exist, however, very good research works that study the trajectory of this task through the time. We refer the reader of this paper, for example, to the book published very recently by Dagan [3].

The methods reported in literature usually perform textual entailment by first extracting one or more of the aforementioned features from the two sentences that we are interested in determining the judgment. Thereafter, they use those features for constructing two feature vectors that may be compared by using basic or more complex similarity metrics, for example, those that involve semantic characteristics. This type of approaches are limited to analyze the presence or absence of common terms, but these elements are not sufficient to obtain high values of precision. A more interesting research line has been the discovering of structural patterns shared by the two sentences. For example, given the

Table 1. Enumeration of different features employed in the RTE task

Feature	Type
ngram of characters and words	Lexical
skipgram of characters and words	Lexical
Jaccard coefficient with synonyms	Lexical
Leacock &Chodorow's word similarity	Knowledge-based
Lesk's word similarity	Knowledge-based
Wu &Palmer's word similarity	Knowledge-based
Resnink's word similarity	Knowledge-based
Lin's word similarity	Knowledge-based
Jiang &Conrath's word similarity	Knowledge-based
Rada Mihalcea's metric using PMI	Corpus-based
Rada Mihalcea's metric using LSA	Corpus-based

following two sentences: *Leonardo Da Vinci painted the Mona Lisa*, and *Mona Lisa is the work of Leonardo da Vinci*, it is possible to find a structural pattern from which we can infer that *X painted Y → Y is the work of X*. This type of patterns may guarantee that the particular type of entailment judgment can be discovered. However, the construction/discovery of this patterns is quite difficult because the major of them are constructed for ad-hoc datasets. Real world patterns are not easy to be discovered automatically, but some efforts have been performed in this direction, in particular, seeking to generate large scale structural patterns [6].

Another research line attempt to generalize the patterns extracted for constructing axioms. This particular logic-based approach [1] seems to be promising, but still a number of problems that need to be solved, because in the major cases these approaches fail when doing the inference process. A huge amount of information is needed in order to have a proper knowledge of the real world. Even if we have properly modeled the logic representation given a set of training data, if we do not have a good knowledge database, we will usually fail to infer the real entailment judgment for the test data.

Other works in literature have tackle the RTE problem from the perspective of question-answering, thus considering that S_1 is a question and S_2 is the answer [5]. So the approaches are limited to determine certain degree of entailment based on the similarity found between the two sentences. Again, this methods may fail because even when there exist a real type of textual entailment between the sentences, some terms may not necessarily be shared by them, i.e., they are not similar from the lexical perspective, but instead they are semantically similar.

Some authors have realized that more than isoled features are needed, therefore, there have been a number of papers describing the construction of lexical resources with the aim of integrating semantic aspects in the process of textual entailment. Most of these authors use tools such as, PoS taggers, dependency parsers and other lexical resources (Wordnet, thesaurus, etc.) in the inference modules that are part of their textual entailment systems [4].

A major problem, from the machine learning perspective is the number of features of the test data that are absent in the training data. Normally, this problem may be overcomed by introducing more data in the training set, or using some kind or smoothing techniques. But, most of the time, the fact that some particular term is not found in the test data is a consequence of a lack of real world knowledge, for example, the term is a hyperonym, hyponym, synonym, etc. Integrating all these features in basic text representation schemas is not so simple, especially if we are interested in conserving the knowledge present in the original sentence. In this way, graph structures are a natural way to represent natural language information, preserving different level of natural language formal description (lexical, syntax, semantics, etc.). There are papers such as: the work [9] that presents statistical machine learning representation of textual entailment via syntactic graphs constituted by tree pairs. They shows that the natural way of representing the syntactic relation between text and hypothesis consists in the huge feature space of all possible syntactic tree fragment pairs, which can only be managed using kernel methods. Moreover the author Okita [10] presents a robust textual entailment system using the principle of just noticeable difference in psychology, which they call a local graph matching-based system with active learning.

Thus, we consider important to use both, graph based structures for preserving the original structure of the sentence and, knowledge databases that may allow our methods to be aware of those lexical relation hold in the terms with the purpose of having more efficient methods for the automatic recognition of textual entailment.

3   A Methodology Based on Graphs Aware of Real World Relation for Textual Entailment

As we mentioned before, we are proposing a method that requires to be aware that some terms are semantically related, a process that require to construct a process for calculating this type of similarity which in fact is much more difficult than calculating lexical similarity. We need, therefore, to use a proper lexical resource for improving the calculation of the degree of semantic similarity between sentences. For this goal, we have constructed a general purpose knowledge database on the basis of the following lexical resources: WordNet[1], ConceptNet5[2] and OpenOffice Thesaurus[3]. The manner we have used this database to infer real-world knowledge with the aim of using semantic information in the textual entailment task, we describe this in the following subsection.

---

[1] http://wordnet.princeton.edu/.
[2] http://conceptnet5.media.mit.edu/.
[3] https://wiki.openoffice.org/wiki/Dictionaries.

3.1   Inference Through Knowledge Databases Using Graph-Based Representation

Sometimes, two terms are semantically related by a given type of relation, let us say, synonym, hyperonym, etc. For example, from the real-world knowledge we know that a "camera" is an "electronic device". In this case, we would like to be able to automatically determine this type of relation. There exist, however other cases in which the relation is not hold directly, but by using graph searching processes we can infer these relation. For example, the terms "head" and "hair" are semantically related, but this relation can not be found directly from our knowledge database. However, we can find and indirect relation between these two words, as shown in Fig. 1, because from the same database we know that "head" is related with "body structure", "body structure" is related with "filament", and "filament" is related with "hair", therefore, both "head" and "hair" are indirectly related by transitivity.

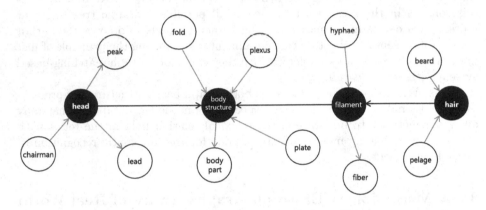

Fig. 1. Concepts network that relates head with hair

This process may be done because we have a knowledge database in which we may execute single queries with the aim of determining whether or not a given pair of concepts are related. Formally, the knowledge database can be seen as a concept graph $G_C = (V, E)$, where V is the set of vertices (concepts), and E is the set of edges (relations). There may be a lot of relations, but for practical purposes we have employed only the following ones: PartOf, Synonym, RelatedTo, WordNetRelatedTo, MemberOf, SimilarTo.

We consider that by adding this process to the methodology we can better accomplish the final purpose of this research work that is to automatically detect the judge of textual entailment for a given pair of sentences.

Having a mechanism for detecting semantic relations, we now propose to employ graph structures for representing sentences through "facts" with the aim of preserving different levels of formal description of language in a single structure. A fact is basically a relation between two language components (Noun phrase, Verb or Adjective); the relation may be one of the following five types: subject, object, qualify, extension or complement. A better description of the facts constructing process follows.

3.2 Building Facts by Sentence Interpretation

In order to build facts, we need to transform the original raw sentence to a one with tags that allow us to interpret the role that each language component plays in the sentence. We propose to start by using the Stanford parser[4]. By using this tool we may obtain PoS tags, the parse and the typed dependencies. The parsing process identify the following chunk tags: Noun Phrases (NP), Verbal Phrases (VP), and Prepositional Phrase (PP), which allow us to generate the following four types of categories[5]:

- *ENTITY*: Nouns in a noun phrase associated to one of the following PoS tags: NN, NNS, NNP, NNPS and NAC.
- *ACTIVITY*: Verbs in a verbal phrase associated to one of the following PoS tags: VB, VBG, VBZ, VBP, VBN, VBD and MD.
- *QUALITY*: Adjectives in a noun phrase associated to one of the following PoS tags: JJ, JJR and JJS.
- *PREPOSITION*: Prepositions in a prepositional phrase associated to one of the following PoS tags: IN and TO.

From the previously detected categories we may discover the following facts:

- *Subject*: This fact is obtained when a given ENTITY is associated to one or more ACTIVITY. We assume that such ENTITY is the "subject" of those activities.
- *Object*: This fact is hold when a given ACTIVITY is associated to one or more ENTITY. In this case, the ENTITY is the "object" of that ACTIVITY.
- *Qualify*: This fact is detected when a given QUALIFY is associated to one ENTITY, thus the QUALITY "qualify" the ENTITY.
- *Extension*: This fact is obtained when a given ENTITY is associated to one PREPOSITION. In this case, we say that the ENTITY has an "extension".
- *Complement*: This fact is given when a given PREPOSITION is associated to one ENTITY. The fact indicate that this ENTITY is a "complement" of a another ENTITY.

The automatic generation of the facts previously mentioned lead us to the rules shown in Table 2, that allow us to extract these facts from the parsed version of the sentences.

We observe this four step process which traduces the original sentence to a set of facts which may be further used in the process of recognizing textual entailment.

---

[4] http://nlp.stanford.edu/software/lex-parser.shtml.
[5] The names have been proposed with the aim of giving some sense to each type of category.

Table 2. Rules for the automatic extraction of facts

Rule	Fact
$X \to ENTITY$ && $X \to ACTIVITY$	subject($ENTITY$, $ACTIVITY$)
$ACTIVITY \to ENTITY$	object($ACTIVITY$, $ENTITY$)
$X \to ENTITY$ && $X \to QUALIFY$	qualify($QUALIFY$, $ENTITY$)
$ENTITY \to PREPOSITION$	extension($ENTITY$, $PREPOSITION$)
$PREPOSITION \to ENTITY$	complement($PREPOSITION$, $ENTITY$)

3.3   Recognition of Textual Entailment

We have extracted facts with the aim of giving an interpretation to the sense of each sentence. However, isolated facts are not as useful as when they are brought together, therefore, we use graph structures for preserving the richness of all these facts. Graphs are a non-linear structure that allow us to relate, in this case, concepts for understanding the manner this relation acts over those concepts. These structures makes it possible to calculate the semantic similarity between pair of sentences, and eventually to describe rules for interpreting those similarities as textual entailment judgment criteria.

Formally, a set of facts is represented as a labeled graph $G = (V, E, L_E, \beta)$, where:

- $V = \{v_i | i = 1, ..., n\}$ is a finite set of vertices, $V \neq \emptyset$, and $n = \#vertices$ in the graph.
- $E \subseteq V \times V$ is the finite set of edges, $E = \{e = \{v_i, v_j\} | v_i, v_j \in V, 1 \leq i, j \leq n\}$.
- L_E, is the fact relation.
- $\beta : E \to L_E$, is a function that assigns a fact relationship to the edge.

Once the graphs are constructed we are able to use graph-based algorithms for finding any type of semantic relation among different sentences. For example, we could find the lexical similarity, or even semantic similarity by using the real-world knowledge database previously constructed, thus enriching the process of detecting the judgement of textual entailment.

We propose an approach in which the textual entailment task is solved analyzing the number of facts shared by the two graphs that represent both the entailing "Text" T, and the entailed "Hypothesis" H. We perform this process by eliminating from both graphs the substructures shared; in this way, the number of graph structures both graphs hold after removing similar substructures allow us to detect the occurrence of textual entailment.

In order to analyze the obtained graph substructures, it is not sufficient to directly compare ENTITIES and ACTIVITIES, because it is needed to detect how two ENTITIES interact given an ACTIVITY, therefore, instead of using a single fact, we use a substructure made up of three concepts linked each pair of concepts by a given relation. We named this type of substructure as an EXPANDED fact (EXP_FACT). The construction of the EXP_FACT are shown in Table 3.

Table 3. The three rules for generating expanded facts

Facts	EXP_FACT
(1) $subject(ENTITY_1, ACTIVITY)$ AND $object(ACTIVITY, ENTITY_2)$	$\rightarrow (ENTITY_1, ACTIVITY, ENTITY_2)$
(2) $qualify(QUALITY, ENTITY)$	$\rightarrow (ENTITY, \text{qualify}, QUALITY)$
(3) $extension(ENTITY_1, PREPOSITION)$ AND $complement(PREPOSITION, ENTITY_2)$	$\rightarrow (ENTITY_1, PREPOSITION, ENTITY_2)$

This manner of analyzing the relation among facts allow us to detect the main ideas that are transmitted in each sentence.

The number of original concepts in the graphs is compared after removing similar EXP_FACT substructures in both graphs (G_1 and G_2). If the proportion of concepts removed is greater than 50 % in the two graphs, then we can say that there exist textual entailment between the two sentences. Otherwise, we can not determine the entailment judgment, so we establish it as "neutral".

4   Experimental Results

In order to have a comparison of the performance of the proposed methodology, we have used the dataset given at the subtask 2 of Task 1 of the SemEval 2014 conference[6]. In that subtask, it is required to solve the problem of textual entailment between two sentences S_1 and S_2 by automatically detecting one of the following judgments:

– *ENTAILMENT*: if S_1 entails S_2
– *CONTRADICTION*: if S_1 contradicts S_2
– *NEUTRAL*: the truth of S_1 cannot be determined on the basis of S_2

The methodology proposed is not able to detect the CONTRADICTION judgment, but only ENTAILMENT or NEUTRAL judgment. Therefore, we propose to use a simple technique based on the existence of cuewords associated to antonyms or negation. Thus, in the pair of sentences S_1 and S_2, if one of this contains negation words or antonyms means that a CONTRADICTION exists.

In the following section we describe the dataset used in the experiments.

4.1   Test Dataset

The test dataset used at SemEVal 2014 was constructed as mentioned in [7]. The description of the number of samples for each type of textual entailment judgment is given in Table 4. Since the dataset is manually annotated, we are able to calculate the performance of the methodology we propose in this paper.

---

[6] International Workshop on Semantic Evaluation.

Table 4. Judgments distributions

Judgment	Test datasetd
NEUTRAL	2793
ENTAILMENT	1414
CONTRADICTION	720

4.2   Obtained Results

It is worth mentioned that the methodology proposed is an unsupervised approach, since the perspective that we do not a training corpus. We only use an external resource (knowledge database) for determining semantic similarities among concepts. In Table 5 we can see the results obtained by other research teams at the SemEval 2014 competition. There we have named **BUAP Graph-Match** to the approach presented in this paper. As we can see, the performance is similar to other approaches submitted to the competition with and accuracy of 79 %. The main difference here is , as we already mentioned, that we do not need to construct a classification model because our approach is unsupervised. We can move to different domains and the performance should not be affected as the target domain can be modeled in some way by the relation stored in the knowledge database.

Table 5. Task1, subtask 2, Semeval 2014 results

Team ID	Accuracy
Illinois-LH_run1	84.575
ECNU_run1	83.641
UNAL-NLP_run1	83.053
SemantiKLUE_run1	82.322
The_Meaning_Factory_run1	81.591
CECL_ALL_run1	79.988
BUAP_run1	79.663
BUAP GraphMatch	**79.072**

5   Conclusions

The methodology proposed for textual entailment aims to use graph structures for representing expanded facts, which contain information of the real world. This unsupervised approach performed well on the textual entailment task with a performance of 79 %. We consider that we have exploited the graph representation by (1) infering transitive relation between pair of concepts, and (2) detecting graph substructures with the aim of determining a textual entailment judgment.

We have noticed that the process of inference helps to improve the performance of the methodology, but a deep analysis over this process need to be done in the future, because we have observed that the semantic similarity between concepts may be degraded when the number of intermediate concepts is too high.

We consider that given the level of complexity of the task carried out in this paper, the methodology is attractive. It should be very interesting to test the same methodology in other natural language task such as summarization, information retrieval, question-answering, etc.

As future work we plan to use other types of graph-based algorithms for calculating similarities. We would also investigate the result of enriching the knowledge database used in the inference process.

References

1. Bos, J.: Is there a place for logic in recognizing textual entailment? Linguist. Issues Lang. Technol. **9**(3), 1–18 (2013)
2. Dagan, I., Glickman, O., Magnini, B.: The pascal recognising textual entailment challenge. In: Quiñonero-Candela, J., Dagan, I., Magnini, B., d'Alché-Buc, F. (eds.) Machine learning Challenges. Evaluating Predictive Uncertainty, Visual Object Classification, and Recognising Tectual Entailment. LNCS, vol. 3944, pp. 177–190. Springer, Heidelberg (2006)
3. Dagan, I., Roth, D., Sammons, M., Zanzotto, F.M.: Recognizing textual entailment: models and applications. Synth. Lect. Hum. Lang. Technol. **6**(4), 1–220 (2013)
4. Day, M.-Y., Tu, C.,Huang, S.-J., Vong, H.C., Wu, S.-W.: IMTKU textual entailment system for recognizing inference in text at NTCIR-10 RITE-2. In: Proceedings of the 10th NTCIR Conference (2013)
5. Harabagiu, S., Hickl, A.: Methods for using textual entailment in open-domain question answering. In: Proceedings of the 21st International Conference on Computational Linguistics and the 44th Annual Meeting of the Association for Computational Linguistics, pp. 905–912. Association for Computational Linguistics (2006)
6. Kouylekov, M., Magnini, B.: Building a large-scale repository of textual entailment rules. In: Proceedings of LREC (2006)
7. Marelli, M., Menini, S., Baroni, M., Bentivogli, L., Bernardi, R., Zamparelli, R.: A sick cure for the evaluation of compositional distributional semantic models. In: Proceedings of LREC (2014)
8. Monz, C., de Rijke, M.: Tequesta: the university of amsterdam's textual question answering system. In: TREC (2001)
9. Moschitti, A., Zanzotto, F.: Encoding tree pair -based graphs in learning algorithms: the textual entailment, pp. 25–32 (2008)
10. Tsuyoshi, O.: Active learning-based local graph matching for textual entailment (2013)

Semi-Supervised Approach to Named Entity Recognition in Spanish Applied to a Real-World Conversational System

Víctor R. Martínez[1]([✉]), Luis Eduardo Pérez[1], Francisco Iacobelli[2],
Salvador Suárez Bojórquez[3], and Víctor M. González[1]

[1] Department of Computer Science, Instituto Tecnologico Autonomo de Mexico, Rio
Hondo #1, Progreso Tizapan, Del. Alvaro Obregon, 01080 Mexico City, Mexico
{victor.martinez,luis.perez.estrada,victor.gonzalez}@itam.mx
[2] Northwestern University, Frances Searle Building 2-343,
2240 Campus Drive, Evanston, IL, USA
f-iacobelli@u.northwestern.edu
[3] BlueMessaging, Ibsen 40, Segundo Piso Col. Polanco, Del. Miguel,
Mexico City, Mexico
info@bluemessaging

Abstract. In this paper, we improve the named-entity recognition
(NER) capabilities for an already existing text-based dialog system (TDS)
in Spanish. Our solution is twofold: first, we developed a hidden Markov
model part-of-speech (POS) tagger trained with the frequencies from
over 120-million words; second, we obtained $2,283$ real-world conversations from the interactions between users and a TDS. All interactions
occurred through a natural-language text-based chat interface. The TDS
was designed to help users decide which product from a well-defined catalog best suited their needs. The conversations were manually tagged using
the classical Penn Treebank tag set, with the addition of an ENTITY tag
for all words relating to a brand or product. The proposed system uses
an hybrid approach to NER: first it looks up each word in a previously
defined catalog. If the word is not found, then it uses the tagger to tag
it with its appropriate POS tag. When tested on an independent conversation set, our solution presented a higher accuracy and higher recall
rates compared to a current development from the industry.

1 Introduction

Text-based Dialog Systems (TDS) help people accomplish a task using written
language [24]. A TDS can provide services and information automatically. For
example, many systems have been developed to provide information and manage
flight ticket booking [22] or train tickets [11]; decreasing the number of automo-
bile ads according to the user preferences [9], or inquiring about the weather

---

Víctor R. Martínez and Luis Eduardo Pérez—These authors contributed equally to
the work.

© Springer International Publishing Switzerland 2015
J.A. Carrasco-Ochoa et al. (Eds.): MCPR 2015, LNCS 9116, pp. 224–235, 2015.
DOI: 10.1007/978-3-319-19264-2_22

report in a certain area [25]. Normally, these kinds of systems are composed by a dialog manager, a component that manages the state of the dialog, and a dialog strategy [12]. In order to provide a satisfactory user experience, the dialog manager is responsible for continuing the conversation appropriately, that is reacting to the user's requests and staying within the subject [16]. To ensure this, the system must be capable of resolving ambiguities in the dialog, a problem often referred to as Named Entity Recognition (NER) [10].

Named Entity Recognition consists in detecting the most salient and informative elements in a text such as names, locations and numbers [14]. The term was first coined for the Sixth Message Understanding Conference (MUC-6) [10], since then most work has been done for the English language (for a survey please refer to [17]). Other well represented languages include German, Spanish and Dutch [17].

Feasible solutions to NER either employ lexicon based approaches or multiple machine learning techniques, with a wide range of success [14,17]. Some of the surveyed algorithms work with Maximum Entropy classifiers [3], AdaBoost [4], Hidden Markov Models [1], and Memory-based Learning [21]. Most of the novel approaches employ deep learning models to improve on accuracy and speed compared to previous taggers [5]. Unfortunately, all these models rely on access to large quantities (in the scale of billions) of labeled examples, which normally are difficult to acquire, specially for the Spanish language.

Furthermore, when considering real world conversations, perfect grammar cannot be expected from the human-participant. Errors such as missing letters, lacking punctuation marks, or wrongly spelled entity names could easily cripple any catalog-based approach to NER. Several algorithms could be employed to circumvent this problems, for example, one could spell-check and automatically replace every error or consider every word that is within certain distance from the correct entity as an negligible error. However, this solutions do not cover the whole range of possible human errors, and can be quite complicated to maintain as the catalog increases in size.

In this work we present a semi-supervised model implementation for tagging entities in natural language conversation. By aggregating the information obtained from a POS-tagger with that obtained from a catalog lookup, we ensure the best accuracy and recall from both approaches. Our solution works as follows: first, we look each word coming from the user in our catalog of entities. If the word is not present, we then use a POS-tagger to find the best possible tag for each word in the user's sentence. We use the POS tag set defined by the Penn Treebank [15] with the addition of an ENTITY tag. We trained our POS-tagger with 120-million tagged words from the Spanish wiki-corpus [20] and thousands of real-world conversations, which we manually tagged.

Our approach solves a specific problem in a rather specific situation. We aimed to develop a tool that is both easy to deploy and easy to understand. In a near future, this model would provide NLP-novice programmers and companies with a module that could be quickly incorporated as an agile enhancement for the development of conversational agents focused on providing information on products or services with a well-defined catalog. Examples of this scenario include the list of movies shown in a theater, or different brands, models and contracts offered by a mobile phone company. To the best of the authors knowledge, no similar solution has been proposed for the Spanish language. Further work could also incorporate our methodology into a Maximum Entropy Classifier, an AdaBoost, or even a DeepLearning technique to further improve on the entity detection.

The paper is structured as follows: in Sect. 2 we present the POS tagging problem and a brief background of POS taggers using hidden Markov models. In Sect. 3 we present the obtained corpus and data used along this work. Section 4 presents the methodology for this work, while Sect. 5 shows our main results. Finally, Sect. 6 describes the real-world implementation and compares our work against the industry's current development.

2   Formal Background

2.1   Part of Speech Tagger

In any spoken language, words can be grouped into equivalence classes called parts of speech (POS) [12]. In English and Spanish some examples of POS are *noun, verb* and *articles*. Part-of-speech tagging is the process of marking up a word in a text as corresponding to a particular part of speech, based on both its definition, as well as its context (*i.e.*, its relationship with adjacent and related words in a phrase or sentence) [12]. A POS tagger is a program that takes as input the set of all possible tags in the language (*e.g.*, noun, verbs, adverbs, etc.) and a sentence. Its output is a single best tag for each word [12]. As aforementioned, ambiguity makes this problem non trivial, hence POS-tagging is a problem of disambiguation.

The first stage of our solution follows the classical solution of using a HMM trained in a previously tagged corpus in order to find the best relation between words and tags. Following are the formal definitions used in our work.

Hidden Markov Models. Loosely speaking, a Hidden Markov Model (HMM) is a Markov chain observed in noise [19]. A discussion on Markov chains is beyond the scope of this work, but the reader is referred to [13,18] for further information. The underlying Markov chain, denoted by $\{X_k\}_{k \geq 0}$ is assumed to take values in a finite set. As the name states, it is *hidden*, that is, it is not observable. What is available to the observer is another stochastic process $\{Y_k\}_{k \geq 0}$ linked to the Markov chain in that X_k governs the distribution of Y_k [19]. A simple definition is given by Capp [19] as follows:

Definition 1. *A hidden Markov Model is a bivariate discrete time process* $\{X_k, Y_k\}_{k \geq 0}$ *where* $\{X_k\}$ *is a Markov chain and, conditional on* $\{X_k\}$, $\{Y_k\}$ *is a sequence of independent random variables. The state space of* $\{X_k\}$ *is denoted by* X, *while the set in which* $\{Y_k\}$ *takes its values is denoted by* Y.

We use this definition as it works for our purpose and has the additional benefit of not overwhelming the reader with overly complex mathematical terms. Following this definition, the problem of assigning a sequence of tags to a sequence of words can be formulated as:

Definition 2. *Let Y be the set of all possible POS tags in the language, and $w_1, w_2, w_3, \ldots, w_n$ a sequence of words. The POS-tagging problem can be expressed as the task of finding for each word w_i, the best tag $\hat{y}_i$ from all the possible tags $y_i \in Y$. That is,*

$$\hat{y}_i = \underset{y_i \in Y}{\mathrm{argmax}}\, P(y_i \mid w_i) \tag{1}$$

Given a sequence of words $w_1, w_2, \ldots, w_T$, an equivalent approach to the POS tagging problem is finding the sequence $y_1, y_2, \ldots, y_T$ such that the joint probability is maximized. That is,

$$\hat{y}_1, \hat{y}_2, \ldots, \hat{y}_T = \underset{y_1, y_2, \ldots, y_T}{\mathrm{argmax}}\, P(w_1, w_2, \ldots, w_T, y_1, y_2, \ldots, y_T) \tag{2}$$

We then can assume that the joint probability function 2 is of the form

$$P(w_1, w_2, \ldots, w_T, y_1, y_2, \ldots, y_T) = \prod_{i=1}^{T+1} q(y_i \mid y_{i-2}, y_{i-1}) \prod_{i=1}^{T} e(w_i \mid y_i) \tag{3}$$

with $y_0 = y_{-1} = *$ and $y_{T+1} = STOP$.

The function q determines the probability of observing the tag y_i after having observed the tags y_{i-1} and y_{i-2}. The function e returns the probability that the word w_i has been assigned the label y_i. By maximizing both functions, the joint probability P is maximized. Now we explain how we can obtain both 4 and 5 from a previously tagged corpus.

Equation 3 captures the concept of using word-level trigrams for tagging [7]. Calculating the function q will depend on the two previous tags. Using the definition of conditional probability, we observe that the function q is the ratio between the number of times y_i is observed after y_{i-2}, y_{i-1} and how many times the bigram y_{i-2}, y_{i-1} was observed in the corpus:

$$q(y_i \mid y_{i-2}, y_{i-1}) = \frac{q(y_i, y_{i-2}, y_{i-1})}{q(y_{i-2}, y_{i-1})} \tag{4}$$

Similarly, the probability that w_i is labelled with y_i can be obtained as the frequentist probability observed in the corpus

$$e(x_i \mid y_i) = \frac{\#\ \text{times } x_i \text{ was seen tagged as } y_i}{\#\ \text{times} x_i \text{ occurs in the corpus}} \tag{5}$$

Viterbi Algorithm. The Viterbi algorithm, proposed in 1966 and published the following year by Andrew J. Viterbi, is a dynamic programming algorithm designed to find the most likely sequence of states, called the Viterbi path, that could produce an observed output [23]. It starts with the result of the process (the sequence of outputs) and conducts a search in reverse, in each step discarding every hypothesis that could not have resulted in the outcome.

The algorithm was originally intended for message coding in electronic signals, ensuring that the message will not be lost if the signal is corrupted, by adding redundancy. This is called an error correcting coding. Today, the algorithm is used in a wide variety of areas and situations [8]. Formally, the Viterbi algorithm is defined as follows

Definition 3. *Let* $\lambda = \{X_k, Y_k\}_{k \geq 0}$ *be an HMM with a sequence of observed inputs* $y_1, y_2, \ldots, y_T$. *The sequence of states* $x_1, x_2, \ldots, x_T$ *that produced the outputs can be found using the following recurrence relation:*

$$V_{1,k} = P(y_1 \mid k) \cdot \pi_k \tag{6}$$

$$V_{t,k} = P(y_t \mid k) \cdot \max_{s \in S}(A_{s,k} \cdot V_{t-1,s}) \tag{7}$$

where $V_{t,k}$ *is the probability of the most likely sequence of states that result in the first t observations and has k as a final state, and* $A_{s,k}$ *is the transition matrix for* $\{X_k\}$

The Viterbi path can be found following backward pointers, which keep a reference to each state s used in the second equation. Let $Ptr(k, t)$ be a function that returns the value of s used in the calculation of $V_{t,k}$ if $t > 1$, or k if $t = 1$. Then:

$$s_T = \operatorname*{argmax}_{s \in S}(V_{T,s}) \tag{8}$$

$$s_{t-1} = Ptr(s_t, t) \tag{9}$$

3   Data

We propose an implementation of the Viterbi algorithm to tag every word in a conversation with a real-world dialog system. The resulting HMM corresponds to an observed output $w_1, w_2, \ldots, w_T$ (the words in a conversation), and the states $y_1, y_2, \ldots, y_T$ (the POS-tags for each word). Some simplifications and assumptions allowed us to implement the Markov model starting from a set of data associating each word (or tuple of words) with one (or more) tags, often called a tagged corpus.

For this work, we used the Spanish Wikicorpus [20], a database of more than 120 million words obtained from Spanish Wikipedia articles, annotated with lemma and part of speech information using the open source library FreeLing. Also, they have been sense annotated with the Word Sense Disambiguation algorithm UKB. The tags used and the quantity of corresponding words in the corpus are given in Table 1. We can note that the least common tag in this corpus is dates (**W**), with no examples in the texts used. We suspect this was due an error on our side while parsing the corpus.

Table 1. Number of words per tag in WikiCorpus [20]

tag	POS	count
A	Adjective	4,688,077
R	Adverb	1,901,136
D	Determinant	10,863,584
N	Noun	21,651,297
V	Verb	8,445,600
P	Pronoun	2,832,306
C	Conjunction	3,413,668
I	Interjection	33,803
S	Preposition	11,773,140
F	Punctuation	5,734
Z	Numeral	1,828,042
W	Date	0

3.1 Rare Words

Even with a corpus of 120 million words, it is quite common that conversations with users contain words that never appeared in the training set. We looked into two possible ways of handling these words: replacing all words with their corresponding tokens, and replacing only *rare* words with their respective tokens. A word is considered *rare* if it's total frequency (number of times it appears in the corpus, regardless of its tag) is less than 5. For this work we only used 5 tokens (Table 2).

Table 2. Translation from words to placeholder tokens for handling words that were not observed in training.

Word	Translation (token)
Number with four digits	_4_DIGITS_
Any other number	_NUMBER_
Word contains a hyphen	_HYPHENATED_
Word is a Roman numeral	_ROMAN_
Word's total frequency < 5	_RARE_

4 Methodology

We downloaded the Spanish WikiCorpus in its entirety, transformed its files into a frequency counts file using a MapReduce scheme [6]. Our aggregated file contains the number of times each word appears, and the number of times each word appears with a certain tag (*e.g.*, the number of times "light" appears as noun). We then applied the rules for rare words, as discussed in Sect. 3.1, which yielded two new corpus and frequency files upon which we trained our HMM.

Each of these two models was tested using five rounds of cross-validation over the whole WikiCorpus of 120 million words. For each round, our program read the frequency count file and calculated the most likely tag for each word in the training set. The result of each round was a confusion matrix. For each tag, we obtained measurements of the model's precision, recall, and F-score over that tag.

5   Results

5.1   Replacing only Rare Words with Tokens (VM1)

For words labeled as nouns or names (N), the algorithm had an average precision of 12.15 % and recall of 46.87 %. From the total of 21, 651, 297 names, it only managed to find 10, 147, 962. From the words the algorithm tagged as nouns or names, only 3 out of every 25 was, in fact, a noun or a name. The left side of Table 3 shows the results for this algorithm.

5.2   HMM Replacing Any Word in Table 2 with a Token (VM2)

This model stands out because of an increase in precision and recall in the classification of all tags. This comes as no surprise since the last rule in our translation Table 2 already considers the whole set of translated words in the previous model. The results are shown on the right side of Table 3.

6   Implementing in a Real-World Scenario

Considering the results described in the previous section, we decided to compare and improve model VM2 in a real world scenario. Such opportunity was presented by BlueMessaging Mexico's text-based dialog systems.

BlueMessaging is a private initiative whose sole objective is to connect business and brands with customers anytime anywhere [2]. It has developed several TDS systems, one of such (named CPG) was designed for helping users decide which product best fits their needs. CPG works with a well-defined product catalog, in a specific domain, and interacts with the user through a natural-language text-based chat interface. For example, CPG could help a user select among hundreds of cell phone options by guiding a conversation with the user about the product's details (see Table 4).

During the course of this work, we collected 2, 283 real-world conversations from CPG. This conversations were manually tagged over the same tag set as the wiki corpus. Words that were part of a product's name or brand were tagged as ENTITY. For example, *Samsung Galaxy S3* were labeled as *Samsung/ENTITY*

Table 3. Classification results for VM1 and VM2

	VM1			VM2		
	Precision	Recall	F-score	Precision	Recall	F-score
Adjective (A)	81.28 %	7.23 %	13.28 %	98.13 %	8.72 %	16.01 %
Conjunction (C)	4.87 %	23.85 %	8.08 %	20.64 %	97.71 %	34.04 %
Determinant (D)	7.74 %	32.38 %	12.48 %	23.18 %	98.53 %	37.49 %
Punctuation (F)	1.18 %	11.10 %	2.13 %	7.71 %	60.83 %	13.56 %
Interjection (I)	2.72 %	3.43 %	3.02 %	13.58 %	18.29 %	15.30 %
Noun (N)	12.15 %	46.87 %	19.27 %	23.40 %	95.68 %	37.56 %
Pronoun (P)	4.13 %	22.04 %	6.95 %	18.11 %	92.32 %	30.25 %
Preposition (S)	8.04 %	33.04 %	12.91 %	23.88 %	99.80 %	38.50 %
Adverb (R)	3.04 %	20.26 %	5.28 %	17.03 %	94.20 %	28.81 %
Date (W)	0.00 %	0.00 %	0.00 %	0.00 %	0.00 %	0.00 %
Verb (V)	6.93 %	31.12 %	11.33 %	21.75 %	93.02 %	35.21 %
Numeral (Z)	5.89 %	27.78 %	9.70 %	23.45 %	99.04 %	37.88 %

Table 4. A translated conversation collected from BlueMessaging Mexico's CPG TDS

User	CPG
Hello	Welcome to CPG. Are you looking for a cell phone?
Do you have the iPhone?	Which iPhone are you looking for? 4s, 5s, 6?
6	Would you like the small or the big screen?

Galaxy/ENTITY S3/ENTITY. After this process, we had a new corpus containing 11,617 words (1,566 unique), from which 1,604 represent known entities.

We then tested three new approaches. The baseline was set by CPG's current catalog approach. We then tested our VM2 model, having scored the best across the tests in the last section. Finally, an hybrid approach was used: using both the POS-tagger and the catalog to identify named entities in the dialog. Each approach was tested with 5 rounds of cross-validation on the conversation corpus. Here we present the results for each of the three.

6.1   Catalog-Based Tag Replacement (DICT)

We tested a simple model consisting only of tag substitution using a previously known catalog of entities. Each entry in this dictionary is a word or phrase that identifies an entity, and an entity can be identified by multiple entries. This is useful for compound product names that could be abbreviated by the user. For example: "galaxy fame", "samsung fame" and "samsung galaxy fame" are all references to the same product. In this case, we used a catalog with 148 entries, identifying 45 unique entities.

To make the substitutions, the sentence to be labeled is compared against every entry in the catalog. If the sentence has an entry as a sub-sequence, the words in that sub-sequence are tagged as ENTITY, overriding any tags they might have had.

Using only this approach to tagging sentences, we reached a precision of 93.56 % and recall of 64.04 % for entities. However, this process does not give any information about any other tag, and does not attempt to label words as anything other than entity (Table 5).

Table 5. Classification results for DICT

	Precision	Recall	F1
ENTITY	93.56 %	64.04 %	76.03 %

6.2 Model VM2 Applied to Real-World Conversations (VM2-CPG)

Having tested this model with another corpus, we can use those results to compare its performance on the corpus of conversations, and see whether the model is too specific to the first corpus. Once again, we tested it using 5 rounds of cross-validation, training with subset of the corpus and leaving the rest as a testing set, for each round. The results for this model are shown on the left part of Table 6. With respect to the previous corpus, the model VM2 in this instance had less recall, but a better precision for nouns, determinants, interjections, pronouns, and adverbs.

6.3 VM3 Model with Catalog-Based Tag Replacement (VM2-DICT)

Lastly, we present the results for the model combining the HMM model, replacing words with the appropriate tokens, and tag replacement using a catalog of known entities as shown on the right hand of Table 6. This method is capable of identifying 24 out of every 25 entities.

7 Comparison with Current System

As a final validation for our models, we compared their performance against the system currently in use by BlueMessaging Mexico's platform. We collected an additional 2, 280 actual user conversations. Again, we manually tagged the sentences according to the wikiCorpus tag set with the inclusion of the ENTITY tag. For both systems, we measured precision and recall. For the VM2-DICT model, the standard error was determined as two standard deviations from the results of the 5-fold cross-validation. Figure 1 shows the results obtained for each system. We found a highly significant difference in precision and recall of the two models (t-test $t = 23.0933$, $p < 0.0001$ and $t = 4.8509$, $p < 0.01$), with VM2-DICT having a better performance in both.

Table 6. Classification results for VM2 tested with the CPG corpus and for VM2 with dictionary replacement

	VM2 over CPG			VM2 with DICT		
	Precision	Recall	F1	Precision	Recall	F1
Adjective (A)	22.83 %	65.30 %	33.84 %	26.21 %	64.83 %	37.33 %
Conjunction (C)	76.22 %	66.75 %	71.17 %	75.68 %	68.19 %	71.74 %
Determinant (D)	63.37 %	43.81 %	51.80 %	64.22 %	45.32 %	53.14 %
Punctuation (F)	0.00 %	0.00 %	0.00 %	0.00 %	0.00 %	0.00 %
Interjection (I)	88.79 %	78.37 %	83.25 %	89.77 %	77.61 %	83.08 %
Noun (N)	77.29 %	60.64 %	67.96 %	77.10 %	60.62 %	67.88 %
Pronoun (P)	84.53 %	52.33 %	64.64 %	84.59 %	53.68 %	65.68 %
Preposition (S)	56.12 %	62.35 %	59.07 %	58.07 %	62.06 %	60.00 %
Adverb (R)	73.40 %	53.86 %	62.13 %	75.63 %	52.45 %	61.94 %
Date (W)	0.00 %	0.00 %	0.00 %	0.00 %	0.00 %	0.00 %
Verb (V)	84.48 %	54.13 %	65.98 %	84.06 %	53.87 %	65.66 %
Numeral (Z)	0.00 %	0.00 %	0.00 %	0.00 %	0.00 %	0.00 %
Entity (ENTITY)	94.66 %	60.47 %	73.80 %	96.53 %	91.31 %	93.85 %

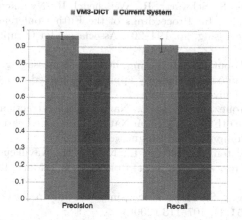

Fig. 1. Comparison between model VM2-dict and BlueMessaging Mexico's current system. The error bars show two standard deviations from the mean.

8 Discussion

We presented a semi-supervised approach to named entity recognition for the Spanish language. We noted that previous works on Spanish entity recognition used either a catalog-based approach or machine learning models with a wide range of success. Our model leverages on both approaches, improving both the accuracy and recall of a real-world implementation.

With a real-world implementation in mind, where solutions are measured by their tangible results and how easily they can be adapted to existing production schemes, we designed our system to be an assemblage of well-studied techniques requiring only minor modifications. We believe that our solution would allow for quick development and deployment of text-based dialog systems in Spanish. In a further work, this assemblage of simple techniques could evolve into more robust solutions, for example by exploring conditional random fields in order to replace some of the hidden Markov model assumptions. Moreover, implementations for deep learning in Spanish language might be possible in a future, as more researchers work in developing tagged corpus.

Acknowledgments. The authors would like to thank the anonymous reviewers for their valuable comments and suggestions. We would also like to thank Consejo Nacional de Ciencia y Tecnología (CONACYT), BlueMessaging Mexico S.A.P.I. de C.V., and to the Asociación Mexicana de Cultura A.C. for all their support. Specially we would like to mention Andrés Rodriguez, Juan Vera and David Jiménez for their wholehearted assistance.

References

1. Bikel, D.M., Miller, S., Schwartz, R., Weischedel, R.: Nymble: a high-performance learning name-finder. In: Proceedings of the Fifth Conference on Applied Natural Language Processing, pp. 194–201. Association for Computational Linguistics (1997)
2. BlueMessaging: About BlueMessaging. http://bluemessaging.com/about/
3. Borthwick, A.: A maximum entropy approach to named entity recognition. Ph.D. thesis, New York University (1999)
4. Carreras, X., Marquez, L., Padró, L.: Named entity extraction using adaboost. In: Proceedings of the 6th Conference on Natural Language Learning, vol. 20, pp. 1–4. Association for Computational Linguistics (2002)
5. Collobert, R., Weston, J., Bottou, L., Karlen, M., Kavukcuoglu, K., Kuksa, P.: Natural language processing (almost) from scratch. J. Mach. Learn. Res. **12**, 2493–2537 (2011)
6. Dean, J., Ghemawat, S.: MapReduce: simplified data processing on large clusters. Commun. ACM **51**(1), 107–113 (2008)
7. Figueira, A.P.F.: 5. the viterbi algorithm for HMMs - part i (2013). Available online at http://www.youtube.com/watch?v=sCO2riwPUTA
8. Forney Jr., G.D.: The viterbi algorithm: a personal history, April 2005. arXiv:cs/0504020, http://arxiv.org/abs/cs/0504020, arXiv: cs/0504020
9. Goddeau, D., Meng, H., Polifroni, J., Seneff, S., Busayapongchai, S.: A form-based dialogue manager for spoken language applications. In: Proceedings of the Fourth International Conference on Spoken Language ICSLP 1996, vol. 2, pp. 701–704. IEEE (1996)
10. Grishman, R., Sundheim, B.: Message understanding conference-6: a brief history. In: Proceedings of the 16th Conference on Computational Linguistics COLING 1996, vol. 1, pp. 466–471. Association for Computational Linguistics, Stroudsburg, PA, USA (1996)

11. Hurtado, L.F., Griol, D., Sanchis, E., Segarra, E.: A stochastic approach to dialog management. In: IEEE Workshop on Automatic Speech Recognition and Understanding 2005, pp. 226–231. IEEE (2005)
12. Jurafsky, D., Martin, J.H.: Speech and Language Processing. Pearson Education India, Noida (2000)
13. Karlin, S., Taylor, H.E.: A First Course in Stochastic Processes. Academic Press, New York (2012)
14. Kozareva, Z.: Bootstrapping named entity recognition with automatically generated gazetteer lists. In: Proceedings of the Eleventh Conference of the European Chapter of the Association for Computational Linguistics: Student Research Workshop, pp. 15–21. Association for Computational Linguistics (2006)
15. Marcus, M.P., Marcinkiewicz, M.A., Santorini, B.: Building large annotated corpus english: the penn treebank. Comput. Linguist. **19**(2), 313–330 (1993)
16. Misu, T., Georgila, K., Leuski, A., Traum, D.: Reinforcement learning of question-answering dialogue policies for virtual museum guides. In: Proceedings of the 13th Annual Meeting of the Special Interest Group on Discourse and Dialogue, p. 8493. Association for Computational Linguistics (2012)
17. Nadeau, D., Sekine, S.: A survey of named entity recognition and classification. Lingvist. Investig. **30**(1), 3–26 (2007)
18. Norris, J.R.: Markov Chains. Cambridge Series in Statistical and Probabilistic Mathematics. Cambridge University Press, New York (1999)
19. Capp, O., Moulines, E., Rydn, T.: Inference in Hidden Markov Models. Springer Series in Statistics. Springer, New York (2005)
20. Reese, S., Boleda, G., Cuadros, M., Padr, L., Rigau, G.: Wikicorpus: a word-sense disambiguated multilingual wikipedia corpus. In: Proceedings of 7th Language Resources and Evaluation Conference (LREC 2010). La Valleta, Malta, May 2010
21. Sang, T.K., Erik, F.: Memory-based named entity recognition. In: Proceedings of the 6th Conference on Natural Language Learning, vol. 20, pp. 1–4. Association for Computational Linguistics (2002)
22. Seneff, S.: Response planning and generation in the mercury flight reservation system. Comput. Speech Lang. **16**(3–4), 283–312 (2002)
23. Viterbi, A.: Error bounds for convolutional codes and an asymptotically optimum decoding algorithm. IEEE Trans. Inf. Theory **13**(2), 260–269 (1967)
24. Williams, J.D., Young, S.: Partially observable Markov decision processes for spoken dialog systems. Comput. Speech Lang. **21**(2), 393–422 (2007)
25. Zue, V., Seneff, S., Glass, J.R., Polifroni, J., Pao, C., Hazen, T.J., Hetherington, L.: Jupiter: a telephone-based conversational interface for weather information. IEEE Trans. Speech Audio Process. **8**(1), 85–96 (2000)

Patterns Used to Identify Relations in Corpus Using Formal Concept Analysis

Mireya Tovar[1,2]([✉]), David Pinto[1], Azucena Montes[2,3], Gabriel Serna[2],
and Darnes Vilariño[1]

[1] Faculty Computer Science, Benemérita Universidad Autónoma de Puebla,
Puebla, Mexico
{mtovar,dpinto,darnes}@cs.buap.mx
[2] Centro Nacional de Investigación y Desarrollo Tecnológico (CENIDET),
Cuernavaca, Mexico
{gabriel,amontes}@cenidet.edu.mx
[3] Engineering Institute, Universidad Nacional Autónoma de Mexico,
Mexico City, Mexico

Abstract. In this paper we present an approach for the automatic identification of relations in ontologies of restricted domain. We use the evidence found in a corpus associated to the same domain of the ontology
for determining the validity of the ontological relations. Our approach
employs formal concept analysis, a method used for the analysis of data,
but in this case used for relations discovery in a corpus of restricted
domain. The approach uses two variants for filling the incidence matrix
that this method employs. The formal concepts are used for evaluating
the ontological relations of two ontologies. The performance obtained
was about 96 for taxonomic relations and 100 % for non-taxonomic relations, in the first ontology. In the second it was about 92 % for taxonomic
relations and 98 % for non-taxonomic relations.

Keywords: Formal concept analysis · Ontology · Semantic relations

1 Introduction

There is a huge amount of information that is uploaded every day to the World
Wide Web, thus arising the need for automatic tools able to understand the
meaning of such information. However, one of the central problems of constructing such tools is that this information remains unstructured nowadays, despite
the effort of different communities for giving a semantic sense to the World
Wide Web. In fact, the Semantic Web research direction attempts to tackle this
problem by incorporating semantic to the web data, so that it can be processed
directly or indirectly by machines in order to transform it into a data network
[1]. For this purpose, it has been proposed to use knowledge structures such as
"ontologies" for giving semantic and structure to unstructured data. An ontology, from the computer science perspective, is "an explicit specification of a
conceptualization" [2].

© Springer International Publishing Switzerland 2015
J.A. Carrasco-Ochoa et al. (Eds.): MCPR 2015, LNCS 9116, pp. 236–245, 2015.
DOI: 10.1007/978-3-319-19264-2_23

Ontologies can be divided into four main categories, according to their generalization levels: generic ontologies, representation ontologies, domain ontologies, and application ontologies. Domain ontologies, or ontologies of restricted domain, specify the knowledge for a particular type of domain, for example: medical, tourism, finance, artificial intelligence, etc. An ontology typically includes the following components: classes, instances, attributes, relations, constraints, rules, events and axioms.

In this paper we are interested in the process of discovering and evaluating ontological relations, thus, we focus our attention on the following two types: taxonomic relations and/or non-taxonomic relations. The first type of relations are normally referred as relations of the type "is-a" (hypernym/hyponymy or subsumption).

There are plenty of research works in literature that addresses the problem of automatic construction of ontologies. The major of those works evaluate manually created ontologies by using a gold standard, which in fact, it is supposed to be manufactured by an expert. By using this approach, it is assumed that the expert has created the ontology in a correct way, however, there is not a guarantee of such thing. Thus, we consider very important to investigate a manner to automatically evaluate the quality of this kind of resources, which are continuously been used in the framework of the semantic web.

Our approach attempts to find evidence of the relations to be evaluated in a reference corpus (associated to the same domain of the ontology) using formal concept analysis. To our knowledge, the use of formal concept analysis in the automatic discovery of ontological relations has nearly been studied in the literature. There are, however, other approaches that may be considered in our state of the art, because they provide mechanisms for discovering ontological relations, usually in the construction of ontologies framework.

In [3], for example, it is presented an approach for the automatic acquisition of taxonomies from text in two domains: tourism and finance. They use different measures for weighting the contribution of each attribute (such as conditional probability and pointwise mutual information (PMI)).

In [4] are presented two experiments for building taxonomies automatically. In the first experiment, the attribute set includes a group of sememes obtained from the HowNet lexicon, whereas in the second the attributes are a basically set of context verbs obtained from a large-scale corpus; all this for building an ontology (taxonomy) of the Information Technology (IT) domain. They use five experts of IT for evaluating the results of the system, reporting a 43.2 % of correct answers for the first experiment, and 56.2 % of correct answers for the second one.

Hele-Mai Haav [5] presents an approach to semi-automatic ontology extraction and design by usign Formal Concept Analysis combined with a rule-based language, such as Horn clauses, for taxonomic relations. The attributes are noun-phrases of a domain-specific text describing a given entity. The non-taxonomic relations are defined by means of predicates and rules using Horn clauses.

In [6] it is presented an approach to derive relevance of "events" from an ontology of the event domain. The ontology of events is constructed using Formal Concept Analysis. The event terms are mapped into objects, and the name entities into attributes. These terms and entities were recovered from an corpus in order to build the incidence matrix.

From the point of view of the evaluation of the ontology, some of the works mentioned above perform an evaluation by means of gold standard [3] in order to determine the level of overlapping between the ontology that has been built automatically and the manually constructed ontology (called gold standard).

Another approach for evaluating ontologies is by means of human experts as it is presented in [4].

In our approach we used a typed dependency parser for determining the verb of a given sentence, which is associated to the ontological concepts of a triple from which the relation component require to be validated through a retrieval system. The ontological concepts together with their associated verbs are introduced, by means of an incidence matrix, to Formal Concept Analysis (FCA) system. The FCA method allow us to find evidence of the ontological relation to be validated by searching the semantic implicit in the data. We use several selection criteria to determine the veracity of the ontological relation.

We do not do ontological creation, but we use formal concept analysis to identify the ontological relation in the corpus and we evaluate it.

In order to validate our approach, we employ a manual evaluation process by means of human experts.

The remaining of this paper is structured as follows: Sect. 2 describes more into detail the theory of formal concept analysis. In Sect. 3 we present the approach proposed in this paper. Section 4 shows and discusses the results obtained by the presented approach. Finally, in Sect. 5 the findings and the future work are given.

2   Formal Concept Analysis

Formal Concept Analysis (FCA) is a method of data analysis that describes relations between a particular set of objects and a particular set of attributes [7]. FCA was firstly introduced by Rudolf Wille in 1992 [8] as an field of research based on a model of set theory to concepts and concept hierarchies which proposes a formal representation of conceptual knowledge [8]. FCA allows data analysis methods for the formal representation of conceptual knowledge. This type of analysis produces two kinds of output from the input data: a concept lattice and a collection of attribute implications. The concept lattice is a collection of formal concepts of the data, which are hierarchically ordered by a subconcept-superconcept relation. The attribute implication describes a valid dependency in the data. FCA can be seen as a conceptual clustering technique that provides intentional descriptions for abstract concepts. From a philosophical point of view, a concept is a unit of thoughts made up of two parts: the extension and the intension [9]. The extension covers all objects or entities beloging to this concept, whereas the intension comprises all the attributes or properties valid for all those objects.

FCA begins with the primitive idea of a context defined as a triple (G, M, I), where G and M are sets, and I is a binary relation between G and M (I is the incidence of the context); the elements of G and M are named objects and attributes, respectively.

A pair (A, B) is a formal concept of (G, M, I), as defined in [3], iff $A \subseteq G$, $B \subseteq M$, $A' = B$ and $A = B'$. In other words, (A, B) is a formal concept if the attribute set shared by the objects of A are identical with those of B; and A is the set of all the objects that have all attributes in B. A is the extension, and B is the intension of the formal concept (A, B).

A' is the set of all attributes common to the objects of A, B' is the set of all objects that have all attributes in B. For $A \subseteq G$, $A' = \{m \in M | \forall g \in A : (g, m) \in I\}$, and dually, for $B \subseteq M$, $B' = \{g \in G | \forall m \in B : (g, m) \in I\}$

The formal concepts of a given context are ordered by the relation of sub-concept - superconcept definided by:

$$(A_1, B_1) \leq (A_2, B_2) \Leftrightarrow A_1 \subseteq A_2 (\Leftrightarrow B_2 \subseteq B_1)$$

FCA is a tool applied to various problems such as: hierarchical taxonomies, information retrieval, data mining, etc., [7]. In this case, we use this tool for identifying ontological relations of restricted domain.

3   Approach for Evaluating Semantic Relations

We employ the theory of FCA to automatically identify ontological relations in a corpus of restricted domain. The approach considers two variants in the selection of properties or attributes for building the incidence matrix that is used by the FCA method for obtaining the formal concepts.

The difference between the two variants is the type of syntactic dependencies parser used in the preprocessing phase for getting the properties.

The first variant uses the minipar tagger [10], whereas the second variant employs the Stanford tagger [11]. For each variant, we selected manually a set of dependency relations in order to extract verbs from each sentence of the corpus that contains an ontology concept. These verbs are then used as properties or attributes in the incidence matrix.

The Stanford dependencies are triples containing the name of the relation, the governor and the dependent. Examples of these triples are shown in Table 1. For the purpose of our research, from each triple we have selected the governor $(p = 1)$, the dependent $(p = 2)$ or both $(p = 1, 2)$ as attributes of the incidence matrix.

In the case of the minipar parser, we use the pattern **C:i:V** for recovering the verbs of the sentence. The grammatical categories that made up the pattern follows: C is a clause, I is an inflectional phrase, and V is a verb or verbal phrase. Some examples of triples recovered from the sentences are shown in Table 2.

Table 1. Dependency relations obtained using the Stanford dependency parser

Relation name	p	Meaning	Example
nsubj	1	Nominal subject	nsubj(specialized, research)
prep	1	Prepositional modifier	prep_into(divided, subfields)
root	2	Root of the sentence	root(ROOT, give)
acomp	1	Adjectival complement	acomp(considered, feasible)
advcl	1,2	Adverbial clause modifier	advcl(need, provide)
agent	1	Agent complement of a passive verb	agent(simulated, machine)
aux	1,2	Auxiliar verb	aux(talked, can)
auxpass	1,2	Passive auxiliar	auxpass(used, is)
cop	1,2	Copula	cop(funded, is)
csubj	2	Clausal subject	csubj(said, having)
csubjpass	1,2	Clausal passive subject	csubjpass(activated, assuming)
dobj	1	Direct object of a verbal phrase	dobj(create, system)
expl	1	Expletive	expl(are, there)
iobj	1	Indirect object	iobj(allows, agent)
nsubjpass	1	Passive nominal subject	nsubjpass(embedded, agent)
parataxis	2	Parataxis	parataxis(Scientist, said)
pcomp	2	Prepositional complement	pcomp(allow, make)
prepc	1	Prepositional clausal modifier	prepc_like(learning, clustering)
prt	1,2	Phrasal verb particle	prt(find, out)
tmod	1	Temporal modifier	tmod(take, years)
vmod	2	Reduced non-finite verbal modifier	vmod(structure, containing)

Table 2. Triples obtained by the Minipar parser

Triples
fin C:i:VBE be
inf C:i:V make
fin C:i:V function

The approach proposed in this paper involves the following three phases:

1. Pre-processing stage. The reference corpus is split into sentences, and all the information (ontology and the sentences) are normalized. In this case, we use the TreeTagger PoS tagger for obtaining the lemmas [12]. An information retrieval system is employed for filtering those sentences containing information referring to the concepts extracted from the ontology. The ontological relations are also extracted from the ontology[1]. Thereafter, we apply the syntactic dependency parser for each sentence associated to the ontology

---

[1] We used Jena for extracting concepts and ontological relations (http://jena.apache.org/).

concepts. In order to extract the verbs from these sentences, we use the patterns shown in Table 3 for each syntactic dependency parser, and each type of ontological relation.

By using this information together with the ontology concepts, we construct the incidence matrix that feed the FCA system.

2. FCA system. We used the sequential version of FCALGS[2] [13]. The input for this system is the incidence matrix with the concepts identified as objects and the verbs identified as attributes. The output is the formal concepts list.

3. Identification of ontological relations. The concepts that made up the triple in which the ontological relation is present are searched in the formal concepts list obtained by the FCA system. The approach assigns a value of 1 (one) if the pair of concepts of the ontological relation exists in the formal concept, otherwise it assigns a zero value. We consider the selection criteria shown in the third column of Table 3 for each type of ontological relation.

As can be seen, in the Stanford approach we have tested three different selection criteria based on the type of verbs to be used. In "stanford$_1$", we only selected the verbs "to be" and "include" that normally exists in lexico-syntactic patterns of taxonomic relations [14]. On the other hand, in "stanford$_3$ we only selected the verbs that exist in the ontological relation.

4. Evaluation. Our approach provides a score for evaluating the ontology by using the accuracy formulae: Accuracy(ontology) $= \frac{|S(R)|}{|R|}$, where $|S(R)|$ is the total number of relations from which our approach considers that exist evidence in the reference corpus, and $|R|$ is the number of semantic relations in the ontology to be evaluated. For measuring this approach, we compare the results obtained by our approach with respect to the results obtained by human experts.

4   Experimental Results

In this section we present the results obtained in the experiments carried out. Firstly, we present the datasets, the results obtained by our approach aforementioned follow; finally, the discussion of these results are given.

4.1   Dataset

We have employed two ontologies, the first is of the Artificial Intelligence (AI) domain and the second is of the standard e-Learning SCORM domain (SCORM)[3] [15] for the experiments executed. In Table 4 we present the number of concepts (C), taxonomic relations (TR) and non-taxonomic relations (NT) of the ontologies evaluated. The characteristics of their reference corpus are also given in the same Table: number of documents (D), number of tokens (T), vocabulary dimensionality (V), and the number of sentences filtered (O) by the information retrieval system (S).

---

[2] http://fcalgs.sourceforge.net/.

[3] The ontologies together with their reference corpus can be downloaded from http://azouaq.athabascau.ca/goldstandards.htm.

Table 3. Patterns used by each variant

Variant	Pattern	Type of selection	Type of relation
minipar	C:i:V *	All verbs recovered	taxonomic, non-taxonomic
stanford$_1$	root(*,*), cop(*,*)	Only the verbs *to be* and *include*	taxonomic
stanford$_2$	nsubj(*,-), prep(*,-), root(*,*), dobj(*,-), acomp(*,-), advcl(*,*), agent(*,-), aux(*,*), auxpass(*,*), cop(*,*), csubj(-,*), csubjpass(*,-), dobj(*,-), expl(*,-), iobj(*,-), cop(*,*), nsubjpass(*,-), parataxis(-,*), pcomp(-,*), prepc(*,-), prt(*,*), tmod(*,-), vmod(-,*)	All verbs recovered	non-taxonomic
stanford$_3$		Only the verbs present in the ontological relations	non-taxonomic

Table 4. Datasets

Domain	Ontology			Reference corpus				
	C	TR	NT	D	T	V	O	S
AI	276	205	61	8	11,370	1,510	475	415
SCORM	1,461	1,038	759	36	34,497	1,325	1,621	1,606

4.2 Obtained Results

As we mentioned above, we validated the ontology relations by means of human expert's judgements. This manual evaluation was carried out in order to determine the performance of our approach, and consequently, the quality of the ontology.

Table 5 shows the results obtained by the approach presented in this paper when the ontologies are evaluated. We used the accuracy criterion for determining the quality of the taxonomic relations. The second column presents two variants for identifying the taxonomic relations. The last three columns indicate the quality (Q) of the system prediction according to three different human experts (E_1, E_2 and E_3). The third column shows the quality obtained by the approach for each type of variant. Table 6 shows the results obtained by the approach when the non-taxonomic relations are evaluated.

Table 5. Accuracy of the ontologies, and quality of the system prediction for taxonomic relations

Domain	Variation	$Accuracy$	$Q(E_1)$	$Q(E_2)$	$Q(E_3)$	$Average$
AI	minipar	0.96	0.90	0.85	0.94	0.90
	stanford$_1$	0.61	0.57	0.56	0.60	0.58
SCORM	minipar	0.89	0.65	0.75	0.64	0.68
	stanford$_1$	0.64	0.49	0.47	0.50	0.49

Table 6. Accuracy of the ontologies and quality of the system prediction for non-taxonomic relations

Domain	Variation	$Accuracy$	$Q(E_1)$	$Q(E_2)$	$Q(E_3)$	$Average$
AI	minipar	0.93	0.80	0.86	0.89	0.85
	stanford$_2$	1.00	0.87	0.92	0.95	0.91
	stanford$_3$	0.95	0.82	0.91	0.91	0.87
SCORM	minipar	0.96	0.85	0.89	0.94	0.89
	stanford$_2$	0.99	0.87	0.89	0.97	0.91
	stanford$_3$	0.90	0.83	0.83	0.90	0.85

Table 7. Accuracy given to the ontologies

Domain	Relation type	Variante	Accuracy
AI	Taxonomic	minipar	96.59 %
		stanford$_1$	73.17 %
	Non-taxonomic	minipar	95.08 %
		stanford$_2$	100.00 %
		stanford$_3$	96.72 %
SCORM	Taxonomic	minipar	92.49 %
		stanford$_1$	64.45 %
	Non-taxonomic	minipar	96.18 %
		stanford$_2$	98.95 %
		stanford$_3$	91.44 %

The results presented here were obtained with a subset of sentences associated to the ontological relations for the AI ontology because of the great effort needed for manually evaluate their validity. In the case of SCORM ontology, we only evaluate the 10 % of the ontological relations and a subset of sentences associated to these. Therefore, in order to have a complete evaluation of the two type of ontological relations, we have calculated their accuracy, but in this case considering all the sentences associated to the relations to be evaluated. Table 7 shows the variantes used for evaluating the ontological relations and the accuracy assigned to each type of relation (*Accuracy*).

As can be seen, the approach obtained a better accuracy for non-taxonomic relations than for taxonomic ones. This result is obtained because the approach is able to associate the verbs that exist in both, the relation and the domain corpus, by means of the FCA method. Therefore, when non-taxonomic relations are evaluated, the approach has more opportunity to find evidence of their validity.

5   Conclusion

In this paper we have presented an approach based on FCA for the evaluation of ontological relations. In summary, we attempted to look up for evidence of the ontological relations to be evaluated in reference corpora (associated to the same domain of the ontology) by using formal concept analysis. The method of data analysis employed was tested by using two types of variants in the selection of properties or attributes for building the incidence matrix needed by the FCA method in order to obtain the formal concepts. The main difference between these two variants is the type of syntactic dependency parser used in the pre-processing phase when obtaining the data properties (Stanford vs. minipar). The Stanford variant was more accurate than the minipar one; actually, the minipar variant obtained a good accuracy for the two types of relations evaluated (taxonomic and non-taxonomic) in AI ontology, whereas the Stanford variant obtained the best results for the non-taxonomic relations. The minipar variant, on the other hand, is quite fast in comparison with the Stanford one.

According to the results presented above, the current approach for evaluating ontological relations obtains an accuracy of 96 % for taxonomic relations, and 100 % for non-taxonomic relations of the AI ontology. In the case of the SCORM ontology, our approach obtains an accuracy of 92 % for taxonomic relations, and 98 % for non-taxonomic relations. Even if these results determine the evidence of the target ontological relations in the corresponding reference corpus, the same results should be seen in terms of the ability of our system for evaluating ontological relations. In other words, the results obtained by the presented approach show, in some way, the quality of the ontologies.

We have observed that the presented approach may have future in the evaluation of ontologies task, but we consider that there still more research that need to be done. For example, as future work, we are interested in analyzing more into detail the reasons for which the approach does not detect 100 % of the ontological relations that have some kind of evidence in the reference corpus.

References

1. Solís, S.: La Web Semántica. Lulu Enterprises Incorporated (2007)
2. Gruber, T.R.: Towards Principles for the Design of Ontologies Used for Knowledge Sharing. In: Guarino, N., Poli, R. (eds.) Formal Ontology in Conceptual Analysis and Knowledge Representation. Kluwer Academic Publishers, Deventer (1993)
3. Cimiano, P., Hotho, A., Staab, S.: Learning concept hierarchies from text corpora using formal concept analysis. J. Artif. Int. Res. **24**(1), 305–339 (2005)

4. Li, S., Lu, Q., Li, W.: Experiments of ontology construction with formal concept analysis. In: ren Huang, C., Calzolari, N., Gangemi, A., Lenci, A., Oltramari, A., Prevot, L., (eds.) Ontology and the Lexicon, pp. 81–97. Cambridge University Press, New york (2010). Cambridge Books Online

5. Haav, H.M.: A semi-automatic method to ontology design by using FCA. In: Snsel, V., Belohlvek, R., (eds.) CLA. CEUR Workshop Proceedings, vol. 110. CEUR-WS.org (2004)

6. Xu, W., Li, W., Wu, M., Li, W., Yuan, C.: Deriving event relevance from the ontology constructed with formal concept analysis. In: Gelbukh, A. (ed.) CICLing 2006. LNCS, vol. 3878, pp. 480–489. Springer, Heidelberg (2006)

7. Belohlávek, R.: Introduction to formal context analysis. Technical report, Department of Computer Science. Palack y University, Olomouk, Czech Republic (2008)

8. Wille, R.: Concept lattices and conceptual knowledge systems. Comput. Math. Appl. **23**(6–9), 493–515 (1992)

9. Wolf, E.K.: A first course in formal concept analysis. In: Faulbaum, F. (ed.) Soft-Stat 1993 Advances in Statistical Software 4, pp. 429–438. Gustav Fischer Verlag (1993)

10. Lin, D.: Dependency-based evaluation of minipar. In: Proceedings of Workshop on the Evaluation of Parsing Systems. Granada (1998)

11. de Marneffe, M.C., MacCartney, B., Manning, C.D.: Generating typed dependency parses from phrase structure trees. In: LREC (2006)

12. Schmid, H.: Probabilistic part-of-speech tagging using decision trees. In: Proceedings of the International Conference on New Methods in Language Processing, Manchester, UK (1994)

13. Krajca, P., Outrata, J., Vychodil, V.: Parallel recursive algorithm for FCA. In: Proceedings of the Sixth International Conference on Concept Lattices and Their Applications, vol. 433, pp. 71–82. CEUR-WS.org, Olomouc (2008)

14. Tovar, M., Pinto, D., Montes, A., González, G., Vilariño, D., Beltrán, B.: Use of lexico syntactic patterns for the evaluation of taxonomic relations. In: Martínez-Trinidad, J.F., Carrasco-Ochoa, J.A., Olvera-Lopez, J.A., Salas-Rodríguez, J., Suen, C.Y. (eds.) MCPR 2014. LNCS, vol. 8495, pp. 331–340. Springer, Heidelberg (2014)

15. Zouaq, A., Gasevic, D., Hatala, M.: Linguistic patterns for information extraction in ontocmaps. In: Blomqvist, E., Gangemi, A., Hammar, K., del Carmen Suárez-Figueroa, M., (eds.) WOP. CEUR Workshop Proceedings, vol. 929. CEUR-WS.org (2012)

Improving Information Retrieval Through a Global Term Weighting Scheme

Daniel Cuellar$^{(\boxtimes)}$, Elva Diaz, and Eunice Ponce-de-Leon-Senti

Computer Sciences Department, Basic Sciences Center, Universidad Autónoma de Aguascalientes (UAA), 940 Universidad Ave., Ciudad Universitaria, 20131 Aguascalientes, Aguascalientes, México
cuellar_garrido@hotmail.com, elvitad@yahoo.com, eponce@correo.uaa.mx

Abstract. The output of an information retrieval system is an ordered list of documents corresponding to the user query, represented by an input list of terms. This output relies on the estimated similarity between each document and the query. This similarity depends in turn on the weighting scheme used for the terms of the document index. Term weighting then plays a big role in the estimation of the aforementioned similarity. This paper proposes a new term weighting approach for information retrieval based on the marginal frequencies. Consisting of the global count of term frequencies over the corpus of documents, while conventional term weighting schemes such as the normalized term frequency takes into account the term frequencies for particular documents. The presented experiment shows the advantages and disadvantages of the proposed retrieval scheme. Performance measures such as precision and recall and F-Score are used over classical benchmarks such as CACM to validate the experimental results.

Keywords: Information retrieval · Indexing · Vector space model · Term weighting · Marginal distribution · Weighting scheme

1 Introduction

Information retrieval is the final step of the process of generating, storing, reviewing and indexing files and documents. Its main objective consists on identifying and retrieving documents containing information relevant to the domain of knowledge of a certain search criteria. For an automatic retrieval system this is basically a list of ordered documents corresponding to a user query. The retrieval effectiveness depends on how this documents are ordered in relation to the relevance of the user's query. The most used method for measuring this retrieval effectiveness has been the precision and recall measure and its derivatives such as F-Score and ROC. *"Recall is the proportion of the relevant documents that have been retrieved, while precision is the proportion of retrieved documents that are relevant"* [1]. Information retrieval task is done over a representation of the document collection (corpus) called index. The most common corpus representation is known as a term document matrix or TDM [2]. A TDM is a

© Springer International Publishing Switzerland 2015
J.A. Carrasco-Ochoa et al. (Eds.): MCPR 2015, LNCS 9116, pp. 246–257, 2015.
DOI: 10.1007/978-3-319-19264-2_24

table that stores the frequency of terms of a thesaurus against the list of documents that contain such terms.

Once the information has been stored in a TDM, it can be compressed or parsed (change the format of the TDM to another form of corpus representation) in order to suit a retrieval system necessities. While the quality of the recovery task depends directly on the information retrieval system and how this system uses its index (the retrieval model), an ordered representation of the information and several indexing techniques used at the indexing phase can improve the recovery effectiveness [3]. The problems with information retrieval models are their inability to discriminate the knowledge domain of the user query, its stiffness to handle typographical errors, the presentation in order of importance of the retrieved documents and the synonymy and polysemy. All these obstacles have been circumvented partially by retrieval systems in some restricted way. The Vector Space Model or VSM for instance [4], one of the most widely used information retrieval algorithms gives exceptional results but lacks in the drawbacks exposed earlier. It assumes term independence since it estimates the similarity between documents and queries by converting the frequency of terms per document and per query into vectors of an orthogonal vector space in the Euclidean space of real numbers, then calculates the cosine of the angle between pairs of them. In such case the similarity between a document and a query is the angle that separates each other, the smaller the angle, higher similarity and vice versa. This information is used then to build the ranked list of documents for each query. The VSM establishes a measure of similarity between documents. Weights establish a way to praise and give more importance to certain terms, which can increase the similarity measure of the retrieval system. Which weighting criteria makes a term improve or reduce its importance is under study and some of the most relevant in the field are discussed in [5]. In the practice, the use of this weighting schemes has improved the accuracy and similarity between the user queries and the documents retrieved [6]. We could see these weights as parameters of the retrieval problem, and by this means, altering or modifying the frequencies of terms in a TDM has proven to affect strongly the outcome of the retrieval model. It is noteworthy that actual weighting schemes in literature [7, 8] applies to the TDM in an individual way, weighting the terms in base of the frequencies presented per document, and not in a global way, taking into account the global count of terms from a corpus. Is recommended in [9, 10] that the terms must be taken by document, because the greater the count of a term in a document, the higher that term represents that document. Also noticing that including the terms of the complete corpus increases the dispersion of the TDM, this is, the quantity of terms whose counting is equal to zero, affecting the disparity weight estimation of the non-zero ones. In contrast, not taking global weightings sets aside the relationships between terms and documents, reinforcing the independence between terms of the index and the "bag of words" concept, which states that *"the terms of a document is represented as the bag (multiset) of its words, disregarding grammar and even word order but keeping multiplicity"* [11]. For that reason we introduce in this paper a new weighting approach by appending a new row to the TDM to include the marginal term sum and using a classic weighting scheme such as the *ntf* [7] on it, applying the same *ntf* measure in a global way, in which the term to appear the most in all the documents represents the document collection and not only a document in question. Section 2 gives a background explanation of the concepts behind weighting schemes, Sect. 3 explains the methodology for this approach and Sect. 4 presents the experimental design in which we successfully demonstrate the viability and effectiveness of the scheme.

2   Conceptual Fundaments on Term Weighting

In 1949, George Kingsley Zipf noted that some words in a language are used more often in comparison with the vast quantity of words over the same language. When he ranked the words in order of use for several languages, a striking pattern emerged. The most used word was always used as twice as often as the second one, three times as often as the third one and so on. He called this the "rank vs. frequency rule" [12], later on was become to be known as the "Zipf Law", an empirical law. Zipf's law states that in a given language, the frequency of occurrence of different words follows a distribution that can be approximated by:

$$P_n \sim \frac{1}{n^a} \tag{1}$$

Where n is the ordered nth word and the exponent a is a parameter of the distribution with a value close to 1. This is the formulation of the "Zipfian distribution" which Zipf himself tried to give an explanation for such peculiar behaviour by proposing "the principle of least effort" which states that *"neither speakers nor hearers using a given language want to work any harder than necessary to reach understanding"*, and such process eventually leads to the disuse of certain words and to the observed Zipf distribution. In 1958, Hans Peter Luhn, a computer scientist ahead of the information retrieval research division at IBM was intending to create a way to index and summarize the content of several scientific documents. He faced the problem of differentiating documents to classify afterwards. He had to ask himself what makes a document different, not only in structure but in content as well. He proposed that the differences reside not only in the appearance of certain words, but in their frequencies as well. He called this "the resolving power of words" and stated that *"the frequency of word occurrence in an article furnishes a useful measurement of word significance"* [13].

His observations lead to the intuitive idea of that common words and rare words have little discriminative power between documents, and that those between them describes better the content of the documents. This seems true in the sense that most common words comprises daily ordinary words such as pronouns, conjunctions and prepositions and rare words could actually stand out a document but not having similar coincidences with other documents for an effective comparison. Given this reasoning, Luhn took the Zipfian distribution to establish upper and lower cut-off points over it for determining terms with a high resolving power. Terms appearing outside these points are considered words with low distinguishing power. In 1968 Claire Kelly Schultz described and popularized Zipf's and Luhn's work in [14], and represented this approach with a Gaussian Bell curve over a Zipfian distribution in which medium frequent terms acquire higher resolving power proportional to the Gaussian curve, being this the first term weighting scheme for resolving the power of the words contained in given documents for a certain language, the curve appearing in [14], is reproduced in Fig. 1. We can see the upper and lower cut-off points which separates the stopwords (words which does not have discriminative resolution in the document retrieval process) from the words that serves to describe and differentiate the documents content, called keywords.

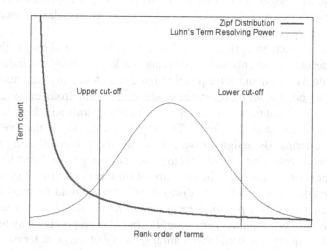

Fig. 1. Luhn's resolving power of terms over a Zipf distribution

In 1959 Maron, Khuns and Ray [15], took the approach started by Luhn and tested a different approach for IR systems over the Boolean Information Retrieval model (BIR) employed extensively in that time. Instead of returning several unranked documents given by the user's query (A combination of terms which relates to the user topics of interest), they took the approach where each document in a collection acquired certain score accordingly to its relevance to a query, weighting the keywords of such documents in the way of Luhn's criteria. This approach outperformed Boolean retrieval as demonstrated in [6], and settled the base for the following term weighting techniques. Term weighting became the de facto practice for ranking the outcome of information retrieval systems, but still presented some complications. The most notorious pointed by Karen Sparck Jones in [16], who exposed the compromise between exhaustivity and specificity.

Exhaustivity refers to the coverage of terms extracted from each document, the more keywords of an index belongs to a specific document, the more chances for that document to match the user's query. While specificity refers to the importance for given keywords of the index. Some terms serves better for matching documents, and due their specific meaning are used more often than other terms for document retrieval. In an information retrieval system, something has to be made in order to balance the compromise between the discriminative and descriptive properties of their index keywords. Sparck Jones saw this compromise and adventured to predict that *"more general terms would be applied to more documents than the separated terms"* and that *"the less specific terms would have a larger collection distribution than the more specific ones"*. Then, the idea of an optimum level between exhaustivity and specificity came thereafter. She transferred those ideas into the information retrieval task and deduced that *"the exhaustivity of a document description is the number of terms it contains, and the specificity of a term is the number of documents to which it pertains"*. So the addition of keywords to an index in relation to their appearance per document increases exhaustivity. And the addition of keywords to an index in relation to their appearance per corpus increases specificity.

The idea is to reach an optimum level of exhaustivity that allows high chances of matching requested documents while reducing the keywords of the index enough to maintain specificity and avoid false positive retrievals. Assessing weights to the index terms appear to be the solution, we can add every term from every document to increase exhaustivity and then assign high value to those terms with high discriminative power in order to increase specificity. The problem with this is that we don't know which are those terms. By assigning weights to this terms just following the Zipf law and Luhn's "word resolving power scheme" not knowing beforehand if they will be used as keywords or not can slant the outcome of information retrieval systems. Sparck Jones also remarked the difficulty to discover if a term should be used for discriminating purposes, to cite *"the proportions of a collection to which a term does and does not belong can only be estimated very roughly"* but also proposed a way to circumvent the problem. If a query has certain matching quantity of frequent terms with a document, and another query has the same amount of matching terms but with non-frequent terms. The document in question would be retrieved in the same manner. We can think that non-frequent terms of an index are more valuable than the frequent ones, but we cannot disregard the latter. The solution seems to point to the relation between the matching values of a term with its collection frequency. This in a global point of view, but can be difficult due the dynamic nature of information retrieval. Documents come and go from the corpus, hindering the calculation of the absolute frequencies of their terms. For that reason, relative frequency seems to be a more appealing solution. Relating the matching values of a term within its document frequency. This is a local approach in which *"the matching value of a term is thus correlated with its specificity and the retrieval level of a document is determined by the sum of the values of its matching terms"*. And by this mean relatively frequent and relatively non-frequent matching terms can retrieve a document in the same manner.

TF-IDF term weighting scheme emerged from the works of Sparck Jones and stands for "Term Frequency – Inverse Document Frequency". Praises terms proportionally to the number of times that appear in each document but offsets the frequency of the word in the corpus. Adjusting its impact of words that appear more frequently in general. It is composed by two components, *tf* or the raw frequency of a term per document:

$$tf(t, d) = |\{t \in d\}| \tag{2}$$

Where tf is the frequency of the term t belonging to the document d. And idf which is a measure that represents the rarity of the term across the corpus:

$$idf(t, D) = \log \frac{|D|}{|\{d \in D : t \in d\}|} \qquad (3)$$

Which is the logarithmically scaled fraction of the total number of documents $|D|$ by the number of documents d belonging to the corpus D containing the term t. In some cases, a 1 is added to the denominator in order to avoid division by 0 when no document in a collection contains a term of the index. But the common practice consists on eliminating such word from the index.

Finally, the TF-IDF factor is the product of both, the TF which maximizes exhaustivity and IDF which maximizes specificity.

$$tfidf(t, d, C) = tf(t, d) \cdot idf(t, C) \qquad (4)$$

Notice that in the TF-IDF weighting scheme, the words that appear the least acquire higher discrimination power and words that appears in every document got disregarded. This in accordance to the Zipf distribution and Luhn's term resolving power. Also, notice that the denominator of the IDF vector contains the sum of documents which contains the term in question, not the sum of a term frequency on the whole corpus.

We can say that TF-IDF is a global weighting scheme. Its value relies on the inclusion of each term per document in the corpus. Including or excluding documents from the corpus affects directly the IDF factor. For that reason it is difficult to implement the TF-IDF in large scale real-time retrieval scenarios [17] such as Web with its dynamic nature. In order to circumvent this problem of weighting an index not depending on the dynamic relationship of their documents, local weighting schemes emerged. Being the *normalized term frequency* or NTF the most common example, it takes the raw frequency of each term of a document divided by the maximum raw frequency of any term in the document. This relative frequency doesn't change over-time but lacks the intrinsic and desired relationship between exhaustivity and specificity of terms noted by Sparck Jones. Here resides the difference between local and global schemes as seen in [18], global schemes implements certain factors taken directly from the corpus, while local schemes does it from the documents. And while global schemes are desired, they are difficult to implement.

3 A Marginal Distribution Weighting Scheme

In this section we present a weighting scheme that improves the results of a NTF weighting scheme. Extrapolating this measure to a new marginal distribution vector that contains the global frequency count of terms. Taking a local weighting scheme and applying it in a global way, making it corpus representative. The advantage of this approach resides in that the weighting method is applied to the marginal distribution vector only and not the whole corpus making it efficient in time each time the weights needs to be recalculated. Also, it is flexible in the sense that can be extrapolated to other local weighting methods as well.

The approach consists on taking the TDM raw frequencies and create an additional vector with the marginal distribution of terms. We calculate the local weighting scheme component on this vector only, instead of the whole corpus and apply the product of it by each term of the TDM to obtain a global weighted TDM from it. By doing this, we get the advantage of the local weighting methods, estimating the term importance of an index relatively to its corpus, and by multiplying for the TDM we increase the term importance proportionally to its document appearance.

This can be achieved treating the original TDM and marginal term frequency vector as a matrix product, but in order to obtain a weighted TDM of the same size of the original. The marginal term frequency vector should be expressed as a diagonal matrix instead of a vector. Let $D = \{d_1, d_2, \ldots, d_n\}$ be a corpus of documents, and let A be the matrix of size $n \times m$ denoted by:

$$A = [a_{ij}]_{n \times m} \tag{5}$$

a_{ij} consist on the number of times that the j_{th} term appears in the i_{th} document of the TDM. Then B is the matrix of size $m \times m$ denoted by:

$$B = [b_{kl}]_{m \times m} \tag{6}$$

Whose b_{kl} are defined as $a_{*j} = \sum_{i=1}^{n} a_{ij}$ when $j = k = l$ and 0 for $k \neq l$. The diagonal of B corresponds to the marginal term frequency vector of the terms in the TDM. This matrix denotes the vector in which we are going to apply the local term scheme, such as NTF and then multiply the matrices A and B. Being C the weighted matrix of a local weighting scheme applied on a global way.

$$
C = A \cdot B = \begin{bmatrix} a_{11} & \cdots & a_{1m} \\ \vdots & \ddots & a_{im} \\ a_{n1} & a_{nj} & a_{nm} \end{bmatrix} \cdot \begin{bmatrix} a_{*1} & 0 & 0 \\ 0 & \ddots & 0 \\ 0 & 0 & a_{*m} \end{bmatrix}
$$
$$
= \begin{bmatrix} a_{11} \times a_{*1} & \cdots & a_{1m} \times a_{*m} \\ \vdots & \ddots & a_{im} \times a_{*m} \\ a_{n1} \times a_{*1} & a_{nj} \times a_{*j} & a_{nm} \times a_{*m} \end{bmatrix} \tag{7}
$$

In the next example we present the explained method over a NTF weighting scheme applied to a TDM. We add the marginal variables to the TDM, in other words, "the marginal variables of a contingency table are the subset of variables found by summing the values in a table along rows or columns, and writing the sum in the margins of the table" [19]. In this case, the "document margin" and the "term margin" of a TDM as shown in the example of the Table 1.

Table 1. TDM with added marginal variables for terms and documents

	Term 1	Term 2	Document margin
Document 1	99	159	258
Document 2	52	105	157
Term Margin	151	264	

We apply the NTF measure to the term margin of our TDM, as if this were a new document to weight. This is shown in Fig. 2. This example applies to two terms only, but we can see the behaviour the NTF has in a document vector, it normalize the most frequent term of the document to 1 and the others below that in function of this one. In normal circumstances, the "ntf" measure should have been applied to each document, but instead, we applied it only to the term margin because we want to know which term of the corpus is more representative to it, and what others follows accordingly.

Term Margin (before ntf)	151	264
$$ntf = \frac{freq_{ij}}{MAX_{k=1}^{m} freq_{ik}}$$	$$ntf = \frac{151}{264}$$	$$ntf = \frac{264}{264}$$
Term Margin (after ntf)	0.57	1

Fig. 2. ntf measure applied to the term margin variable

In [2], is stated that *"a document or zone that mentions a query term more often has more to do with that query and therefore should receive a higher score the most representative word of a document"* this is true in a document perspective but in a corpus perspective, the most representative term of the document collection does not serve to discriminate because the majority of the documents in the collection contains that term and is the same case as the "stop words list" of a retrieval system, a list used to exclude from the TDM all the common terms shared by all the documents of the corpus in the indexing phase. Also, Gerard Salton in [7], states *"this implies that the best terms should have high term frequencies but low overall collection frequencies"*. Another weighting schemes such as the term frequency-inverted document frequency or "tf-idf" [16], addresses this problem in a different way from our approach, in the sense that the aforementioned multiplies the "tf" by an "idf" factor which varies inversely with the number of documents n to which a term is assigned in a collection of N documents, being the "idf" factor a statistical descriptive coefficient meanwhile our approach tries to use the information given by the global term frequency distribution to soften the impact of the highly frequent words in a corpus. This, and further discussion on good term criteria for discrimination can be found in [18]. For that reason we want now to invert the weights estimated by "ntf", this is achieved easily dividing 1 by each weight of the term margin variable, and with this the less frequent terms of a corpus become more representative and the most frequent, less representative. Now we can test both approaches and according to the present experiment, inverting the importance of the most representative terms tend to give better results, suggesting that queries with less frequent terms improves the specificity of the search. This operation is shown in Fig. 3.

Term Margin (before)	151	264
$\dfrac{1}{ntf} = \dfrac{MAX_{k=1}^{m} freq_{ik}}{freq_{ij}}$	$\dfrac{1}{ntf} = \dfrac{264}{151}$	$\dfrac{1}{ntf} = \dfrac{264}{264}$
Term Margin (after ntf)	1.75	1

Fig. 3. Inverse of ntf measure applied to the term margin variable

Next we continue multiplying each term of the TDM by its corresponding term marginal variable, this is, the global weight of a term by the quantity of terms per document. By now we can discard the document margin variable in spite of that we are not using it. The result for the "inverse marginal term frequency" or "gobal normalized term frequency" (gntf) as we call it, can be seen in Table 2.

Table 2. TDM with "gntf" scheme applied

	Term 1	Term 2	Document margin
Document 1	173.25	159	
Document 2	91	105	
Term Margin	1.75	1	

4   Experimental Design

The extents of the previous example are not visible at simple sight, but this is a case of a TDM of 2 × 2 dimensions, then we present an experimental design for testing the "gntf" scheme in it, we are trying to see if extrapolating classic weighting schemes globally applying them to the corpus instead of each document has actually impact in the results of a retrieval system and if these method is viable to use in large collections of documents. The benchmark selected was the CACM dataset consisting of 3204 documents and 5763 terms in its thesaurus, giving us a TDM of 5763 × 3204 dimensions. This Benchmark was selected in order to test the efficiency of the scheme in spite the dispersion of the TDM, being a collection which includes only the titles and some abstracts of 3204 scientific papers, the majority of the frequencies of his TDM are zeroes. And in the literature [20, 21] it can be seen that almost all retrieval models develops poorly on it. Also we implemented a VSM retrieval program [4], for the task because of his proved and steady performance, in [20], the authors question and test the VSM against LSI and GVSM, showing that the efficiency of each one is benchmark dependant, reinforcing the intuition of the "no free lunch theorem" as well [22], and noticing that the VSM is more consistent in its results, regardless the benchmark.

For testing and comparing results, we implemented a program which reads the results given by the VSM and builds a report in precision and recall [23, 24], for certain list of queries provided beforehand, the precision-recall measure takes into account the quantity of documents retrieved (retrieval), against the quantity of retrieved documents which are actually relevant to a query in question (precision). The higher the recall and the higher the precision, the better result, but these are conflicting objectives and represent a multi objective problem, increasing the documents retrieved, the recall rises but the precision tends to fall and vice versa. For this motive, optimizing the result of a precision-recall measure is achieved approximating the front in the like of a pareto chart. The list of queries given in the CACM benchmark consist in 52 queries with its respective list of results of relevant and non-relevant documents for the query in question. Finally is to be mentioned that the charts showing these results were built accordingly to specifications given for TREC benchmarks [25], which is the de facto test in the literature and is provided by one of the longest and more confident resources for information retrieval in the web. The first step in our experiment consisted on creating a TDM in the form of a csv file to contain the raw frequencies of the CACM corpus, this step was necessary because the CACM benchmark provides the TDM in a compressed form, and a parser was necessary to build a table in which the marginal variables could be estimated.

Once we calculated and estimated the "gntf" weighting described on the CACM TDM, we entered the table and the list of queries to the VSM program and asked to get 100 incremental replicas for the 52 queries given in the benchmark, each replica is incremental in the sense that the 1st one retrieve only one document for each query, the 2nd two documents for each query and so on until retrieving 100 documents for each query. Each pair of replica-query then reports a recall-precision measure which is stored and served to interpolate and build the following shown in Fig. 4.

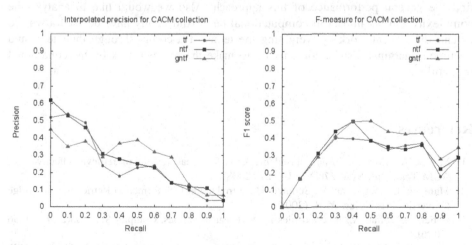

Fig. 4. Precision-Recall and F1 score for "f", "ntf" and "gntf" weighting schemes on CACM dataset

The results shown in Fig. 4 summarizes and compares three different schemes for weighting terms in the CACM TDM, the "f" measure which represents the interpolated precision of the raw frequencies (the TDM without weighting), the "ntf" measure which represents the interpolated precision of a classical term frequency weighting scheme and lastly the "gntf" measure which represents the interpolated precision of "ntf" inverted and interpolated to a corpus level. All the three graphs are shown at 11 recall levels in accordance to the guidelines given in [25]. We can see form the chart that even the high dispersion level of the CACM TDM benchmark, applying a weighting scheme such as "ntf" to being corpus representative does really has impact in the outcome of the VSM and generally speaking, a slightly better than their counterpart at middle and higher levels of recall. The only drawback is at low levels of recall (30 % and below) which corresponds to the VSM retrieving low quantities of documents, but notice that is at this same point when all the measures recovers from a high tendency to loose precision and is the "gntf" which does it best, even reverting the loose completely, something that the others could not do.

5   Conclusions and Further Work

We can say from our experiments, that using a weighting scheme such as the "ntf" applied to the term frequency margin count like the case of our "gntf" does improves the performance over its classical counterpart. Also, a program for porting other measures to a corpus level via the marginal distribution of their term frequencies is done. We have proven that a global weighting scheme is viable and can be done, contrary to the common belief argued in [2, 9, 10], reinforced by other weighting schemes with a global approach applied successfully like the one in [18], and both in accordance and sustained by [26].

In the future we will continue to test more benchmarks and schemes like Okapi and TF-IDF, porting them to a corpus level via its marginal term distribution, in order to find the general performance of this approach. Also we would like to analyse the complexity and his impact in computational time. And finally have devised a way to extend the present work by weighting the terms of a corpus through their term and document marginal distributions, trying to make the retrieval task by his conditional probabilities.

References

1. Moffat, A., Zobel, J.: Rank-biased precision for measurement of retrieval effectiveness. ACM Trans. Inf. Syst. **27**(1), 2:1–2:27 (2008)
2. Manning, C., Raghavan, P., Schütze, H.: Introduction to Information Retrieval. Cambridge University Press, New York (2008)
3. Zobel, J., Moffat, A.: Inverted files for text search engines. Comput. Surv. **38**(2), 6.1–6.56 (2006)
4. Salton, G., McGill, M.: Introduction to Modern Information Retrieval. McGraw-Hill, New York (1986)

5. Lan, M., Sung, S.-Y., Low, H.-B., Tan, C.-L.: A comparative study on term weighting schemes for text categorization. In: Proceedings of the International Joint Conference on Neural Networks vol. 1, pp. 546–551 (2005)
6. Spärck, K.: Information Retrieval Experiment. Butterworth-Heinemann, London (1981)
7. Salton, G., Buckley, C.: Term-weighting approaches in automatic text retrieval. Inf. Process. Manag. **54**(5), 513–523 (1988)
8. Reed, J., Jiao, Y., Potok, T., Klump, B., Elmore, M., Hurson, A.: TF-ICF: A new term weighting scheme for clustering dynamic data streams. In: Proceedings of the 5th International Conference on Machine Learning and Applications, pp. 258–263 (2006)
9. Baeza, R., Ribeiro, B.: Modern Information Retrieval. Addison Wesley, New York (2011)
10. Rajendra, A., Pawan, L.: Building an Intelligent Web, Theory and Practice. Jones and Bartlett Publishers, Sudbury (2008)
11. Harris, Z.: Distributional Struct. Word **10**, 146–162 (1954)
12. Zipf, G.: Human behavior and the principle of least effort. J. Clin. Psychol. **6**(3), 306 (1950)
13. Luhn, H.: The Automatic creation of literature abstracts. IBM J. Res. Dev. **2**, 159–165 (1958)
14. Schultz, C., Luhn, H.P.: Pioneer of Information Science Selected Works. Macmillan, London (1968)
15. Maron, M., Kuhns, J., Ray, L.: Probabilistic Indexing: A Statistical Technique for Document Identification and Retrieval. Thompson Ramo Wooldridge Inc., Los Angeles (1959). Data Systems Project Office, Technical Memorandum 3
16. Sparck, K.: A statistical interpretation of term specificity and its application in retrieval. J. Documentation **28**, 11–21 (1972)
17. Hui, K., He, B., Luo, T., Wang, B.: Relevance weighting using within-document term statistics. In: Proceedings of the 20th ACM International Conference on Information and Knowledge Management, pp. 99–104 (2011)
18. Cummins, R., O'Riordan, C.: Evolving local and global weighting schemes in information retrieval. Inf. Retrieval **9**, 311–330 (2006)
19. Trumpler, R., Weaver, H.: Statistical Astronomy. University of California Press, Berkeley (1953)
20. Kumar, A., Srinivas, S.: On the performance of latent semantic indexing-based information retrieval. J. Comput. Inf. Technol. **17**, 259–264 (2009)
21. Hofmann, T.: Probabilistic latent semantic indexing. In: Proceedings of the Twenty-Second Annual International Sigir Conference on Research and Development in Information Retrieval, pp. 50–57 (1999)
22. Wolpert, D., Macready, W.: No free lunch theorems for optimization. IEEE Trans. Comput. **1**, 67–82 (1997)
23. Raghavan, V., Bollmann, P., Jung, G.: A critical investigation of recall and precision as measures of retrieval system performance. ACM Trans. Inf. Syst. **7**, 205–229 (1989)
24. Ishioka, T.: Evaluation of criteria for information retrieval. Syst. Comput. Japan **35**(6), 42–49 (2004)
25. Text Retrieval Conference (TREC): Common Evaluation Measures. http://trec.nist.gov/pubs/trec15/appendices/CE.MEASURES06.pdf
26. Warren, G.: A theory of term weighting based on exploratory data analysis. In: Proceedings of the 21st International ACM SIGIR Conference on Research and Development in Information Retrieval, pp. 11–19 (1998)

Sentiment Groups as Features of a Classification Model Using a Spanish Sentiment Lexicon: A Hybrid Approach

Ernesto Gutiérrez[✉], Ofelia Cervantes, David Báez-López,
and J. Alfredo Sánchez

Universidad de las Americas, Cholula, Mexico
{ernesto.gutierrezca,ofelia.cervantes,david.baez,
alfredo.sanchez}@udlap.mx

Abstract. Discovering people's subjective opinion about a topic of interest has become more relevant with the explosion in the use of social networks, microblogs, forums and e-commerce pages all over the Internet. Sentiment analysis techniques aim to identify polarity of opinions by analyzing explicit and implicit features within the text. This paper presents a hybrid approach to extract features from Spanish sentiment sentences in order to create a model based on support vector machines and determine polarity of opinions. In addition to this, a Spanish Sentiment Lexicon has been constructed. Accuracy of the model is evaluated against two previously tagged corpora and results are discussed.

Keywords: Sentiment analysis · Opinion mining · Polarity identification · Spanish lexicon · Sentiment lexicon · Lexicon-based sentiment identification · Support vector machines · Classification model

1 Introduction

Explosion in the use of social networks, microblogs, forums and e-commerce pages has generated an increased interest in discovering opinions, feelings and sentiment of authors regarding such information. The essence of this information is subjectivity, even though sentiment could exist related to objective text, many decisions people make are influenced by others' opinions. One clear example is stock prices that are subject not only to objective parameters but also to speculation. This subjective information also raises interest to discover trends and changes in society's opinions.

Sentiment analysis, also known as opinion mining, tries to cope with this subjectivity problem. Sentiment analysis defines the subjectivity problem as a polarity problem, where the main goal is to determine the polarity of text. There is a wide variety of uses and applications of sentiment analysis such as: marketing trends, satisfaction of end users surveys, political opinion trends, among others. In this manner, detection of sentiment polarity facilitates understanding such subjective information.

In Spanish texts, as in other languages, sentiment analysis poses challenges for Natural Language Processing (NLP) due to the inherent nature of the language, the ambiguity in the use of words, misspelling, grammatical mistakes, missing or misused punctuation, use of slang, and the lack of standardization while tagging corpora.

© Springer International Publishing Switzerland 2015
J.A. Carrasco-Ochoa et al. (Eds.): MCPR 2015, LNCS 9116, pp. 258–268, 2015.
DOI: 10.1007/978-3-319-19264-2_25

This paper presents a hybrid approach to extract features from a corpus consisting of Spanish opinion sentences. After feature extraction, we construct a model using support vector machines (SVM) and a Spanish Sentiment Lexicon created semi-automatically. In order to test the accuracy of the model, two previously annotated corpora were used achieving 87.9 % accuracy in a 5-fold cross-validation, 52.6 % accuracy for 3-class TASS 2014 corpus and 65.4 % for SFU Reviews corpus.

The rest of the paper is organized as follows: In Sect. 2 a brief synthesis of related work about sentiment analysis on Spanish texts is presented; in Sect. 3 the process of supervised classification to construct the model is explained; in Sect. 4 implementation details of a sentiment analysis system are described; in Sect. 5 the accuracy of the model is evaluated; and finally, in Sect. 6 conclusions and future work are outlined.

2 Related Work

Sentiment analysis has many interesting applications such as: determining customer satisfaction, identifying opinion trends, the market sentiment,[1] political opinion trends, determining some entity's reputation and more important listening the voice of the majority respect to some topic of interest. Excellent work has been done regarding sentiment analysis in English. However, it is relevant to increase the efforts focused on other languages.

Specifically in Spanish, more work is needed on freely available lexicons in order to improve algorithms to detect sentiment polarity. In this regard, a few works are open and freely available such as the work of Molina-Gonzales et al. [10] where a Spanish Opinion Lexicon (SOL) was constructed using machine translation over the lexicon constructed by Hue and Liu in [8]. In addition to machine translation, the SOL lexicon was improved manually and enhanced with domain information from a movies reviews corpus.[2] Another example of freely available lexicon is the work of Perez-Rosas et al. [13] where two corpora were obtained using Latent Semantic Analysis over WordNet and SentiWordNet: full-strength and medium-strength lexicons. Open lexicons only for academic and research purposes are also in the scope: Redondo et al. [14] adapted the ANEW corpus previously done in English by Bradley and Lang [3] using manual translation of 1034 words; and Sidorov et al. [15] manually annotated the Spanish Emotions Lexicon (SEL) with the six basic human emotions: anger, fear, joy, disgust, surprise, and sadness.

Other works aim to extract automatically lexicons from annotated corpora, such as the work of Gutierrez, et al. [7], in which graph-based algorithms have been used to annotate extracted words from previously annotated corpora. In their approach, every word is either potentially positive or negative if it appears in a phrase tagged as positive or negative respectively and semantic relations are captured with a graph-based representation. Montejo-Raez et al. [11] implemented a strategy to obtain a set of polarity

---

[1] http://www.investopedia.com/terms/m/marketsentiment.asp.

[2] http://www.muchocine.net.

words from twitter by using the query *"me siento"* (I feel) and then manually tagging the polarity of words.

Regarding lexicon based methods for sentiment analysis, Taboada et al. [16] presented a heuristic Semantic Orientation Calculator (SO-CAL) which uses annotated dictionaries of polarity words (adjectives, adverbs, verbs and nouns), negation words and intensifiers. Moreno-Ortiz et al. [12] also applied a heuristic calculation to obtain what they call Global Sentiment Value (GSV), however, their GSV formula has poor accuracy.

Machine Learning approaches are often used to solve sentiment polarity classification problems. Martinez-Camara et al. [9] performed several tests with different features as Term Frequency-Inverse Document Frequency (TF/IDF) and Binary Term Occurrence (BTO) with SVM and Naive Bayes classifiers. In the work of Anta et al. [1] a series of tests using N-grams obtained through a combination of preprocessing tasks (lemmatization, stemming, spell-checking) with Bayesian classifiers and decision tree learners is also performed.

Our hybrid approach is similar to [5, 18] in the sense that it combines a lexicon-based approach with a supervised-learning approach. Del-Hoyo et al. [5] formed feature vectors concatenating TFIDF features and the result of their Semantic Tool for detecting affect in texts. Vilares et al. [18] used part-of-speech tags with syntactic dependencies. Nevertheless, our work is different from [5, 18] in the sense that no bag of words is needed to construct the feature vector. Therefore, feature vector dimensionality is reduced.

3 Classification Model

The work presented in this paper aims to cope with challenges in sentence-level sentiment analysis. The methodology that has been followed is depicted in Fig. 1: A supervised-learning approach was used to determine sentiment polarity of sentences. First, a process collects sentences from twitter and several e-commerce pages to construct a corpus. The Sentiment Groups features were extracted from this corpus. Furthermore, a sentiment lexicon was obtained from corpora using the most representative polarity words. Using support vector machines (SVM) three models were obtained with a radial basis kernel for binary classification. Then, 5-fold cross-validation was used to obtain accuracy and the F1-measure, although similar results were obtained using 6, 8 and 10-fold-cross-validation.

3.1 Tweets + Reviews Corpus

The Tweets + Reviews (TR) Corpus was obtained downloading tweets[3] and a subset of the reviews[4] from e-commerce pages available in the work of Dubiau and Ale [6]. Once downloaded, tweets and reviews were split into sentences. A total of 6687

---

[3] Tweets obtained with search query *"futbol mexico holanda"* from twitter.com.

[4] https://github.com/ldubiau/sentiment_classifier.

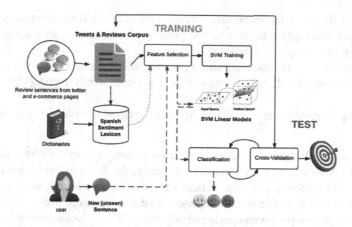

Fig. 1. Supervised-learning in sentiment analysis

sentences were obtained. Each sentence was tagged accordingly to its polarity as: terrible (N+), bad (N), neutral (NEU), good (P) and excellent (P+).

To ensure quality of corpus, tagging and selection of sentences was done in a three-step process:

(i) Manual tagging: A group of five trained colleagues tagged each sentence, and each sentence was reviewed by at least two different taggers.

(ii) Sentiment polarity identification: A heuristic sentiment algorithm was used to tag automatically the sentences. This heuristic calculator determines sentiment polarity using basic linguistic rules and polarity words and is explained in detail in Sect. 4.2.

(iii) Corpus construction: Sentences that matched the heuristic sentiment calculator and manual tagging were selected to construct the corpus.

With this procedure we obtained 3084 sentences as shown in Table 1.

Table 1. Tweets & reviews corpus

Class	No. of observations
P+	419
P	1000
NEU	245
N	1281
N+	140

3.2 Spanish Sentiment Lexicon

Our Spanish Sentiment Lexicon (SSL) is one of the major components of the sentiment analysis system. For this work, the SSL was constructed semi-automatically:

1. Extracting the most representative (frequent) words selected statistically from the TR Corpus and tagging them with the polarity of the sentence they came from and with their part-of-speech (PoS) tag.
2. Adding to the SSL the most common polarity adjectives and adverbs coming from on-line dictionaries. It was necessary to tag those words coming from external sources.
3. Validating manually the Sentiment Spanish Lexicon in order to increase lexicon quality.

Even though it has been an iterative and time consuming process, we have constructed a consistent Spanish Sentiment Lexicon. Currently, the Spanish Sentiment Lexicon is formed by 4583 words divided into adjectives, adverbs, nouns, negation adverbs, and intensifier words. Intensifier words depends upon the language and could be quantitative adverbs or comparative/superlative adjectives. This is summarized in Table 2.

Table 2. Spanish sentiment lexicon

Category	Positive	Negative	Total
Adjective	635	724	1359
Adverb	20	23	43
Noun	138	279	417
Verb	1027	1690	2717
Negation adverbs	–	–	10
Intensifier words	–	–	37
All	1820	2796	4583

Labeling of words proceeded as follows: -2 for N+, -1 for N, 0 for NEU, 1 for P and 2 for P+. Negative and positive labels are defined accordingly to intentionality rather than interpretation. For instance, there are words that are interpreted with negative feeling such as: "government", "politicians", "acne", among others, but in fact, those words are not indeed negative, even though many times they are used in contexts where negative feelings are towards them. In this sense, words that are part of our lexicon are those whose intentionality is clearly positive or negative. Other words of ambiguous intentionality or context dependent like: ambitious, small, big, long, among others are tagged as context-dependent. The Spanish Sentiment Lexicon is freely available.[5]

3.3 Feature Selection: Sentiment Groups

In sentiment analysis, commonly selected features are: frequency or occurrence of terms, frequency or occurrence of n-grams (especially bigrams and trigrams) and part-of-speech (PoS) tags. In this paper the notion of Sentiment Groups as features for sentiment analysis classification is introduced.

---

[5] SSL is freely available contacting authors.

A Sentiment Group is a group of related words. These words can be polarity words (*bueno*-good, *malo*-bad, etc.), intensity words (*poco*-little, *mucho*-much, quantitative adverbs in case of spanish language) or negation adverbs (*no*-not, *nada*-nothing, etc.), a special case in spanish is double negation, and we identify it and treat double negation as intensification of single negation (*No me gustó nada la película*, I did not like the movie at all). Neutral and objective words as well as conjunctions and prepositions are not part of a Sentiment Group. However, they are essential for delimiting Sentiment Groups. Several Sentiment Groups can occur within a sentence. This situation occurs often when there is incorrect punctuation use, which is common in informal text such as tweets and reviews.

A Sentiment Group is defined by these rules:

- All words within Sentiment Group are at most 2 words from distance.
- A Sentiment Group can contain double or triple negation (because of Spanish nature, other languages may not contain neither double nor triple negation)
- Sentiment group separators are conjunctions (*y*-and, *o*-or, etc.) and punctuation: comma, semicolon.

Examples of sentiment groups are depicted in Table 3. In addition to grouping, words that belong to Sentiment Groups are tagged with its PoS tag and its polarity (+ for positive and − for negative).

Table 3. Examples of sentiment groups

Sentence	Polarity, intensity or negation words	Sentiment groups
I like the new iPhone, what I do not like its price *Me gusta el nuevo iPhone, lo que no me agrada es el precio*	*gusta, no, agrada*	G1(gusta) → G1(+VB) G2(no-, agrada) → G2 (NEG, +VB)
Sound quality is good but esthetically is horrible *La calidad del sonido es buena pero estéticamente es horrible*	*calidad, buena, horrible*	G1(calidad, buena) → G1 (+NN, +JJ) G2(horrible) → G2(-JJ)
It takes too long to start, it is pretty slow *Tarda mucho en arrancar, es muy lento*	*Tarda, mucho, muy, lento*	G1(tarda, mucho) → G1(- VB, INT) G2(muy, lento) → G2 (INT, -JJ)
I don't like it and I don't want it *No me gusta y no lo quiero*	*no, gusta, no, quiero*	G1(no, gusta) → G1 (NEG, +VB) G2(no, quiero) → G2 (NEG, +VB)

A Sentiment Group contains basically atomic units of sentiment independent from each other. In this manner, we can extract several atomic units of sentiment and use those characteristics to build the feature vector. A Sentiment Group can be seen as a fast-heuristic approach to a syntactic dependency parser for informal text.

3.4   Feature Vector

We performed some tests on the corpus and discovered that average length of sentences was 9 words with standard deviation of 8. However, we also found some anomalous cases where length of sentences extended up to 45 words. Taking into account maximum expected length of sentence as 20 (a little more than average + standard deviation) words and with no punctuation and a terrible use of grammar rules a sentence can contain up to ten Sentiment Groups. Hence, we decided to construct the feature vector using ten Sentiment Groups. In the feature vector each Sentiment Group is characterized by all polarity, intensity and negation words and their PoS tags. Feature vector is then formed as (1).

$$F = \{NEG_n, DNEG_n, INT_n, +NN_n, -NN_n, +JJ_n, -JJ_n, +VB_n, -VB_n, +RB_n, -RB_n | 1 \leq n \leq 10\}$$
(1)

It is important to note that order matters while constructing the feature vector, so each word belongs to the corresponding Sentiment Group. Double negation is taken into account. Table 4 shows the interpretation for each Sentiment Group feature.

Table 4. Features in each sentiment group within feature vector

Tag	Description
NEG	Describes if there is a negation within the Sentiment Group
DNEG	Describes if there is a double or triple negation in SG
INT	Describes if there is an intensifier within SG
(SIGN)NN	It is the addition of all positive or negative nouns
(SIGN)JJ	It is the addition of all positive or negative adjectives
(SIGN)VB	It is the addition of all positive or negative verbs
(SIGN)RB	It is the addition of all positive or negative adverbs

3.5   Support Vector Machines

Support Vector Machines (SVM) are widely used as classifiers in sentiment analysis related tasks [1, 5, 9, 18]. Due to the inherent characteristic of SVM as linear classifiers, it was decided to construct three binary models instead of one multiclass model but still using non -linear kernel for better performance, specifically radial basis function kernel. In this manner, three balanced corpus were obtained using undersampling and were used to train three SVM models: good-bad (P, N) model with 2839 sentences, excellent-good (P+,P) model with 840 sentences and bad-terrible (N,N+) model with 282 sentences.

4   Tinga Sentiment Analysis System

In order to test our proposed model we implemented a testbed that we refer to as Tinga,[6] which is part of a Scala library for Natural Language Preprocessing and it also includes modules for text preprocessing, tokenizing, part-of-speech tagging, basic text features extraction, and a module that wraps a Java Support Vector Machine library.[7]

4.1   Text Preprocessing

Much has being said about text preprocessing regarding sentiment analysis, some approaches clean text by getting rid of punctuation, stopwords, diacritics and stemming or lemmatizing words [1, 9]. However, in our proposal we use minimal classic preprocessing as in [2] and only follow the next preprocessing steps to normalize informal text (such as tweets):

- Emoji[8]-emoticons identification: A regex was used to identify emojis and emoticons present in informal text (like tweets). Each emoji and emoticon was previously classified as positive or negative and it is replaced in text using polarity words (*excelente, buen, neutro, malo, terrible*).
- Hashtag split: Many hashtags are formed by several words. We implemented an algorithm to split hashtags into several words.
- Repetition of characters: In Spanish only a few words allow repetition of characters, for instance, consonants c, l, n and r are the only ones that are allowed to be repeated. Although all vowels can be repeated to form words, neither polarity nor intensity words are within this set of words. Therefore we detect and erase the repetition of characters including vowels, non $\{c,l,n,r\}$ consonants and punctuation.
- Upper-case words: In chat slang, upper case means yelling.
- Adversative conjunctions detection: In addition to determination of negation, determination of adversative conjunctions changes the sentiment of opinion.
- Special characters and punctuation: All Spanish characters are allowed but only basic punctuation marks are allowed.
- Spell checking: Tinga has implemented a Bayes theorem based spell checker.
- Tokenizing and PoS tagging: A sentence is split into valid word tokens and tagged according to its grammatical category (PoS tag).

4.2   Polarity Identification

Cascading SVM Classifiers. With the three models obtained from the training phase it is possible to classify into two classes (P, N) or into four classes (P+, P, N, N+) by simply cascading SVM models as shown in Fig. 2. Neutral (NEU) and none (NONE)

---

[6] https://github.com/PhotonServices/tinga.

[7] http://www.csie.ntu.edu.tw/~cjlin/libsvm/.

[8] http://emojipedia.org/.

classes are discarded from classifiers but taken into account in our heuristic sentiment calculator.

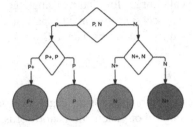

Fig. 2. Cascading SVM classifiers

Heuristic Sentiment Calculator. A basic Heuristic Sentiment Calculator (HSC) was implemented to match manual tagging. HSC uses some basic rules:

- Split the sentences into Sentiment Groups (SG)
- Obtain polarity of each SG by multiplying all polarity and intensity words within each SG taking into account negation, double negation.
- Obtain sentence level sentiment using weighted sum of all SG present

For a more formal calculator using syntactic dependencies, see [17].

5 Results

Our model was validated using 5, 6, 8 and 10-fold cross validation over balanced corpora (see Sect. 3.5) and also tested against the TASS 2014 corpus and the SFU Reviews Corpus. Cross-validation gives good results, however, relying only on cross-validation is not sufficient to tackle the sentiment analysis problem. Therefore, we test our approach against TASS 2014 (1 k corpus) [19] and SFU Review Corpus [16]. For TASS 2014 our accuracy was 52.6 % for 3-class problem and 35.2 % for 5-class problem. For the SFU Review Corpus our accuracy was 65.4 % as shown in Table 5.

Table 5. Accuracy and F-1 measure of proposed hybrid approach

Corpus	Accuracy	F-1 Measure
TR 5-fold cross validation (2 classes)	87.9 %	87.6 %
TASS 2014 (3 classes)	52.6 %	–
TASS 2014 (5 classes)	35.2 %	–
SFU Review Corpus (2 classes)	65.4 %	64.2 %

Testing the TASS corpus is a challenging task because of the nature of its tagging, for instance, we found some examples clearly neutral but tagged as positive. We did better in the SFU Review corpus, but still with low accuracy because of difference of structure. The TR corpus consist of sentences while SFU Review corpus consist of long texts. While analyzing long texts we need to tackle context in order to weight most relevant sentences.

6 Conclusions and Future Work

We have presented a hybrid method to classify polarity of Spanish comments with support vector machines trained with a vector formed with lexical-syntactic features: part-of-speech tags and polarity valence of words. A Spanish Sentiment Lexicon was constructed to be the reference of polarity valence of words. In addition to this, we implemented Tinga, our sentiment analysis system, to test our proposed model. Finally, to test accuracy and F-1 measure TASS 2014 and SFU Reviews corpus were used. Our approach use cascading classifiers, but in future work an SVM with polynomial kernel will be tested in order to make classification in one single step.

A Spanish Sentiment Lexicon is the result of a semi-automatic process and more formalization is needed by evaluating reliability using kappa agreement. Also, it is necessary to increment its potential by adding context information to polarity words. In this manner, the polarity of the words will be influenced also by its context, giving better results. We are also working on a graph based approach [4] to tackle both: sentiment analysis classification and automatic lexicon annotation.

Limitations of this work are given mainly due to the heuristic approach. More work on validation of lexicon and feature extraction is needed to improve the robustness to the model.

Acknowledgements. This work has received support from the CONACYT-OSEO project no. 192321 and from the CONACYT-PROINNOVA project no. 198881.

References

1. Anta, A.F., Chiroque, L.N., Morere, P., Santos, A.: Sentiment analysis and topic detection of spanish tweets: a comparative study of NLP techniques. Procesamiento del Lenguaje Nat. **50**, 45–52 (2013)
2. Balahur, A.: Sentiment analysis in social media texts. In: WASSA 2013: 4th Workshop on Computational Approaches to Subjectivity, Sentiment and Social Media Analysis (2013)
3. Bradley, M.M., Lang, P.J., Bradley, M.M., Lang, P.J.: Affective Norms for English Words (ANEW): Instruction manual and affective ratings. Technical Report C-1, The Center for Research in Psychophysiology, University of Florida, pp. 1–45 (1999)
4. del-Hoyo, R., Hupont, I., Lacueva, F.J., Abadía, D.: Hybrid text affect sensing system for emotional language analysis. In: Proceedings of the International Workshop on Affective-Aware Virtual Agents and Social Robots, AFFINE 2009, pp. 3:1–3:4. ACM, New York, NY, USA (2009)
5. Castillo, E., Cervantes, O., Vilariño, D., Báez-Lopez, D., Sánchez, J.A.: Udlap: sentiment analysis using a graph based representation. In: Proceedings of the 8th International Workshop on Semantic Evaluation (2015)

6. Dubiau, L., Ale, J.M.: Análisis de sentimientos sobre un corpus en espanol: experimentación con un caso de estudio. In: Proceedings of the 14th Argentine Symposium on Artificial Intelligence, ASAI, pp. 36–47 (2013)
7. Gutiérrez, Y., González, A., Orquín, A.F., Montoyo, A., Muñoz, R.: RA-SR: Using a ranking algorithm to automatically building resources for subjectivity analysis over annotated corpora. WASSA 2013, 94 (2013)
8. Hu, M., Liu, B.: Mining and summarizing customer reviews. In: Proceedings of the Tenth ACM SIGKDD International Conference on Knowledge Discovery and Data Mining, KDD 2004. ACM, New York, NY, USA, pp. 168–177 (2004)
9. Martínez-Cámara, E., Martín-Valdivia, M.T., Ureña-López, L.A.: Opinion classification techniques applied to a spanish corpus. In: Muñoz, R., Montoyo, A., Métais, E. (eds.) NLDB 2011. LNCS, vol. 6716, pp. 169–176. Springer, Heidelberg (2011)
10. Molina-González, M.D., Martínez-Cámara, E., Martín-Valdivia, M.-T., Perea-Ortega, J.M.: Semantic orientation for polarity classification in Spanish reviews. Expert Syst. Appl. 40, 7250–7257 (2013)
11. Montejo-Ráez, A., Díaz-Galiano, M.C., Perea-Ortega, J.M., Ureña-López, L.A.: Spanish knowledge base generation for polarity classification from masses. In: Proceedings of the 22nd International Conference on World Wide Web Companion, WWW 2013 Companion. International World Wide Web Conferences Steering Committee, pp. 571–578. Republic and Canton of Geneva, Switzerland (2013)
12. Moreno-Ortiz, A., Hernández, C.P.: Lexicon-based sentiment analysis of twitter messages in spanish. Procesamiento del Lenguaje Nat. 50, 93–100 (2013)
13. Perez-Rosas, V., Banea, C., Mihalcea, R.: Learning sentiment lexicons in spanish. In: Proceedings of the Eight International Conference on Language Resources and Evaluation (LREC 2012). European Language Resources Association (ELRA), Istanbul, Turkey (2012)
14. Redondo, J., Fraga, I., Padrón, I., Comesaña, M.: The Spanish adaptation of ANEW (Affective Norms for English Words). Behav. Res. Methods 39, 600–605 (2007)
15. Sidorov, G., Miranda-Jiménez, S., Viveros-Jiménez, F., Gelbukh, A., Castro-Sánchez, N., Velásquez, F., Díaz-Rangel, I., Suárez-Guerra, S., Treviño, A., Gordon, J.: Empirical study of machine learning based approach for opinion mining in tweets. In: Batyrshin, I., Mendoza, M.G. (eds.) Advances in Artificial Intelligence. Lecture Notes in Computer Science, pp. 1–14. Springer, Heidelberg (2013)
16. Taboada, M., Brooke, J., Tofiloski, M., Voll, K., Stede, M.: Lexicon-based methods for sentiment analysis. Comput. Linguist. 37, 267–307 (2011)
17. Vilares, D., Alonso, M.A., Gómez-Rodríguez, C.: Clasificación de polaridad en textos con opiniones en español mediante análisis sintáctico de dependencias. Procesamiento del Lenguaje Nat. 50, 13–20 (2013)
18. Vilares, D., Alonso, M.Á., Gómez-Rodríguez, C.: Supervised polarity classification of spanish tweets based on linguistic knowledge. In: Proceedings of the 2013 ACM Symposium on Document Engineering, DocEng 2013, pp. 169–172. ACM, New York, NY, USA (2013b)
19. Villena-Roman, J., García-Morera, J., Sánchez, C.P., García Cumbreras, M.A., Martínez-Cámara, E., Ureña-López, L.A., Martín-Valdivia, M.T.: TASS 2014 - workshop on sentiment analysis at SEPLN - overview. In: Proceedings of the TASS Workshop at SEPLN 2014 (2014)

Applications of Pattern Recognition

Modified Binary Inertial Particle Swarm Optimization for Gene Selection in DNA Microarray Data

Carlos Garibay[1], Gildardo Sanchez-Ante[1(✉)], Luis E. Falcon-Morales[1], and Humberto Sossa[2]

[1] Data Visualization and Pattern Recognition Lab, Tecnológico de Monterrey, Campus Guadalajara, Av. Gral. Ramon Corona 2514, Col. Nvo. Mexico, 45201 Zapopan, JAL, Mexico
gildardo@itesm.mx
[2] Instituto Politécnico Nacional-Centro de Investigación En Computación, Av. Juan de Dios Batiz S/N, Gustavo A. Madero, 07738 México, D.F., México
hsossa@cic.ipn.mx

Abstract. DNA microarrays are being used to characterize the genetic expression of several illnesses, such as cancer. There has been interest in developing automated methods to classify the data generated by those microarrays. The problem is complex due to the availability of just a few samples to train the classifiers, and the fact that each sample may contain several thousands of features. One possibility is to select a reduced set of features (genes). In this work we propose a wrapper method that is a modified version of the Inertial Geometric Particle Swarm Optimization. We name it MIGPSO. We compare MIGPSO with other approaches. The results are promising. MIGPSO obtained an increase in accuracy of about 4 %. The number of genes selected is also competitive.

Keywords: DNA microarray · PSO · Feature selection · Wrapper method

1 Introduction

Genetic data can be used to diagnose, treat, prevent and cure many illnesses. Among some new technologies developed in this area, one that has been growing fast in the recent years is the DNA microarrays. A DNA microarray is a platform that enable the execution of several thousands of experiments simultaneously. The microarray is a matrix of nucleic acid probes (genes) arranged on a solid surface. The probes contain also certain fluorescent nucleotides. When the tissue from a subject is put in contact with the microarray, some of the genes from the microarray may bind with genetic material from the tissue. This process is called hybridization. When that happens, there is a fluorescent expression that can be measured. Such information is stored as a vector $x = [x_1, x_2, \ldots, x_m]$ where each $x_i \in x$ represents the *level of expression* of a certain gene. For illnesses such

© Springer International Publishing Switzerland 2015
J.A. Carrasco-Ochoa et al. (Eds.): MCPR 2015, LNCS 9116, pp. 271–281, 2015.
DOI: 10.1007/978-3-319-19264-2_26

as cancer, the size of the vector usually takes several thousands. The samples from microarrays are labeled using other techniques such as biopsy. Having the vector and the label, a classfier can be used to learn the patterns behind that information.

Training a classifier with a small number of samples, and a big number of features, becomes a challenge. Another complication happens when the datasets are imbalanced. One possibility to deal with those complications consists in retaining only those genes that are really relevant to characterize the pattern for a given disease. This is called *feature reduction*.

In this paper, we propose an improved binary geometric particle swarm optimization to perform feature reduction in DNA microarray datasets. The method we propose is compared with other alternatives reported in the literature. Our method allows an increase in accuracy of about 3 % over the best previous result reported.

The rest of the paper is organized as follows: Sect. 2 presents an overview of previous works, Sect. 3 describes Particle Swarm Optimization, Sect. 4 introduces our proposed method, Sect. 5 describes the experiments and results and finally, Sect. 6 presents the conclusions and future work.

2   Previous Work

The methods to perform feature selection can be classified as: filter methods, wrapper methods and embedded methods [1,2]. In the *filter* approach, simple statistics are computed to rank features and select the best ones. Examples of such methods are: information gain, Euclidean distance, t-test and Chi-square [3]. Filters are fast, can be applied to high-dimensional datasets, and can be used independently from the classifier. Their major disadvantages are: (a) they consider each feature independently, ignoring possible correlations; (b) that they ignore the interaction with the classifier [4].

In the *wrapper* approach, the feature selection is performed using classifiers. A feature subset is chosen, the classifier is trained and evaluated. The performance of the classifier is associated with the subset used. The process is repeated until a "good" subset is found [1]. In general, this approach provides a better classification accuracy, since the features are selected taking into consideration their contribution to the classification accuracy of the classifiers. The disadvantage of the wrapper approach is its computational requirement when combined with classifiers such as neural networks and support vector machines.

Authors such as [5] report a hybrid method that requires three steps. In the first one they use unconditional univariate mixture modeling, then they rank features using information gain and in the third phase they use a Markov blanket filter. This method works well to eliminate redundant features. The experiments included three classifiers: k-NN, Gaussian generative model and logistic regression. Guyon et al. [6] present a complete analysis on gene selection for cancer identification. In their work, they show that selected features are more important than the classifier, and that ranking coefficients for features can be used as classifier weights. Their proposal consists in a Support Vector Machine (SVM) with Recursive Feature Elimination (RFE) as gene selector. The method is tested in several public databases. In [7], a hybrid model is introduced. This model consists of two steps. The first one is a filter to rank the genes. The second one evaluates the subsets with a wrapper. The experiments included four public cancer datasets. In [8], Ant Colony Optimization (ACO) is used in gene selection. The experiments included information from prostate tumor and lung carcinomas. The paper is not clear on how the subsets are evaluated, but they show some results using SVM and Multi-Layer Perceptron (MLP). In [9] a wrapper approach is used, combining ACO with SVM. Each ant constructs a subset of features that is evaluated using an SVM. The classifier accuracy is used to update pheromone. The experiments include five datasets.

Regarding Particle Swarm Optimization (PSO), in [10] the authors propose a method named IBPSO, which stands for Improved Binary Particle Swarm Optimization. The information on which genes will be considered at any given time is provided through a binary set, where 1 means that the corresponding gene will be used and 0 otherwise. The evaluation of subsets is given by the accuracy obtained by a 1-NN (Nearest Neighbor). The experiments included eleven datasets. In nine of those cases the technique outperformed other approaches. In [11], a fairly complex process is introduced. The authors proposed an Integer-Coded Genetic Algorithm (ICGA) to perform gene selection. Then, a PSO driven Extreme Learning Machine (ELM) algorithm is used to deal with problems related to sparse/imbalanced data. Vieira [12], on the other hand, presents a modified binary PSO (MBPSO) for feature selection with the simultaneous optimization of SVM kernel parameter setting. The algorithm is applied to the analysis of mortality in septic patients.

In [13] authors propose a two steps process. In the first one a filter is applied to retain the genes with minimum redundancy and maximum relevance (MRMR). The second step uses a genetic algorithm to identify the highly discriminating genes. The experiments were run on five different datasets. In a recent work, Chen et al. [14], combine a PSO algorithm with the C4.5 classifier. The important genes were proposed using PSO, and then C4.5 was employed as a fitness function.

3 Particle Swarm Optimization

Particle Swarm Optimization (PSO) is an optimization technique that was originally developed by Kennedy and Eberhart [15] based on studies of social models of animals or insects. Since then, it has gained popularity, and many applications have been reported [16]. Particle swarm optimization is similar to a genetic algorithm in the sense that both work with a population of possible solutions.

In PSO the potential solutions (particles), are encoded by means of a position vector x. Each particle has also a velocity vector v that is used to guide the exploration of the search space. The velocity of each particle is modified iteratively by its personal best position p (i.e., the position giving the best fitness value so far), and g, the best position of the entire swarm. In this way, each particle combines its own knowledge (local information) with the collective knowledge (global information). At each time step t, the velocity is updated and the particle is moved to a new position. The new position $x(t+1)$ is given by Eq. 1, where $x(t)$ is the previous position and $v(t+1)$ the new velocity.

$$x(t+1) = x(t) + v(t+1) \qquad (1)$$

The new velocity is computed by Eq. 2:

$$v(t+1) = v(t) + U(0, \phi_1)(p(t) - x(t)) + U(0, \phi_2)(g(t) - x(t)) \qquad (2)$$

where $U(a, b)$ is a uniformly distributed number in $[a, b]$. ϕ_1 and ϕ_2 are often called acceleration coefficients. They determine the significance of $p(t)$ and $g(t)$, respectively [17]. The process of updating position and velocity of the particles continues until a certain stopping criteria is met [18]. In the original version of the algorithm, the velocity was kept within a range $[-V_{max}, +V_{max}]$.

3.1   Inertial Geometric PSO

The original PSO algorithm has been extended in several ways. For instance, in [19], the authors propose a modification of Eq. 2, by introducing a parameter ω, called *inertia weight*. This parameter reduces the importance of V_{max} and it is interpreted as the *fluidity* of the medium in which particles are moving [18]. Thus, the equation to update the velocity of particles is given by Eq. 3:

$$v(t+1) = \omega v(t) + U(0, \phi_1)(p(t) - x(t)) + U(0, \phi_2)(g(t) - x(t)) \qquad (3)$$

PSO was conceived as a method for the optimization of continuous nonlinear functions [15]. However, Moraglio et al. [20] proposed a very interesting way of using PSO for Euclidean, Manhattan and Hamming spaces. This is relevant to us since in the particular case of the DNA microarray features selection, we will be searching on a combinatorial space. By defining a geometric framework, the authors can generalize the operation of PSO to any search space endowed with a distance. This approach is called Geometric PSO (GPSO). In a later work by the same group, they extend GPSO to include the inertia weight [21], giving place to Inertial GPSO (IGPSO). Algorithm 1 shows the pseudocode of the IGPSO [21]. The definition of the convex combinations and the extension ray recombinations are included in the same reference. Note that GPSO and IGPSO do not compute explicitly the velocity of the particles, in difference with PSO. They only compute the position at different times.

4   Methodology

We tried to use the original IGPSO algorithm to work with DNA microarray data. However, in some preliminary experiments we found that the performance of the method was not good in our problem. A further analysis of the causes allowed us to propose a modification that improves the behavior of the method. Here we describe our methodology. It consists of four steps: (1) encoding of DNA microarray feature selection, (2) initialization of the population, (3) definition of a fitness function and (4) adaptation of IGPSO to update the particles.

Algorithm 1. IGPSO
1: Initialize position x_i at random
2: **while** Termination condition not reached **do**
3:　　**for** Each particle i **do**
4:　　　　Update personal best p_i
5:　　　　Update global best g
6:　　**end for**
7:　　**for** Each particle i **do**
8:　　　　Update position using a randomized convex combination
9:　　　　Update position using weighted extension ray
10:　　**end for**
11: **end while**

4.1 Encoding of DNA Microarray Data

A solution for the gene selection problem in DNA microarray data basically consists in a list of genes that are used to train a classifier. A common way to encode a solution i is by using a binary vector $x_i = [x_{i1}, x_{i2}, \ldots, x_{im}]$ where each bit x_{ij} takes a binary value. If gene j is selected in solution i, $x_{ij} \leftarrow 1$, otherwise $x_{ij} \leftarrow 0$. m represents the total amount of genes. Such particles could be represented on a Hamming space.

4.2 Population Initialization

The population is divided into four groups. The first one contains 10 % of the total population. Particles in this set will be initialized with N genes chosen at random. The second group contains 20 % of the population. Those particles will be initialized with $2N$ genes being selected. The third group is the 30 % of the population and its elements will be initialized with $3N$ genes selected. The rest of the particles (40 %) will be initialized with $(1/2)m$ of the genes selected. Following the advice of [22], we define $N \leftarrow 4$.

4.3 Fitness Evaluation

Since the method that we propose is a wrapper, we need to train and test a classifier with each combination of genes given by PSO. The accuracy of the classifier is used as the fitness function. In this work, we report results for a multi-layer perceptron (MLP), although different classifiers can be used. Equation 4 shows how to compute the fitness.

$$f(x_i) = \alpha(acc) + \beta \left(\frac{m - r}{m - 1} \right) \tag{4}$$

where α, β are numbers in the interval $[0, 1]$, such that $\alpha = 1 - \beta$, acc is the accuracy given by the classifier, m is the number of genes in the microarray, and r is the reduced number of genes selected. α and β represent the importance given to accuracy and number of selected genes, respectively.

4.4   Particle Position Updating

Our proposal is based on IGPSO [21], with two improvements. We noticed that sometimes the IGPSO was getting stuck in local minima. To avoid that, we propose considering a threshold value T, along with a counter k. k is increased if during an iteration, the global best g is not changed. If the counter reaches T, then the global best is reset with zeros. This will induce the exploration of the space, allowing to escape from the local minima.

The position update is performed in steps 8 and 9 of Algorithm 1. Step 8 consists in a 3-parental uniform crossover for binary strings. In our work, we take that process as defined in [20]. It is important to mention that step 8 includes the use of Eq. 3, which requires three parameters: ω, ϕ_1 and ϕ_2.

The step 9 in Algorithm 1 is performed by means of an *extension ray* in Hamming spaces [21]. The extension ray C is defined as $C \in ER(A, B)$ iff $C \in [A, B]$ or $B \in [A, C]$. Where A, B and C are points in the space.

For the case of feature selection, the effect of the operations in steps 8 and 9 as originally defined, is that at each iteration the particles will select less genes. Thus, the distance between A and B will become very small compared with the maximum distance m. Moreover, the positions in which A and B will be equal will be mostly in positions occupied by 0's. That will in fact provoke that those bits be flipped to 1's. This means that more genes will be selected, which is the oposite to what we are looking for. In our exploratory experiments, this not only increased the time required to compute the fitness of each particle, but also decreased the accuracy.

Thus, we propose a modification to operator ER, that we call MER. The idea is the following. Once the probability P is computed, a vector $R = [r_1, r_2, \ldots, r_n]$, of n random numbers in $(0, 1)$ is generated and we count how many $r_i \leq P$. That amount (q) represents the number of bits that are equal in A and B and that will therefore change from 1 to 0, or from 0 to 1 in C. To avoid that the operator choses only those bits in which A and B are 0, only $q/2$ bits where A and B are 0 will be randomly chosen to be changed to 1, and similarly, $q/2$ bits will be selected from the ones in which A and B are 1 to be changed to 0. Algorithm 2 describes this steps in detail.

5   Experiments and Results

5.1   Experimental Setup

All our experiments were run on a HP Proliant G8 server with 2 Intel processors, 8 cores each and 256 Gb RAM. The algorithms are evaluated on two public datasets. One is the colon cancer dataset, which contains the expression levels of 2000 genes in 22 normal colon tissues and 40 tumor [23]. The other dataset is the prostate dataset (GEMS), which involves the expression levels of 10,509 genes in 52 normal prostate tissue, and 50 tumor [24].

Algorithm 2. Modified Extension Ray algorithm

1: **function** MER($A, B, w_{ab}, w_{b,c}$)
2:     **Input:** Two particles $A = [a_1, a_2, \ldots, a_m]$ and $B = [b_1, b_2, \ldots, b_m]$, $w_{a,b}, w_{b,c}$
3:     **Output:** $C = [c_1, c_2, \ldots, c_m]$
4:     $d(B, C) = d(A, B) \cdot w_{bc}/w_{ab}$
5:     $n \leftarrow m - d(A, B)$
6:     Compute probability $P = d(B, C)/(m - d(A, B))$
7:     Generate a vector $R = [r_1, r_2, \ldots, r_n]$ of random numbers in $(0, 1)$
8:     Let $q \leftarrow 0$
9:     **for** Each gene i **do**
10:         **if** $(r_i \leq P)$ **then**
11:             $q \leftarrow q + 1$
12:         **end if**
13:     **end for**
14:     Let Z be the list of gene indices where $(a_i \vee b_i) = 0$
15:     Let O be the list of gene indices where $(a_i \wedge b_i) = 1$
16:     Choose at random $q/2$ indices from Z and make their positions 1 in C.
17:     Choose at random $q/2$ indices from O and make their positions 0 in C.
18: **end function**

5.2   Parameter Tuning

The algorithm requires the determination of parameters w, ϕ_1 and ϕ_2. A design of experiments was performed considering three factors (the parameters), three levels for the first factor and six levels for the other two. A factorial experiment required a total of 47 different combinations of values, with 10 repetitions for each combination. For colon dataset, the best values were: $w = 0.3, \phi_1 = 2.2, \phi_2 = 2.2$. For prostate cancer, $w = 0.3, \phi_1 = 1.4, \phi_2 = 1.4$.

5.3   Comparison with Other Methods

In this section we compare the results of MIGPSO against other methods reported in the literature. Table 1 shows time, accuracy and number of selected genes. As it is possible to observe, MIGPSO performs better in all. It improves an average of 4 % the accuracy, while time is reduced to almost half. The number of genes selected is reduced to almost one third.

In Table 2, we show a comparison for the colon dataset with other wrapper methods. The classifiers considered are: Naive Bayes (NB), C4.5, and Support Vector Machines (SVM). Six different methods for feature selection were considered [7]. The best wrapper reported for this dataset is SVM + Hybrid PSO/GA with 92 % accuracy. MIGPSO with an MLP reach an accuracy of 96 %. That is an increase of almost 4 % in average. But it is worth noting that the amount of genes required is smaller in MIGPSO. Only 6 genes, compared with 18 in the other case. In perspective with the total amount of genes, that means a reduction of three orders of magnitude.

Table 1. Comparison between the original IGPSO and MIGPSO. For prostate, IGPSO was allowed to run 13 hours and it did not gave an answer.

Dataset Method	Feature Selection (%)	Accuracy	Time	Number of selected genes
Colon	IGPSO	92.76	15823	21
	MIGPSO	95.8	9218	6
Prostate	IGPSO	NA	NA	NA
	MIGPSO	98.04 %	27538	84

Table 2. Comparison with several classifiers and feature selection methods for colon cancer dataset.

Classifier	Feature Selection Method	Colon dataset	
		Accuracy (%)	Number of genes selected
NB	BIRSw	85.5	3.5
NB	FCBF	77.6	14.6
NB	SFw	84.1	5.9
NB	CSFsf	82.6	22.1
NB	FOCUSsf	77.1	4.6
NB	Hybrid PSO/GA	85.5	12.9
C4.5	BIRSw	83.8	2.9
C4.5	FCBF	88.3	14.6
C4.5	SFw	80.7	3.3
C4.5	CSFsf	86.9	22.1
C4.5	FOCUSsf	79.1	4.6
C4.5	Hybrid PSO/GA	83.9	15.1
SVM	Hybrid PSO/GA	91.9	18.0
MLP	MIGPSO	**95.8**	6.1

Table 3 shows a similar comparison for prostate dataset. The classifiers considered are: SVM, Self-Organizing Map (SOM), BPNN (back propagation neural network), Naïve Bayes, Decision Trees (CART), Artificial Immune Recognition System (AIRS). PSO with C4.5 Decision Tree (PSODT). All those experiments use PSO as feature selection. The best method reported for this dataset is PSODT with 94.31 % accuracy. Our MIGPSO with an MLP goes to 98.04 % accuracy. Almost 4 % above the best reported.

Table 3. Comparison for prostate cancer dataset. The first seven results are taken from [14]. The number of genes for some of the methods was not available.

Classifier	Feature Selection	Prostate dataset	
		Accuracy (%)	Number of genes
SVM	PSO	87.66	NA
SOM	PSO	67.24	NA
BPNN	PSO	56.72	NA
NB	PSO	62.35	NA
CART	PSO	82.55	NA
AIRS	PSO	52.51	NA
PSODT	PSO	**94.31**	126
MLP	MIGPSO	**98.04**	87

6   Conclusions and Future Work

New technologies such as DNA microarrays are generating information that can be of relevance to detect, treat and prevent illnesses in a better way. However, the generated data presents challenges for current computational methods. The difference between the number of attributes and samples can be of several orders of magnitude. Even more, most datasets of microarray information are imbalanced. There are many more samples for healthy than from diseased subjects. The main concerns for automated processing and classification of new data are on the one hand the reduction of attributes analyzed and on the other hand, the use of robust classifiers. Our experiments suggest that performing a good attribute selection is worth when classifying. Having less attributes reduces the processing time, but also may increase the correct classification of new data. In this work we show that a modification of a geometric version of PSO allows a Multi-Layer perceptron to classify with high accuracy.

There are of course some issues that might be worth to explore in future work, such as: (1) running more experiments with other classifiers in the wrapper, (2) running experiments with more datasets, particularly with multi-class problems.

Acknowledgments. The authors thank Tecnológico de Monterrey, Campus Guadalajara, as well as IPN-CIC under project SIP 20151187, and CONACYT under project 155014 for the economical support to carry out this research.

References

1. Koller, D., Sahami, M.: Toward optimal feature selection. Technical report, Stanford InfoLab, Stanford University (1996)
2. Kohavi, R., John, G.H.: Wrappers for feature subset selection. J. Artif. Intel. **97**(1), 273–324 (1997)

3. Guyon, I., Elisseeff, A.: An introduction to variable and feature selection. J. Mach. Learn. Res. **3**, 1157–1182 (2003)
4. Saeys, Y., Inza, I., Larrañaga, P.: A review of feature selection techniques in bioinformatics. J. Bioinf. **23**(19), 2507–2517 (2007)
5. Xing, E.P., Jordan, M.I., Karp, R.M., et al.: Feature selection for high-dimensional genomic microarray data. In: ICML, vol. 1, pp. 601–608. Citeseer (2001)
6. Guyon, I., Weston, J., Barnhill, S., Vapnik, V.: Gene selection for cancer classification using support vector machines. J. Mach. Learn. **446**(1–3), 389–422 (2002)
7. Ruiz, R., Riquelme, J.C., Aguilar-Ruiz, J.S.: Incremental wrapper-based gene selection from microarray data for cancer classification. J. Patt. Recog. **39**(12), 2383–2392 (2006)
8. Chiang, Y.M., Chiang, H.M., Lin, S.Y.: The application of ant colony optimization for gene selection in microarray-based cancer classification. Int. Conf. Mach. Learn. Cybern. **7**, 4001–4006 (2008)
9. Yu, H., Gu, G., Liu, H., Shen, J., Zhao, J.: A modified ant colony optimization algorithm for tumor marker gene selection. Genom. Proteom. Bioinf. **7**(4), 200–208 (2009)
10. Chuang, L.Y., Chang, H.W., Tu, C.J., Yang, C.H.: Improved binary PSO for feature selection using gene expression data. Comp. Biol. Chem. **32**(1), 29–38 (2008)
11. Saraswathi, S., Sundaram, S., Sundararajan, N., Zimmermann, M., Nilsen-Hamilton, M.: ICGA-PSO-ELM approach for accurate multiclass cancer classification resulting in reduced gene sets in which genes encoding secreted proteins are highly represented. T. Comput. Biol. Bioinf. **8**(2), 452–463 (2011)
12. Vieira, S.M., Mendonça, L.F., Farinha, G.J., Sousa, J.M.: Modified binary PSO for feature selection using SVM applied to mortality prediction of septic patients. J. Appl. Soft Comput. **13**(8), 3494–3504 (2013)
13. El Akadi, A., Amine, A., El Ouardighi, A., Aboutajdine, D.: A two-stage gene selection scheme utilizing MRMR filter and GA wrapper. Know. Inf. Syst. **26**(3), 487–500 (2011)
14. Chen, K.H., Wang, K.J., Tsai, M.L., Wang, K.M., Adrian, A.M., Cheng, W.C., Yang, T.S., Teng, N.C., Tan, K.P., Chang, K.S.: Gene selection for cancer identification: a decision tree model empowered by particle swarm optimization algorithm. BMC Bioinf. **15**(1), 49 (2014)
15. Kennedy, J., Eberhart, R.: Particle swarm optimization. IEEE Int. Conf. Neural Networks. **4**, 1942–1948 (1995)
16. García-Gonzalo, E., Fernández-Martínez, J.: A brief historical review of particle swarm optimization (PSO). J. Bioinf. Intel. Cont. **1**(1), 3–16 (2012)
17. Yang, X.S.: Engineering Optimization: An Introduction With Metaheuristic Applications. Wiley, New York (2010)
18. Poli, R., Kennedy, J., Blackwell, T.: Particle swarm optimization. Swarm Intell. **1**(1), 33–57 (2007)
19. Shi, Y., Eberhart, R.: A modified particle swarm optimizer. In: IEEE Conference Evolutionary Computation., pp. 69–73 (1998)
20. Moraglio, A., Di Chio, C., Togelius, J., Poli, R.: Geometric particle swarm optimization. J. Artif. Evol. Appl. **11**, 247–250 (2008)
21. Moraglio, A., Togelius, J.: Inertial geometric particle swarm optimization. In: IEEE Conference on Evolutionary Computation, pp. 1973–1980 (2009)
22. Alba, E., García-Nieto, J., Jourdan, L., Talbi, E.G.: Gene selection in cancer classification using PSO/SVM and GA/SVM hybrid algorithms. In: IEEE Conference on Evolutionary Computation, pp. 284–290 (2007)

23. Alon, U., Barkai, N., Notterman, D.A., Gish, K., Ybarra, S., Mack, D., Levine, A.J.: Broad patterns of gene expression revealed by clustering analysis of tumor and normal colon tissues probed by oligonucleotide arrays. Proc. Nat. Acad. Sci. **96**(12), 6745–6750 (1999)
24. Singh, D., Febbo, P.G., Ross, K., Jackson, D.G., Manola, J., Ladd, C., Tamayo, P., Renshaw, A.A., D'Amico, A.V., Richie, J.P., et al.: Gene expression correlates of clinical prostate cancer behavior. Cancer cell **1**(2), 203–209 (2002)

Encoding Polysomnographic Signals into Spike Firing Rate for Sleep Staging

Sergio Valadez[1]([✉]), Humberto Sossa[1], Raúl Santiago-Montero[2],
and Elizabeth Guevara[1]

[1] Instituto Politécnico Nacional, CIC, Av. Juan de Dios Batiz, S/N, Col. Nva.
Industrial Vallejo, 07738 Mexico, DF, Mexico
svaladezg1400@alumno.ipn.mx, hsossa@cic.ipn.mx, eli.guevara@gmail.com
[2] División de Estudios de Posgrado e Investigación, Instituto Tecnológico de León,
Av. Tecnológico S/N, León, Guanajuato, Mexico
rsantiago66@gmail.com

Abstract. In this work, an encoding of polysomnographic signals into a
spike firing rate, based on the BSA algorithm, is used as a discriminant
feature for sleep stage classification. This proposal obtains a better sleep
staging compared with the mean power signals frequency. Furthermore,
a comparison of classification results obtained by different algorithms -
such as Support Vector Machines, Multilayer Perceptron, Radial Basis
Function Network, Naïve Bayes, K-Nearest Neighbors and the decision
tree algorithm C4.5 - is reported, demonstrating that Multilayer Perceptron has the best performance.

Keywords: Sleep stages · EEG · EOG · EMG · BSA algorithm · FIR
filter · Encoding · Polysomnogram · Spike train · Spike firing rate ·
Classification

1 Introduction

Sleep is a basic human need -such as drinking and eating- that is an essential part
of any individuals' life. However, even though precise functions of sleep remain
unknown, sleep disturbances can cause physical and mental health problems,
injuries, weak productivity performance, and even a greater risk of death [1,2].
To diagnose any sleep deficiency, sleep studies are needed to identify any disorder.
There exist different kinds of sleep studies; one of the most used is the polysomnog-
raphy [1]. This study records different physiological changes that occur during
sleep, including electroencephalogram (EEG), electrooculogram (EOG) and elec-
tromyogram (EMG). By placing electrodes on the scalp and face, EEG measures
small brain voltages, EOG records the electrical activity of eye movements and
EMG obtains the electric changes in a muscle, usually the chin [2].

Sleep stage identification is a first step in the diagnosis of any sleep deficiency
[3]. Therefore, to achieve that goal it is necessary to understand how sleep functions. Generally, sleep is divided into two major states: Rapid Eye Movement

© Springer International Publishing Switzerland 2015
J.A. Carrasco-Ochoa et al. (Eds.): MCPR 2015, LNCS 9116, pp. 282–291, 2015.
DOI: 10.1007/978-3-319-19264-2_27

(REM) and Non-Rapid Eye Movement (NREM) [2]. During the REM stage, EEG waves are fast and not synchronized, meanwhile in NREM are slower. Considering wave amplitude and frequency, NREM is divided into three stages named N1, N2, and N3. Due to the physiological changes in the body during sleep, EEG, EMG and EOG patterns change several times. Thus, polysomnographic recordings could help to identify the sleep stages.

In the last decade several works have tackled the problem of automated sleep staging, see [3] for a review of different approaches. Most of these studies extract a large number of polysomnographic signal features using spectral information through Fourier [4,5] and Wavelet [6–8] Transforms. However, very few researches have used a feature considering the conversion of polysomnographic signals into spike trains (see [9] and [10]).

In this work, we propose the conversion of somnographic signals into spike firing rates by first obtaining spike trains from the polysomnograms. A spike train is generated using the Bens Spiker Algorithm (BSA) [11]. Later, it is converted into a spike firing rate proposed by the Nobel laureate E. Adrian [12]. With only one feature per signal and without cleaning it, the spike firing rate allows classifying sleep stages with better results than the mean power frequency [13] of the same signal. The mean power frequency was chosen for comparison for two reasons. First because the frequency domain shows the better performance representing polysomnographic signals for sleep staging [3]. Second because the mean power frequency is an average value, as the spike firing rate, but in different domains.

To demonstrate the advantages of the spike firing rate encoding for the sleep staging, a comparison with the mean power frequency of the same signals was performed. These experiments showed that with the spike firing rate a better classification was obtained.

The remainder of the paper is organized as follows: Sect. 2 describes the experimental data and methodology used. Section 3 shows the results of two experiments. Finally, Sect. 4 presents the conclusions and future work.

2   Methods

In this section, a description of the somnographic data used for the experiments is presented, also the steps to characterize a signal into spike firing rate and mean power frequency and the algorithms used for sleep classification.

2.1   Polysomnographic Data

The data was obtained by Dr. Mary Carskadon, professor in the Department of Psychiatry and Human Behavior at the Warren Alpert Medical School at Brown University. It contains recordings of sleep stages from four people and two files for each person. The first one contains night recordings baseline (BSL) of rested sleep and the second contains recordings of recovery following a sleep deprivation night (REC). The structure for each file is the following:

- DATA: physiological characteristics recorded within approximately 10.5 h.
- stages: researcher classified stages for each 30-s epoch.
- srate: sampling rate for the data.

The physiological characteristics recorded in DATA are: EEG, EOG and EMG (tone chin muscle) recordings. DATA contains recordings of nine channels in total. Channel 1 (C3/A2), Channel 2 (O2/A1), Channel 8 (C4/A1) and Channel 9 (O1/A2) correspond to EEG. These channels are named as "recording electrode location" /"reference electrode location" and placed over the scalp according to the International 10–20 system [14] (see Fig. 1). Channel 3 (ROC/A2) and Channel 4 (LOC/A1) are from EOG, where ROC and LOC are the right and left outer canthus (corner of the eye) electrodes positioned according to [15]. Channel 5 (chin EMG 1), Channel 6 (chin EMG 2) and Channel 7 (chin EMG 3) correspond to EMG. An overnight recording is divided into 30-s epochs. A sleep stage is assigned to each epoch manually by a researcher according to [16] and [17]. The stages in the dataset are eight: Awake (W), N1, N2, N3, N4, REM sleep (R), Movement time and Unscored. The sampling rate for the data is 128 samples/s.

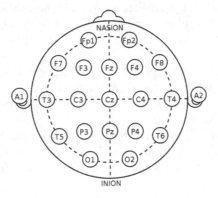

Fig. 1. Location of scalp electrodes according to the international 10–20 system in the context of an EEG recording [14].

For simplicity and considering that the signals for the same somnogram are similar, only the BSL recording for one person and one channel of each kind of signal was used: EEG from Channel 1, EOG from Channel 3 and EMG from Channel 7. These channels are denoted as EEG, EOG and EMG respectively in the rest of the article. Usually, N3 and N4 stages are merged as one. Thus, this combination was named as N3. Movement time stage was not considered due to it not containing any useful information. Therefore, the number of stages are six: N1, N2, N3 and R for sleep stages; W for awake; and Unscored signals. Figure 2 presents some polysomnographic recordings of each sleep stage.

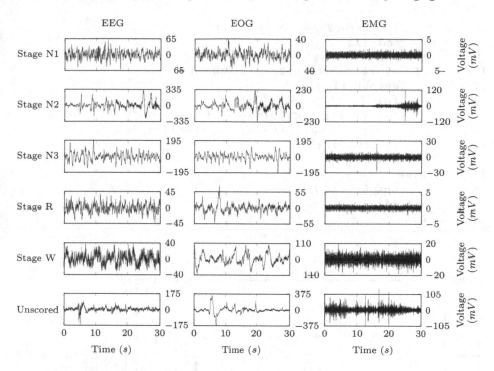

Fig. 2. Polysomnographic recordings -EEG, EOG and EMG- of the sleep stages (N1, N2, N3, R), awake (W) and Unscored signals. The data was obtained from four people by Dr. Mary Carskadon in her laboratory at Brown University.

2.2 Encoding Polysomnographic Signals into Spike Firing Rates

The Ben Spiker Algorithm (BSA) (see Pseudocode 1) is a technique for converting analog values (such as polysomnographic signals) into a spike train. The BSA was introduced in [11] and tries to obtain a reverse convolution of the analog signal, also assumes the use of a Finite Impulse Response (FIR) reconstruction filter, a threshold and no external parameters.

Using the BSA algorithm, the spike trains for EEG and EOG epochs were generated applying FIR filter 1 (see Fig. 3) with the threshold equal to 0.9550, as in [11]. Due to the EMG signal being smaller in amplitude than EEG and EOG, the FIR filter 1 was modified in an attempt to increase its response in amplitude. This new filter was named FIR filter 2 (see Fig. 3).

Pseudocode 1. BSA pseudocode.

```
1   for i = 1 to size(input)
2       error1 = 0;
3       error2 = 0;
4       for j = 1 to size(filter)
5           if i+j−1 <= size(input)
6               error1 += abs(input(i+j−1) − filter(j));
```

```
7                    error2 += abs(input(i+j-1));
8            end if;
9        end for;
10       if error1 <= (error2 - threshold)
11           output(i)=1;
12           for j = 1 to size(filter)
13                if i+j-1 <= size(input)
14                    input(i+j-1) -= filter(j);
15                end if;
16           end for;
17       else
18           output(i)=0;
19       end if;
20   end for;
```

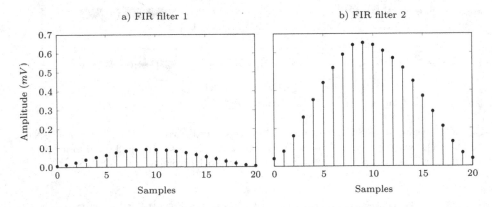

Fig. 3. Finite Impulse Response (FIR) filters to encode a polysomnographic epoch into a spike train. (a) Original filter found by Schrauwen and Campenhout [11]. This filter was used to encode each epoch of the EEG and EOG into a spike train. (b) An amplitude extended filter for the EMG.

The classification feature was generated applying BSA over the signals and obtaining a spike train for each epoch of duration $T = 30$s. An example of an EEG epoch encoding into a spike trains is presented in Fig. 4. The spike firing rate per epoch was calculated as the total number of spikes that occurs in a time window T divided by T [12]:

$$spike\,firing\,rate = \frac{number\,of\,spikes}{T} \qquad (1)$$

2.3   Encoding Polysomnographic Signals into Mean Power Frequencies

To transform the polysomnographic signals in the time domain to a signal in the frequency domain, the Fast Fourier Transform was applied. The correlation

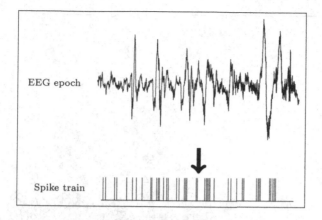

Fig. 4. Encoding of an EEG epoch into a spike train using FIR filter 1 (see Fig. 3).

signal function was used to obtain the power spectral density (PSD). The PSD was calculated using one of the most popular methods [18]. Later, the mean power frequency is calculated as the product sum of the PSD and the frequency; divided by the total sum of PSD [13]:

$$mean\ power\ frequency = \frac{\sum_{i=1}^{M} f_i P_i}{\sum_{i=1}^{M} P_i} \tag{2}$$

where f_i is the PSD frequency value, P_i is the power spectrum and M is the size of the frequency bin i.

2.4   Classification Algorithms

The classification algorithms used were: Multilayer Perceptrons (MLP) [19,20], Support Vector Machines (SVM) [21], K-nearest neighbors (KNN) [22], C4.5 decision tree algorithm [23], Naïve Bayes [22] and Radial Basis Function Networks (RBFN) [24]. These classifiers were applied using the software WEKA v3.6.10 [25] with the default configurations for each of the algorithms and a 10-fold cross-validation [26].

3   Results

For validating the advantages of the proposed encoding, a comparison of classification results between different algorithms was performed. Each somnogram epoch was encoding into spike firing rate and mean power frequency as described in Sects. 2.2 and 2.3 respectively. For the comparison, the algorithms cited in Sect. 2.4 were used. The distribution of the BSL epochs based in the spike firing rate encoding for the EEG and EMG of a person is shown in Fig. 5. As can be appreciated, the problem of separating perfectly all epochs is difficult to solve

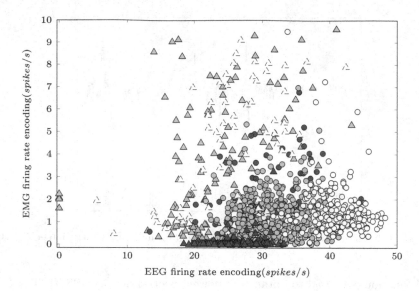

Fig. 5. Distribution of the epochs based in the spike firing rate encoding (*spikes*/s) of the EEG and EMG of a person in various stages: ● N1, ◉ N2, ○ N3, ▲ R, △ W and △ the Unscored signal. The FIR filter 1 was used to encode each EEG epoch into a spike train and FIR filter 2 for the EMG epochs. Subsequently, spike trains were used to calculate its firing rates encoding.

due its overlapping. However, R, N1 and N2 epochs are clearly clustered. Thus, a first test was made considering the separation of the R epochs from the rest of the epochs. Later, a second experiment was performed taking into account all the six sleep stages.

The results of the first experiment are shown in Table 1. As can be seen, the best classification percentage (94.78 %) was achieved with MLP considering the spike firing rate. Using the mean power frequency as a feature was only 85.14 % with C4.5.

The results for the second test are shown in Table 2. Considering the spike firing rate, a 74.54 % of classification efficiency was achieved again with MLP and using the mean power frequency the percentage was only 59.83 % with RBF. The percentage of accuracy decreases because the overlapping between epochs.

Table 1. Classification accuracy (%) when separating R stage from all others: N1, N2, N3, W and the Unscored signal.

Encoding	SVM	NB	RBF	1NN	C4.5	MLP
Mean power frequency	80.79	67.11	84.35	84.82	**85.14**	79.21
Spike firing rate	81.58	87.67	90.75	93.68	93.91	**94.78**

Table 2. Classification accuracy (%) when separating all sleep stages: N1, N2, N3, R, W and the Unscored signal.

Encoding	NB	1NN	SVM	C4.5	RBF	MLP
Mean power frequency	53.43	51.86	41.11	57.86	**59.83**	52.65
Spike firing rate	68.62	69.33	69.56	71.93	73.36	**74.54**

However, the classification percentage remains the best with the spike firing rate encoding.

4 Conclusions

These experimental results demonstrate that the spike firing rate can be used as a feature for sleep staging. It was proved with a comparison between the spike firing rate and the mean power frequency. This comparison showed that the proposed encoding has a better classification accuracy to separate the R sleep stage from the rest of other stages. Also, the spike firing rate allows separating all the sleep stages better than the mean power frequency, but due to it being a complex problem, the classification accuracies are much lower in this case. Therefore, the results suggest that the firing rate codification is efficient to classify sleep stages. In both cases, the classification algorithm with the best performance was the MLP.

An aspect that must be considered as further research is to optimize the sampling and amplitude filter parameters through evolutionary algorithms in order to improve encoding.

Acknowledgments. S. Valadez would like to thank CONACYT for the scholarship granted in pursuit of his doctoral studies, Dr. M. Carskadon for providing the database of sleep stages, M. Linden, D. Sheinberg and Brown University for the *Exploring Neural Data* course at www.coursera.org. H. Sossa would like to thank SIP-IPN and CONA-CYT under grants 20151187 and 155014 for the economical support to carry out this research. R. Santiago-Montero would like to thank Instituto Tecnológico de León and Tecnológico Nacional de México. E. Guevara thanks CONACYT for the scholarship granted for her doctoral studies.

References

1. National Heart, Lung, and Blood Institute: (2014). http://www.nhlbi.nih.gov/
2. National Heart, Lung, and Blood Institute: Sleep, sleep disorders, and biological rhythms. NIH Curriculum supplement series grades sleep (2003)
3. Şen, B., Peker, M., Çavuşoğlu, A., Çelebi, F.V.: A comparative study on classification of sleep stage based on EEG signals using feature selection and classification algorithms. J. Med. Syst. **38**(3), 18–21 (2014)

4. Zoubek, L., S., C., Lesecq, S., Buguet, A., Chapotot, F.: A two-steps sleep/wake stages classifier taking into account artefacts in the polysomnographic signals. In: Proceedings of the 17th IFAC World Congress, pp. 5227–5532 (2008)
5. Doroshenkov, L.G., Konyshev, V.A., Selishchev, S.V.: Classification of human sleep stages based on EEG processing using hidden markov models. Biomed. Eng. **41**(1), 25–28 (2007)
6. Ebrahimi, F., Mikaeili, M., Estrada, E., Nazeran, H.: Automatic sleep stage classification based on EEG signals by using neural networks and wavelet packet coefficients. In: Proceedings of the 30th Annual International Conference of the IEEE Engineering in Medicine and Biology Society, EMBS 2008, pp. 1151–1154 (2008)
7. Subasi, A., Erçelebi, E.: Classification of EEG signals using neural network and logistic regression. Comput. Methods Programs Biomed. **78**, 87–99 (2005)
8. Oropesa, E., Cycon, H.L., Jobert, M.: Sleep stage classification using wavelet transform and neural network. ICSI technical report TR-99-008, pp. 1–7 (1999)
9. Nuntalid, N., Dhoble, K., Kasabov, N.: EEG classification with BSA spike encoding algorithm and evolving probabilistic spiking neural network. In: Lu, B.-L., Zhang, L., Kwok, J. (eds.) ICONIP 2011, Part I. LNCS, vol. 7062, pp. 451–460. Springer, Heidelberg (2011)
10. Kasabov, N.K.: NeuCube: a spiking neural network architecture for mapping, learning and understanding of spatio-temporal brain data. Neural Netw. **52**, 62–76 (2014)
11. Schrauwen, B., Van Campenhout, J.: BSA, a fast and accurate spike train encoding scheme. In: Proceedings of the International Joint Conference on Neural Networks, vol. 4(4), pp. 2825–2830 (2003)
12. Adrian, E.D.: The impulses produced by sensory nerve endings. J. Physiol. **61**, 49–72 (1926)
13. Phinyomark, A., Thongpanja, S., Hu, H., Phukpattaranont, P., Limsakul, C.: The usefulness of mean and median frequencies in electromyography analysis. In: Naik, G.R., (ed.) Computational Intelligence in Electromyography Analysis- A Perspective on Current Applications and Future Challenges, pp. 195–220. Intech (2012)
14. Jasper, H.: Report of the committee on methods of clinical examination in electroencephalography: 1957. Electroencephalogr. Clin. Neurophysiol. **10**(2), 370–375 (1958)
15. Marmor, M.F., Brigell, M.G., McCulloch, D.L., Westall, C.A., Bach, M.: ISCEV standard for clinical electro-oculography (2010 update). Doc. Ophthalmol. **122**, 1–7 (2011)
16. Rechtschaffen, A., Kales, A.: A Manual of Standardized Terminology, Techniques and Scoring System for Sleep Stages of Human Subjects, vol. 204, Bethesda, Md., U.S. National Institute of Neurological Diseases and Blindness, Neurological Information Network (1968)
17. Iber, C., Ancoli-Israel, S., Chesson, A., Quan, S.: The AASM Manual for the Scoring of Sleep and Associated Events: Rules, Terminology and Technical Specifications. American Academy of Sleep Medicine, Westchester (2007)
18. Welch, P.D.: The use of fast fourier transform for the estimation of power spectra: a method based on time averaging over short, modified periodograms. IEEE Trans. Audio Electroacoust. **15**, 70–73 (1967)
19. McCulloch, W.S., Pitts, W.: A logical calculus of the ideas immanent in nervous activity. Bull. Math. Biophys. **5**, 115–133 (1943)
20. Rosenblatt, F.: The perceptron: a probabilistic model for information storage and organization in the brain. Psychol. Rev. **65**(6), 386–408 (1958)

21. Vapnik, V., Lerner, A.: Pattern recognition using generalized portrait method. Autom. Remote Control **24**(6), 709–715 (1963)
22. Duda, R., Hart, P., Stork, D.: Pattern Classification, 2nd edn. Wiley Interscience, New York (2001)
23. Quinlan, J.: C4.5: Programs for Machine Learning. Morgan Kaufmann Publishers, San mateo (1993)
24. Theodoridis, S., Koutroumbas, K.: Pattern Recognition, 4th edn. Academic Press, San Diego (2009)
25. Hall, M., Frank, E., Holmes, G., Pfahringer, B., Reutemann, P., Witten, I.H.: The weka data mining software: an update. SIGKDD Explor. **11**(1), 10–18 (2009)
26. Kohavi, R.: A study of cross-validation and bootstrap for accuracy estimation and model selection. In: Proceedings of the 14th International Joint Conference on Artificial Intelligence (IJCAI), vol. 2, pp. 1137–1143 (1995)

Improving Bayesian Networks Breast Mass Diagnosis by Using Clinical Data

Verónica Rodríguez-López[✉] and Raúl Cruz-Barbosa

Universidad Tecnológica de la Mixteca,
69000 Huajuapan de León, Oaxaca, Mexico
{veromix,rcruz}@mixteco.utm.mx

Abstract. Nowadays, breast cancer is considered a significant health problem in Mexico. Mammogram is an effective study for early detecting signs of this disease. One of the most important findings in this study is a mass, which is the main indicator of malignancy. However, mass detection and diagnosis are difficult. In this study, the impact of the inclusion of seven clinical features on the performance of Bayesian Networks models for mass diagnosis is presented. Here, Naïve Bayes, Tree Augmented Naïve Bayes, K-dependence Bayesian classifier, and Forest Augmented Naïve Bayes models with eight image features nodes were augmented with several clinical features subsets. These models were trained with a data set extracted from the public BCDR-F01 database. The experimental results have shown that the Bayesian networks models augmented with a subset of three clinical features have improved their performance up to 0.82 in accuracy, 0.80 in sensitivity, and 0.83 in specificity. Therefore, these augmented models are considered as suitable and promising methods for mass classification.

Keywords: Breast cancer · Bayesian networks · Clinical data · Mass diagnosis

1 Introduction

Breast cancer is a disease in which malignant cells are formed in the breast tissues [17]. Nowadays, it is the most common cancer among women in the world. Although the highest rates of incidence for this disease are found in developed countries, the lowest survival rates occur in less developed countries [26]. As an example of this situation, Mexico is found at an intermediate level, with incidence rates four times lower than the highest ones, however, since 2006 breast cancer is the primary cause of death from malignant tumors among women [4].

A mammogram is an X-ray film of the breast that has shown to be an effective tool for early detection of breast cancer. One of the most important findings in this study is a mass, which is one of the main signs for detection of this disease. The American College of Radiology (ACR) defines a mass as a three-dimensional structure demonstrating convex outward borders, usually evident on two orthogonal views. These lesions are diagnosed from their characteristics of size,

© Springer International Publishing Switzerland 2015
J.A. Carrasco-Ochoa et al. (Eds.): MCPR 2015, LNCS 9116, pp. 292–301, 2015.
DOI: 10.1007/978-3-319-19264-2_28

shape, margins and density [1]. Margin refers to the mass border, and density is the amount of fat tissue in the mass compared with the surrounding breast tissue. Fatty masses with round shaped and well defined margins usually indicate benign cellular changes, while irregular shaped masses with high density and ill-defined margins, have a high probability of malignancy [8].

The diagnostic of masses is difficult, because they often are hidden by complex or similar breast tissue, which make them hard to distinguish from their characteristics. Moreover, the wide range of mass characteristics add more complexity to this, which becomes a more difficult, tedious and time-consuming task. For this, Computer Aided Diagnosis Systems (CADx) have been developed in order to help in the mammogram analysis. These systems have shown to improve the performance in mass diagnosis from 75 % to 82 % [6,24,25].

Bayesian networks are probabilistic models that have been applied on breast diagnosis. There are several reports about the performance of Bayesian networks for breast diagnosis when they are trained with clinical data and mammogram findings provided by radiologists [2,10,13]. However, few works have been discussed about their performance when they are trained from automatically obtained features [18,25]. In this study, the performance comparison of Bayesian networks models for breast mass diagnosis is presented. The main aim of this paper is to analyze the impact of the inclusion of seven clinical features on the performance of Bayesian Networks models as Naïve Bayes, Tree Augmented Naïve Bayes, K-dependence Bayesian classifier, and Forest Augmented Naïve Bayes. These models are trained with automatically obtained image features and are augmented with several clinical features subsets.

The paper is organized as follows. In Sect. 2, we explain the image features used for mass description. A brief review of Bayesian networks classifiers is presented in Sect. 3. In Sect. 4, the data sets used in our experiments is described, and the experimental results are presented in Sect. 5. Finally, conclusions and future work are given in Sect. 6.

2 Image Features

In this study, eight image features extracted from mammogram images are used to characterize the shape, margins and density of masses. This set of image descriptors were selected as in [22].

The mass shape can be round, oval, lobular, or irregular [1]. Round and convex masses are probably benign, whereas the malignant ones have irregular shape [8]. The compactness feature is used to describe the mass shape.

The most important predictor of malignancy is the mass margin. A mass can appear with circumscribed (well-defined), microlobulated, obscured (by density of adjacent tissue), indistinct (ill-defined), or spiculated margins. Benign masses usually have very well defined margins, while the poorly defined are associated with malignant ones [8]. The mean, standard deviation and entropy of the normalized radial length (NRL) are calculated and used as margins descriptors. The NRL is defined as the normalized Euclidean distance from a point on the mass boundary to the mass centroid [9].

The mass density can be fatty, or also be lower, higher or isodense than the surrounding glandular breast tissue. Fatty or isodense masses have less probability of malignancy than the ones with higher density [8]. To describe this density, three intensity features and energy of gray-level co-occurrence matrices (GLCM) are used. From the intensity of pixels belonging to the mass region, three basic statistics are calculated: median, skewness, and kurtosis [6,9]. The energy is obtained from a sample of 32×32 pixels of the mass center for the four directions $\{0°, 45°, 90°, 135°\}$, and $d = 1$. The range, in the four directions, is taken as density descriptor.

3   Bayesian Networks Classifiers

A Bayesian Network (BN) is a probabilistic graphical model where the nodes represent variables, and the arcs, dependence among variables. BN are recognized as powerful tools for knowledge representation and inference under conditions of uncertainty [7].

A formal definition of BN is as follows: A Bayesian network is a pair (D, P), where D is a directed acyclic graph, and $P = \{p(x_1|\pi_1), p(x_2|\pi_2), ..., p(x_n|\pi_n)\}$ is a set of n conditional probability distributions, one for each variable, and Π_i is the parent set of node X_i in D. The set P defines the associated joint probability distributions as [5],

$$p(x_1, x_2, ..., x_n) = \Pi_{i=1}^{n} p(x_i|\pi_i) \qquad (1)$$

Several types of BN models have been proposed as classifiers, some of them are: Naïve Bayes (NB), Tree Augmented Naïve Bayes (TAN), K-dependence Bayesian classifier (KDB), and Forest Augmented Naïve Bayes (FAN).

3.1   Naïve Bayes

Naïve Bayes model is the simplest form of a Bayesian network, in which the root node of a tree-like structure corresponds to a class variable. Also, the class node is the only parent for each attribute variable. The key assumption of a NB model is that all attributes are independent given the value of the class variable [7].

3.2   Tree Augmented Naïve Bayes

A TAN classifier is an extension of the Naïve-Bayes model, and also has a tree-like structure. In this model is allowed that each variable has at most two parents: the class variable, and other attribute. The Friedman method [11] can be applied to learn the structure of a TAN model.

3.3   K-Dependence Bayesian Classifier

A KDB classifier is a Bayesian Network where each attribute variable has at most k parents. This BN model is also considered as an extension of Naïve-Bayes, and its structure can be learned with the algorithm proposed by Sahami [23].

Table 1. Clinical features included in the BCDR-F01 database.

Feature	Description	Type
Age	The age of the patient at the time of the study	Nominal (range [28,82])
Breast density	The density of the breast at the time of the study according to the BI-RADS standard	Ordinal (range [1–4])
Mammography nodule	The lesion contains a mass	Binary
Mammography calcification	Calcifications were detected in the lesion	Binary
Mammography microcalcification	Microcalcifications were detected in the lesion	Binary
Mammography axillary adenopathy	Axillary adenopathy detected	Binary
Mammography architectural distortion	Signs of architectural distortion	Binary
Mammography stroma distortion	Signs of stroma distortion	Binary

3.4 Forest Augmented Naïve Bayes

This BN classifier is a variant of TAN where the attribute variables form a forest graph. One of the advantages of this classifier is that eliminates unnecessary relations among attributes. A method to learn the structure of FAN classifiers is proposed by Lucas [15].

4 Datasets

In this paper, a dataset extracted from the public BCDR-F01 (Film Mammography dataset number 1) database [20] was used. This database is the first public released dataset of the Breast Cancer Digital Repository (BCDR) which contains craniocaudal and mediolateral oblique mammograms of 190 patients. The mammograms were digitized with a resolution of 720 × 1167 pixels, using 256 grey levels. For each mammography, the coordinates for the lesion contours, and numerical anonymous identifiers for linking instances and lesions are provided. In addition to this, information including clinical and image-based descriptors of each lesion, is also available. The summary of the clinical descriptors available for this dataset is shown in Table 1.

In order to form the dataset used in our experiments, the 224 mammogram images with mass lesion that include all clinical descriptors were selected from the BCDR-F01 database. Next, the eight image features explained in Sect. 2 were extracted from each smallest bounding box containing a mass (region of interest-ROI, see Fig. 1). The ROIs were obtained with the help of ImageJ program [21]. In addition to the image features, for each mass the clinical information about the

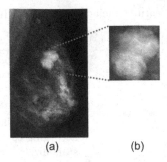

(a)   (b)

Fig. 1. An example of a BCDR image used in this study: (a) original image and (b) the corresponding ROI containing a mass.

age of patient, breast density, calcification, microcalcification, axillary adenopathy, arquitectural distortion, and stroma distortion were included. In summary, this dataset includes 224 mass cases: 112 benigns and 112 malignants where each mass is represented by eight image descriptors and seven clinical features.

5   Experiments and Results

The goal of the experiments is to analyze whether the Bayesian networks performance on mass diagnosis can be improved with a combination of image and clinical features. The performance of the BN models is evaluated with the Leave-one-out cross validation technique. The performance measurements used to report the results are accuracy, sensitivity, and specificity. Classification accuracy is the proportion of masses that are correctly classified by the model. The ratio of malignant masses that are correctly identified is the sensitivity; and specificity is the ratio of benign masses that are correctly identified [24]. All Bayesian networks models are trained and tested using the Matlab®software with help of the Bayes Net toolbox [16] and the BNT Structure Learning Package [14].

We start our experiments with four types of Bayesian networks models: Naïve Bayes, Tree Augmented Naïve Bayes, K-dependence Bayesian classifier (for $K = 2$), and Forest Augmented Naïve Bayes. All models were trained with the values of the eigth image features explained in Sect. 2, which were selected as in [22]. The corresponding network topology for these models are presented in Fig. 2.

5.1   Bayesian Networks Models Using the Complete Set of Clinical Features

In our first experiment, the impact of inclusion of the complete set of clinical features on the initial Bayesian networks models (shown in Fig. 2) is evaluated.

The seven clinical features were added to each Bayesian network model: NB, TAN, KDB, and FAN. Each clinical feature was added taking into account its

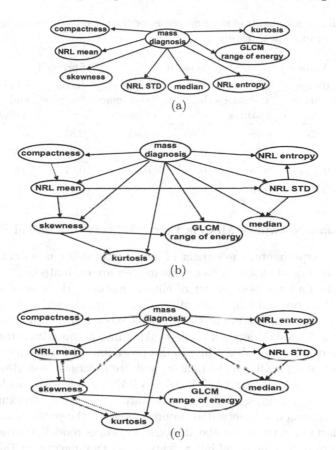

Fig. 2. Topology for the initial Bayesian network models trained with the image features: (a) NB, (b) TAN, (c) KDB and FAN, the former is represented with solid arc lines and the latter is the same structure except where the dotted arcs appear.

causal relationship with the diagnosis node or with the mass descriptor nodes. According to medical literature, the age of patient impact on breast diagnosis; breast density impact on shape, margins, and density of masses; presence of calcification, microcalcification, axillary adenopathy, arquitectural distortion, and stroma distortion, are associated with breast disease [3,8].

The classification results for this experiment are presented in Table 2. From this table, it can be seen that the addition of all clinical features do not help to improve the previous performance on the initial BN models. Only, the extended NB model showed a significant improvement that can be explained by its simple topology. A decrease in sensitivity with a important improvement for the specificity are observed in the other models.

Table 2. Performance results of mass diagnosis for the Bayesian networks models using the complete set of clinical features.

Model	Accuracy		Sensitivity		Specificity	
	Image features	Clinical and image features	Image features	Clinical and image features	Image features	Clinical and image features
NB	0.73	0.78	0.65	0.69	0.81	0.88
TAN	0.79	0.78	0.80	0.70	0.77	0.85
KDB	0.79	0.78	0.80	0.70	0.77	0.85
FAN	0.79	0.78	0.80	0.70	0.77	0.85

5.2   Bayesian Networks Models Using Subsets of Clinical Features

In our second experiment, the impact of inclusion of different subset of clinical features on the initial Bayesian Networks models are evaluated.

In order to find the best subset of clinical features, the sequential forward selection method over all four types of Bayesian Network models is applied [19]. Following the same convention as the previous experiment, each clinical feature is added in the Bayesian network models taking into account its causal relationship with the other nodes. According to this procedure, the best subset is formed by the calcification, axillary adenopathy, and arquitectural distortion features. Other possible topologies for TAN, KDB and FAN models with this best combination of clinical and image features were searched by using structure learning algorithms, but was not observed an improvement on the performance results.

The performance results of the Bayesian networks models trained with the best combination of clinical and image features are summarized in Table 3. From this table, it can be seen that all models trained with the best combination of clinical and image features outperformed those obtained with only image features. For all models, this combination significantly improves the specificity. That is, the results show that the inclusion of clinical features helps to improve particularly the benign mass diagnosis of our dataset.

Table 3. Performance results of mass diagnosis for the Bayesian networks models using the best combination of clinical and image features.

Model	Accuracy		Sensitivity		Specificity	
	Image features	Clinical and image features	Image features	Clinical and image features	Image features	Clinical and image features
NB	0.73	0.76	0.65	0.67	0.81	0.85
TAN	0.79	0.82	0.80	0.80	0.77	0.83
KDB	0.79	0.82	0.80	0.80	0.77	0.83
FAN	0.79	0.82	0.80	0.80	0.77	0.83

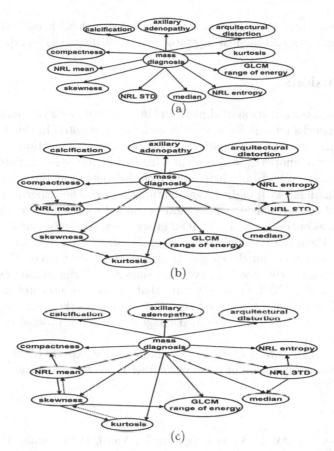

Fig. 3. Topology for the final Bayesian network models using the best combination of clinical features: (a) NB, (b) TAN, (c) KDB and FAN, the former is represented with solid arc lines and the latter is the same structure except where the dotted arcs appear.

Also from Table 3, it can be seen that the best obtained models are TAN, KDB, and FAN. These models, which include dependence among image features, show a good accuracy performance that is comparable with the average performance of radiologists [6] and with the average performance (0.82) of experimentals CADx [24]. In summary, these results indicate that they are more suitable for identification of both mass type than the ones only using image features.

The topologies presented in Fig. 3 for TAN, KDB, and FAN using the best combination of clinical and image features, show a structure that is easy to interpret and validate by experts. They suggest that the clinical factors like the presence of calcification, axillary adenopathy, and arquitectural distortion; as well as, shape, margins, and density of masses (see Sect. 2 for image descriptors) are important factors for mass diagnosis. Moreover, the dependences among image features captured by the extended models, reveal that the shape of a mass has influence on margins, and this impact on the density attributes. These findings, factors and their relationships, are consistent with both the analysis followed by experts and medical literature for mass diagnosis [8,12].

6   Conclusions

The impact of a combination of clinical and image features on the performance of Bayesian Networks models for mass diagnosis was presented in this work. Seven clinical features about age of patient, breast density, calcification, microcalcification, axillary adenopathy, arquitectural distortion, and stroma distortion were analyzed. Several subsets of them were included on initial Bayesian Networks classifiers combined with eight image features nodes. The experimental results have shown that Bayesian networks models trained with a combination of three clinical features outperformed with 0.82 in accuracy, 0.80 in sensitivity, and 0.83 in specificity, those obtained with only image features. This improvement, particularly has more impact on the diagnosis of benign masses than on the malignant ones. Furthermore, the results have also shown that the augmented models: TAN, KDB, and FAN, have a structure that is easy to interpret and validate by experts. For this, it can be said that Bayesian networks with a combination of clinical and automatically obtained image features are promising models for mass classification.

As future work, interpretation and validation by experts of the best found models and the comparision of them with other type of classifiers are considered.

References

1. Balleyguier, C., Ayadi, S., Van Nguyen, K., Vanel, D., Dromain, C., Sigal, R.: BIRADS^{TM} classification in mammography. Eur. J. Radiol. **61**(2), 192–194 (2007)
2. Burnside, E.S., Davis, J., Chhatwal, J., Alagoz, O., Lindstrom, M.J., Geller, B.M., Littenberg, B., Shaffer, K.A., Kahn Jr, C.E., Page, C.D.: Probabilistic computer model developed from clinical data in national mammography database format to classify mammographic findings 1. Radiology **251**(3), 663–672 (2009)
3. Burnside, E.S., Rubin, D.L., Fine, J.P., Shachter, R.D., Sisney, G.A., Leung, W.K.: Bayesian network to predict breast cancer risk of mammographic microcalcifications and reduce number of benign biopsy results: initial experience 1. Radiology **240**(3), 666–673 (2006)
4. Cárdenas, J.S., Bargalló, E.R., Erazo, A.V., Maafs, E.M., Poitevin, A.C.: Consenso mexicano sobre diagnóstico y tratamiento del cáncer mamario: quinta revisión. Elsevier Masson Doyma México (2013)
5. Castillo, E.: Expert Systems and Probabilistic Network Models. Monographs in Computer Science. Springer-Verlag, New York (1997)
6. Cheng, H., Shi, X., Min, R., Hu, L., Cai, X., Du, H.: Approaches for automated detection and classification of masses in mammograms. Pattern Recogn. **39**(4), 646–668 (2006)
7. Cheng, J., Greiner, R.: Learning Bayesian belief network classifiers: algorithms and system. In: Stroulia, E., Matwin, S. (eds.) Canadian AI 2001. LNCS (LNAI), vol. 2056, p. 141. Springer, Heidelberg (2001)
8. De Paredes, E.S.: Atlas of Mammography. Lippincott Williams & Wilkins, Philadelphia (2007)

9. Delogu, P., Evelina Fantacci, M., Kasae, P., Retico, A.: Characterization of mammographic masses using a gradient-based segmentation algorithm and a neural classifier. Comput. Biol. Med. **37**(10), 1479–1491 (2007)

10. Fischer, E., Lo, J., Markey, M.: Bayesian networks of BI-RADS descriptors for breast lesion classification. In: Proceedings of the 26th Annual International Conference of the IEEE Engineering in Medicine and Biology Society (IEMBS 2004). vol. 2, pp. 3031–3034 (2004)

11. Friedman, N., Geiger, D., Goldszmidt, M.: Bayesian network classifiers. Mach. Learn. **29**(2–3), 131–163 (1997)

12. Jackson, V., Dines, K., Bassett, L., Gold, R., Reynolds, H.: Diagnostic importance of the radiographic density of noncalcified breast masses: analysis of 91 lesions. Am. J. Roentgenol. **157**(1), 25–28 (1991)

13. Kahn Jr, C.E., Roberts, L.M., Shaffer, K.A., Haddawy, P.: Construction of a bayesian network for mammographic diagnosis of breast cancer. Comput. Biol. Med. **27**(1), 19–29 (1997)

14. Leray, P., Francois, O.: BNTStructure Learning Package. Technical report FRE CNRS 2645, Laboratoire PSI-INSA Rouen-FRE CNRS, France (2004)

15. Lucas, P.J.S.: Restricted bayesian network structure learning. In: Gamez, J.A., Moral, S., Salmeron, A. (eds.) Advances in Bayesian Networks. STUDFUZZ, vol. 146, pp. 217–234. Springer, Heidelberg (2004)

16. Murphy, K.: How to use Bayes net toolbox (2004). http://www.ai.mit.edu/murphyk/Software/BNT/bnt.html

17. National Cancer Institute, NCI: General information about breast cancer. http://www.cancer.gov/cancertopics/pdq/treatment/breast/Patient/page1.   Accessed Jan 2015

18. Patrocinio, A.C., Schiabel, H., Romero, R.A.: Evaluation of Bayesian network to classify clustered microcalcifications. In: Proceedings of the SPIE Medical Imaging 2004. pp. 1026–1033. International Society for Optics and Photonics (2004)

19. Pazzani, M.: Searching for dependencies in Bayesian classifiers. In: Fisher, D., Lenz, H.-J. (eds.) Learning from Data. LNS, pp. 239–248. Springer, New York (1996)

20. Ramos-Pollán, R., Guevara-López, M.A., Suárez-Ortega, C., Díaz-Herrero, G., Franco-Valiente, J.M., Rubio-del Solar, M., González-de Posada, N., Vaz, M.A.P., Loureiro, J., Ramos, I.: Discovering mammography-based machine learning classifiers for breast cancer diagnosis. J. Med. Syst. **36**(4), 2259–2269 (2012)

21. Rasband, W.: ImageJ: Image processing and analysis in Java. Astrophysics Source Code Library (2012)

22. Rodríguez-López, V., Cruz-Barbosa, R.: On the breast mass diagnosis using Bayesian networks. In: Gelbukh, A., Espinoza, F., Galicia-Haro, S. (eds.) MICAI 2014, Part II. LNCS, vol. 8857, pp. 474–485. Springer, Heidelberg (2014)

23. Sahami, M.: Learning limited dependence Bayesian classifiers. In: Proceedings of the Second International Conference on Knowledge Discovery and Data Mining. vol. 96, pp. 335–338. AAAI Press (1996)

24. Sampat, M., Markey, M., Bovik, A.: Computer-aided detection and diagnosis in mammography. In: Handbook of Image and Video Processing. ch. 10.4, pp. 1195–1217. Elsevier Academic Press (2005)

25. Velikova, M., Lucas, P.J., Samulski, M., Karssemeijer, N.: A probabilistic framework for image information fusion with an application to mammographic analysis. Med. Image Anal. **16**(4), 865–875 (2012)

26. World Health Organization: Breast cancer prevention and control. http://www.who.int/cancer/detection/breastcancer/en/index1.html. Accessed Jan 2015

Patrolling Routes Optimization
Using Ant Colonies

Hiram Calvo[✉], Salvador Godoy-Calderon,
Marco A. Moreno-Armendáriz, and
Victor Manuel Martínez-Hernández

Centro de Investigación en Computación, Instituto Politécnico Nacional,
Av. Juan de Dios Bátiz e/M.O. de Mendizábal s/n, 07738 Mexico City, Mexico
{hcalvo,sgodoyc,marco_moreno}@cic.ipn.mx,
mhernandezb07@sagitario.cic.ipn.mx

Abstract. In general, route optimization by using ant colony algorithms has been widely used with good results so far. This work presents a novel method within this kind of techniques for optimizing patrolling routes for personnel working in public security. Our algorithm can be used in all places with this kind of activities, allowing to allocate an optimal number of human and material resources for patrolling. We present a case study based on data from the municipality of Cuautitlán Izcalli, in Mexico. For three different patrolling requirements, we were always able to find optimal routes in relatively short time (around 50 algorithm iterations).

Keywords: Patrolling routes · Public security · Ant-colony system

1 Introduction

Nowadays public security and crime fighting are one of the most important social priorities in all great cities of the world. Despite the enormous quantity of human and material resources that governments assign for this matter, it is still evident the need for alternative mechanisms that allow to increase the effectiveness and efficiency of police forces [4]. One of the main variables that limit this effectiveness is the response time to crime events. Particularly, immediate-reaction events show that the marginal improvements obtained in this matter are not enough to reduce the general criminal incidence of the zone, as well as to substantially modify the perception of insecurity between citizens [7].

A better perspective of this situation can be achieved if the problem is translated to the sphere of prevention instead of reaction. If public forces were capable to anticipate when and where the criminal activity of a specific kind might be increased, a double benefit could be achieved. On one hand, it would be possible to concentrate resources and logistic activity necessary to fight that specific kind of criminal activity in the

The authors wish to thank the support of the Mexican Government (CONACYT, SNI, SIP-IPN, COFAA-IPN, and BEIFI-IPN).

© Springer International Publishing Switzerland 2015
J.A. Carrasco-Ochoa et al. (Eds.): MCPR 2015, LNCS 9116, pp. 302–312, 2015.
DOI: 10.1007/978-3-319-19264-2_29

anticipated place and time. On the other hand, it could be possible to establish dynamically and with solid foundations several of the common parameters of everyday work in public security, such as the specific design of surveillance rounds, the distribution of forces in time and space, and, of course, the development of security operations, or even information and prevention campaigns through massive communication media [10].

Unfortunately, careful creation and planning of patrolling routes covering the highest criminal incidence is usually not carried out, resulting in personnel being established in a zone of responsibility, where a patrol chief decides under his own criteria where to patrol, without a specific order or previous planning. As a consequence, we can foresee two important problems: the first one is that not all major criminal incidence areas are covered, causing lack of security in certain regions; the second problem is the slow reaction time of patrols. When they receive a call for help, they decide their routes based mostly on the personnel's experience in traversing short, or less transited roads.

Several methods have been developed for tackling the problem of route optimization. The field of multi-robot cooperative tasks provides an interesting set of examples; see [1] for a thorough compendium of several models. Within this approach, we found two major drawbacks. The first one is that some of them are designed for small devices [12], and the second one is that they are designed for automatic execution, and usually they do not allow incorporating certain restrictions pertaining to real world human driving and wide area sectorization. Other approaches are based on workload balancing models [13], local search techniques [16], and agents [2]. However, to our knowledge, ant colony systems, while being known to be effective for finding optimal routes [8, 15], have not been applied to police patrol route planning considering three real needs[1]: (a) finding the optimal route for a patrol to attend an emergency call; (b) finding the optimal route between the current location of a patrol and a set of nearby streets that require surveillance; and finally (c) to find the optimal route for a patrol, so that it can survey different points of major criminal incidence in a specified neighborhood. In the next section, we will present a short introduction to ant colony systems, then in Sect. 3 we describe our proposed method; in Sect. 4 we present experiments with a case study, and finally in Sect. 5 we draw our conclusions.

2 Ant Colony Optimization

Ant Colony Optimization algorithms are models inspired in real ant colonies. Studies show how animals that are almost blind, such as ants, are capable of following the shortest path to their supplies (food) [3]. This is due to the ability ants have to exchange information, since each one of them, while moving, leaves a trace of a substance called pheromone along their path. Thus, while an isolated ant moves essentially in a random way, *agents* of an ant colony detect the pheromone trace left by other ants, and tend to follow such trace. These ants, in turn, leave their own pheromone along the travelled

[1] Direct communication by the personnel of the Emergency Central C4 of the municipality of Cuautitlan Izcalli.

path, making it more attractive, since the pheromone trace has been reinforced. With time, the pheromone evaporates, causing the trace to weaken. In short, it could be say that the process is characterized by a positive feedback, in which the probability for an ant to choose a path increases with the number of ants that previously have chosen the same path. One of the first known applications of the ant colony system was the travelling salesman problem (TSP) [6], obtaining favorable results. From that algorithm several heuristics have been developed to improve the original algorithm, and have been applied to other problems such as the vehicle routing problem (VRP) [5] and the Quadratic Assignment Problem (QAP) [11].

In this paper we present result of a heuristic based on an improved version of the ant colony optimization (ACO) algorithm called MMAS (Max Min Ant System) [14].

The ACO algorithms are iterative processes. In each iteration, a colony of m ants is deployed, and each one of the ants constitutes a solution to the problem. Ants build solutions in a probabilistic way, being guided by a trace of artificial pheromone, and by information calculated *a priori* in a heuristic way. The probabilistic rule for traversing nodes on a graph is:

$$p_{ij}^k(t) = \frac{[\tau_{ij}(t)]^\alpha \cdot [\eta_{ij}]^\beta}{\sum_{l \in N_i^k} [\tau_{il}(t)]^\alpha \cdot [\eta_{lj}]^\beta} \tag{1}$$

where $p_{ij}^k(t)$ is the probability, in a t iteration of the algorithm, the k ant currently situated in city i, chooses city j as the next stop. N is the set of cities not yet visited by the ant k. $\tau_{ij}(t)$ is the amount of pheromone accumulated on the arc (i,j) of the network at the t iteration. η_{ij} is the heuristic information for which, in the case of TSP, the inverse of the distance between i and j cities. α and β are parameters of the algorithm to be adjusted.

When all ants have built a solution, pheromone must be updated on each arc. The formula for this is:

$$\tau_{ij}(t+1) = (1 - \rho) \cdot \tau_{ij}(t) + \cdot \tau_{ij}^{best},$$

$$\Delta\tau_{ij}^{best} = \begin{cases} \frac{1}{L^{best}} & \text{if the arc } (i,j) \text{ belongs to } T^{best} \\ 0 & \text{otherwise} \end{cases} \tag{2}$$

Where ρ is the pheromone evaporation coefficient. T^{best} can be the best solution found at the moment, or the best solution found in the current iteration. The level of pheromone should be in a range $[T_{min}, T_{max}]$. These limits are established in order to avoid stagnation in the search of solutions. All pheromone is initialized with T_{max}. After updating the pheromone, a new iteration can be started. The final result is the best solution found over all iterations.

This gives us a global view of the MMAS algorithm. In the next section we will present its application to the problem of human and material resources for patrolling routes. We aim to a three-folded purpose: (a) To find the optimal route between a patrol's current location, and a point where a call for help has been raised. (b) To find optimal routes for patrolling a small set of nearby streets in a neighborhood, and finally (c) To find optimal routes for patrolling different points of major criminal incidence in a specified neighborhood.

3   Proposed Methodology

We will illustrate our methodology with the example case of a neighborhood of the municipality of Cuautitlán Izcalli, México. This neighborhood was selected considering the current geographic level for assignment of patrolling routes. In Fig. 1 the structure at street level can be seen. Patrols must cover the points considered as the most important ones.

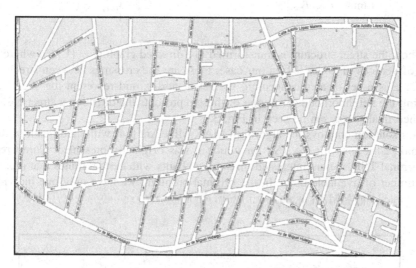

Fig. 1. Structure showing streets of a neighborhood in Cuautitlán Izcalli, México

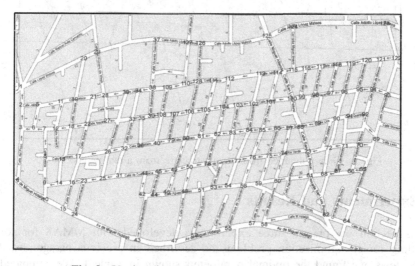

Fig. 2. Vertices located on each intersection of streets

Table 1. Parameters used in the MMAS algorithm

Parameter	Value
Number of vertices of the generated graph	128
Number of undirected graphs	26
Number of directed arcs	38
Number of ants generated in each iteration	20
Number of iterations	50
Pheromone evaporation constant	0.98
Limits $[T_{min}, T_{max}]$	$\left[0.0078, \frac{1}{p}\right]$

Then, the street structure is transformed to a directed graph $G = (V, E)$, where V is a set of vertices or nodes [9]. In our case, those are the crossings between streets. See Fig. 2. E is a set of arcs connecting the set of nodes, and represent the streets conforming the neighborhood. Each one of these represents the direction a street has. The obtained final graph can be seen in Fig. 3.

Our solution employs the algorithm MAX-MIN Ant [14] with modifications to the original restrictions for the TSP for which it was originally presented. Compared to the original TSP, we are interested on having N ants with certain routes that represent the number of available units. In the original problem, we have only one individual. We adapted the MMAS as shown in Fig. 4.

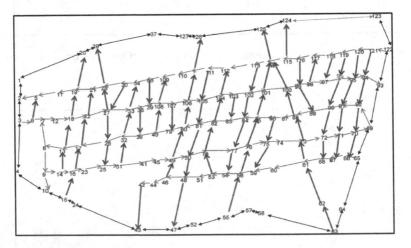

Fig. 3. Graph obtained from street structure from a real neighborhood

4   Experiments and Results

In this section we present three experiments developed with MMAS for solving patrolling routes optimization problems as described in previous paragraphs. After several tests, we found the optimal parameters shown in Table 1. We compared our results against a random walk baseline, which consists basically on using the algorithm shown in Fig. 4 without using formulas (1) and (2), *i.e.*, using a plain random roulette with equal probabilities, and not using pheromones at all.

```
Let G be the graph of all street intersections
Let Aᵢ be a set of n ants (i = 0, … n) // Each ant simulates a patrol
Let R be a subset of G consisting in all points required for a certain
    route.
Repeat m times:
    Set Aᵢ.pos=initial vertex for all n ants
    Do:
        For each ant Aᵢ:
            * Compute next Aᵢ.pos as a random roulette based on the
                probabilities according to formula (1)
            Add Aᵢ.pos to the set Aᵢ.visited
            if Aᵢ.pos ∈ R then add Aᵢ.pos to the list Aᵢ.route
            if |Aᵢ.route| == |R|
                Add Aᵢ to the set of Solutions
                Reset Aᵢ
            if |Aᵢ.visited| > |R|:
                Reset Aᵢ
        while |Solutions| < required number of solutions
        For each ant Sᵢ in Solutions:
            Calculate cost of Sᵢ.route
        Tᵇᵉˢᵗ = Sᵢ with less expensive Sᵢ.route
        * Update pheromone level according to formula (2)
    End
```

Fig. 4. Pseudocode for the MMAS algorithm. Asterisks show the steps to be modified for the random walk baseline comparison

4.1 Goal A: Target Route Optimization

The goal of this experiment is to optimize routes that were created as a preventive perimeter given an alert call in a point or specific street. For this purpose, 5 points in the map's neighborhood were randomly selected, as well as a common starting point. See Fig. 5 and Table 2.

We can see in all cases that the optimal route was automatically found by the ant colony algorithm, while the random walk algorithm did not converge to the optimal solution after the same number of iterations (50), except for Alert 2. For Alerts 1 to 4

Table 2. Results obtained from experiment A

Alert #	Start point	End point	Ant colony MMAS			Random walk baseline		
			Cost	Time	Optimal?	Cost	Time	Optimal?
1	1	31	716	23	Yes	819	32	No
2	1	109	674	35	Yes	674	35	Yes
2	1	105	780	35	Yes	1432	43	No
4	1	89	1197	37	Yes	1781	57	No
5	1	64	1964	96	Yes	–	–	N/A

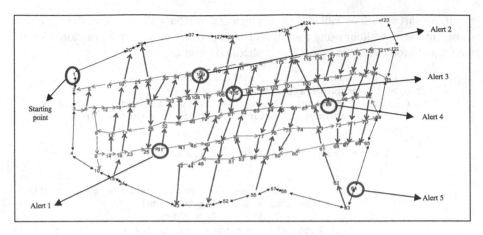

Fig. 5. Selected alerts in experiment A

the random walk algorithm obtained a route, but for Alert 5 the maximum number of iterations was reached without finding a route to the target node.

4.2 Goal B: Route Optimization

The goal of this experiment is to find optimal routes for patrolling a small set of nearby streets in a neighborhood. This kind of routes is generally assigned to individual patrols.

For this experiment, three different surveillance areas were selected, each one with 6 nearby points, located randomly in the studied neighborhood, as well as a common point, see Fig. 6. It is important to note that the selected areas were managed independently, and that this experiment aims to illustrate the optimal route from a specific point to a particular area, and so, it does not model interaction with other areas. Optimal routes are shown in Table 3. Our Ant Colony algorithm was able to find all optimal routes for this experiment, whereas the random walk baseline found routes for Alerts 2 and 3, but they were not optimal. See Table 4.

Table 3. Optimal routes for experiment B

#	Start point	Rute points	Route cost	Time	Optimal route
1	1	59-61-73-75-86-88	2,387	66	1-2-3-6-7-12-18-22-27-26-25-31-41-45-49-50-78-77-76-75-74-73-88-87-86-75-74-73-72-67-68-61-60-59
2	1	31-32-33-39-40-45	1,190	148	1-2-3-6-7-12-18-22-27-33-35-39-108-109-38-34-30-27-33-35-36-32-26-25-31-41-45-49-80-79-40
3	1	91-92-93-95-120-122	2,136	62	1-20-29-37-127-126-125-124-123-122-121-120-95-94-93-69-92-91

Table 4. MMAS vs. Random walk routes for experiment B

Alert #	Ant colony MMAS			Random walk baseline		
	Cost	Time	Optimal?	Cost	Time	Optimal?
1	2,387	66	Yes	–	–	N/A
2	1,190	148	Yes	3,215	217	No
3	2,136	62	Yes	4,384	250	No

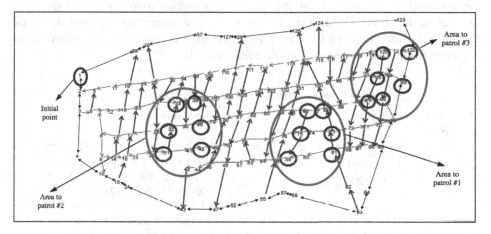

Fig. 6. Selected areas in experiment B from the sample neighborhood

Found routes are ready to be implemented in a real patrolling scenario. Routes like these were calculated for all neighborhoods of the municipality of Cuautitlán Izcalli, always finding optimal routes.

4.3 Goal C: Diverse Patrolling Areas Optimization

The goal of this experiment is to find optimal routes for patrolling diverse areas that are distributed throughout the whole neighborhood. For this experiment, the algorithm was executed to find two optimal routes. Each one of them must pass through three different surveillance areas. Each area is integrated with 4 nearby points, randomly selected from the studied neighborhood. All routes depart from a common initial point. See Fig. 7.

All optimal routes shown in Table 5 were found by our Ant Colony Algorithm, while the Random walk algorithm was not able to find a route covering the requested route points within the specified number of iterations. In general, several routes were calculated for all neighborhoods in the municipality of Cuautitlán Izcalli, always finding optimal routes, implying the proposed algorithm is a reliable way of calculating patrolling routes given important points to be covered. These points can be obtained from daily operation of patrolling routes planning sessions.

5 Conclusions and Future Work

This work has shown the advantage of using evolutionary techniques in the optimization of patrolling routes, obtaining promising results. Based on the carried experiments, we found that ant colony based algorithms are efficient and effective for optimizing several kinds of patrolling routes. In all cases we were able to find an optimal route within a limited number of iterations, while the random walk algorithm found an optimal route in only a few cases. For Patrolling Area Optimization, the random walk algorithm was not able to find a patrolling route within the specified number of iterations. These experiments show that computing the probability of transition for an ant based on a pheromone component, improves the ability of an exploration algorithm to find a feasible solution in short time.

Table 5. Optimal routes for experiment C

#	Start point	Rute points	Route cost	Time	Optimal route
1	1	59-61-86-88	4,546	3,421	1-2-3-4-10-15-16-23-25-31-32-33-35-39-
2		32-33-39-40			108-107-106-105-104-103-102-101-
3		93-95-120-122			100-99-98-97-96-95-94-93-122-121-
					120-95-94-93-122-121-120-119-96-90-
					89-88-87-86-75-74-73-72-68-61-60-59-
					56-54-53-78-82-81-80-79-40
4	1	11-18-21-22	4,126	990	1-2-3-6-7-12-18-22-27-28-21-19-11-5-2-
5		47-48-52-53			1-20-29-37-127-126-127-37-127-126-
6		113-115-124-125			125-124-123-122-121-120-119-118-
					117-116-115-114-113-102-84-77-54-
					53-51-48-47-52

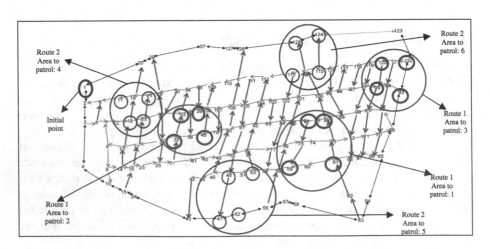

Fig. 7. Areas to be covered by the routes sought in Experiment C

The problems tackled in our experiments are extendable to cover many problems arising currently in great urban zones of the world. A drawback that we found is that processing time increases considerably as the number of points to be visited is greater, so that we would need to explore other improvements if we needed to consider a wider area patrolling planning. Fortunately, this is not the case in most real problems, since large cities tend to divide in sectors their patrolling forces. Around 50 iterations were needed to find an optimal route with our method. The baseline method was not able to find an optimal route within this number of iterations. Further experimentation with the baseline algorithm for finding the needed number of iterations to obtain an optimal route (if possible) has been left as future work.

Also as future work, we plan considering traffic factors affecting patrolling maneuvers, as well as considering other factors impeding free vehicular transit and thus, affect the response time of a patrol.

References

1. Agmon, N.: Multi-robot patrolling and other multi-robot cooperative tasks: an algorithmic approach. Dissertion, Barllan University (2009)
2. Calvo, R., de Oliveira, J.R., Figueiredo, M., Romero, R.A.: Parametric investigation of a distributed strategy for multiple agents systems applied to cooperative tasks. In: Proceedings of the 29th Annual ACM Symposium on Applied Computing. ACM (2014)
3. Colorni, A., Dorigo, M., Maniezzo, V: Distributed Optimization by Ant Colonies, actes de la première conférence européenne sur la vie artificielle, pp. 134–142. Elsevier Publishing, Paris, France (1991)
4. Creed, W.E.D., Miles, R.E.: Trust in organizations. In: Kraemer, K.L., Tyler, T.R. (eds.) Trust in organizations: frontiers of theory and research, pp. 16–38. Sage, Thousand Oaks (1996)
5. Dantzig, G.B., Ramser, J.H.: The truck dispatching problem. Manag. Sci. 6(1), 80–91 (1959)
6. Dorigo, M., Gambardella, L.M.: Ant colony system: a cooperative learning approach to the traveling salesman problem. IEEE Trans. Evol. Comput. 1, 53–66 (1997)
7. Eck, J.E., Maguire, E.: Have changes in policing reduced violent crime? an assessment of the evidence. In: Blumstein, A., Wallman, J. (eds.) The Crime Drop in America, pp. 207–228. Cambridge University Press, New York (2000)
8. Fard, E.S., Monfaredi, K., Nadimi, M.H.: Application methods of ant colony algorithm. Am. J. Softw. Eng. Appl. 3(2), 12–20 (2014)
9. Jiang, B., Claramunt, C.: A structural approach to the model generalization of an urban street network. GeoInformatica 8(2), 157–171 (2004)
10. Liu, L. (ed.): Artificial Crime Analysis Systems: Using Computer Simulations and Geographic Information Systems: Using Computer Simulations and Geographic Information Systems. IGI Global, Hershey (2008)
11. Maniezzo, V., Colorni, A.: The ant system applied to the quadratic assignment problem. IEEE Trans. Knowl. Data Eng. 11(5), 769–778 (1999)
12. Portugal, D., Rocha, R.P.: Cooperative multi-robot patrol in an indoor infrastructure. In: Spagnolom, P., Mazzeo, P.L., Distante, C. (eds.) Human Behavior Understanding in Networked Sensing, pp. 339–358. Springer International Publishing, Switzerland (2014)

13. Shafahi, A., Haghani, A.: Balanced routing of patrolling vehicles focusing on areas with historical crime. In: Transportation Research Board 95th Annual Meeting, 15–4387 (2015)
14. Stützle, T., Hoos, H.H.: MAX–MIN ant system. Future Gener. Comput. Syst. **16**(8), 889–914 (2000)
15. Toklu, N.E., Gambardella, L.M., Montemanni, R.: A multiple ant colony system for a vehicle routing problem with time windows and uncertain travel times. J. Traffic Logistics Eng. **2**(1), 52–58 (2014)
16. Watanabe, T., Takamiya, M.: Police patrol routing on network voronoi diagram. In: Proceedings of the 8th International Conference on Ubiquitous Information Management and Communication. ACM (2014)

Author Index

Printed in the United States
By Bookmasters

Printed in the United States
By Bookmasters